GRÖBNER BASES IN RING THEORY

GRÖBNER BASES IN RING THEORY

Huishi Li
Hainan University, China

NEW JERSEY • LONDON • SINGAPORE • BEIJING • SHANGHAI • HONG KONG • TAIPEI • CHENNAI

Published by

World Scientific Publishing Co. Pte. Ltd.

5 Toh Tuck Link, Singapore 596224

USA office: 27 Warren Street, Suite 401-402, Hackensack, NJ 07601

UK office: 57 Shelton Street, Covent Garden, London WC2H 9HE

British Library Cataloguing-in-Publication Data
A catalogue record for this book is available from the British Library.

GRÖBNER BASES IN RING THEORY

ISBN-13 978-981-4365-13-0
ISBN-10 981-4365-13-0

Printed in Singapore.

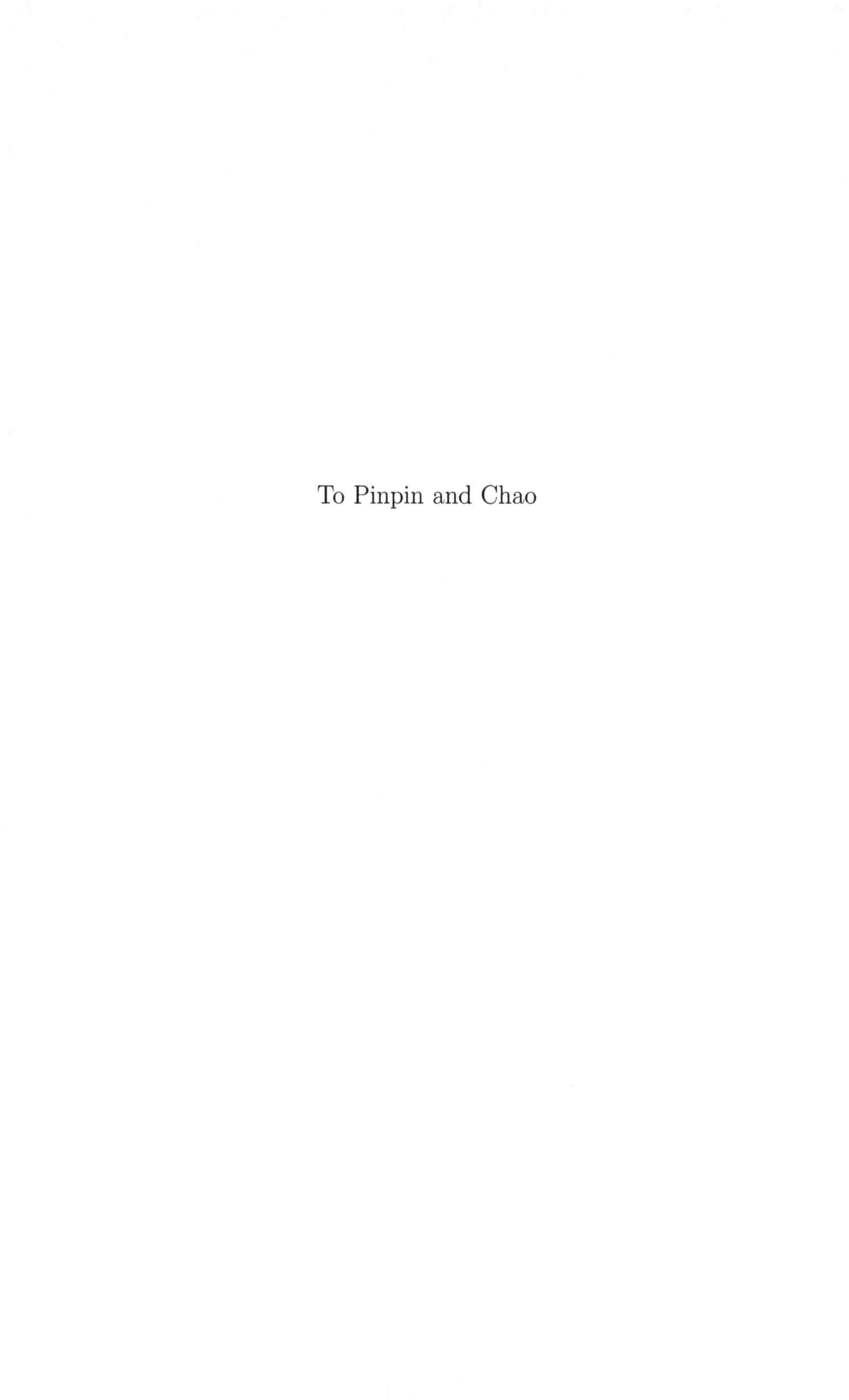

To Pinpin and Chao

Preface

The Gröbner basis theory, which is at the heart of modern computational (computer) algebra, has evolved to be a well-defined subdiscipline of the theory of rings (algebras) and their modules (representations), of which the core idea is to computationally study the structural properties of rings (algebras) and their modules defined by relations that form a Gröbner basis. Focusing on *noncommutative associative algebras* over a field K, this monograph attempts to introduce the mathematical fundamentals of using Gröbner bases in ring theory at a modest level. Such an idea is realized through covering several featured topics studied in recent years, such as, for a given Γ-filtered algebra A, the realization of lifting structural properties from the associated Γ-graded algebra $G^{\Gamma}(A)$ to A via the Γ-leading homogeneous algebra $A^{\Gamma}_{\mathbf{LH}}$; the construction of Gröbner bases in more extensive but widely studied algebras, especially the structural foundation of calculating non-homogeneous Gröbner bases by passing to the homogeneous case; the establishment of a constructive PBW theory via Gröbner bases; and how to give full scope to the associated monomial algebra $A^{\mathcal{B}}_{\mathbf{LH}}$ of an algebra A defined by a Gröbner basis $\mathcal{G}$ in the structure theory of A and certain (graded) algebras closely related to A.

Our emphasis in this book is on opening the way of applying certain computational methods (in terms of Gröbner bases) to recognizing and determining various structural properties of algebras, which is illustrated by numerous illuminating examples.

Except for presenting the latest developments on the topics involved, we endeavor to make this volume self-contained, for instance, the material of Chapters 1 and 3 may be used as lecture notes for an introduction to (noncommutative) Gröbner basis theory. Moreover, some extended examples and problems related to the topics studied in separate chapters (though not explicitly labeled as exercises) are left, as exercises, to the interested reader.

The National Natural Science Foundation of China has my heartfelt thanks for the support of the research (10571038, 10971044) that went into this work.

During the preparation of this book, Professor J. Okninski, Dr. Jiangfeng Zhang, and Dr. Yuchun Wu provided many interesting references to me; and the Department of Literature Transfer in the Library of Hainan University transferred many more references to me from other libraries in China. I wish to express my gratitude for their help.

Two anonymous referees gave valuable suggestions and comments on improving the manuscript of this book; Professor Yu Kiang Leong read the entire manuscript very carefully, corrected many misused English words, and suggested improvements to my English style and grammar, in almost every page. I would very much like to thank them for their patience and kindness.

This is my second time publishing a book with World Scientific Publishing Co. I am very grateful to the publisher for their help and cooperation during the preparation of this book. Moreover, my thanks also go to Ms Lai Fun Kwong for her timely and friendly communications concerning the preparation and publication of this book in about four years.

March, 2011
Haikou, China

Huishi Li

Contents

Chapter 0

Introduction

Let $R = \oplus_{\gamma\in\Gamma}R_\gamma$ be a Γ-graded K-algebra over a field K, where Γ is a totally ordered semigroup, and suppose that R has a Gröbner basis theory for two-sided ideals with respect to an admissible system $(\mathcal{B}, \prec)$, where $\mathcal{B}$ is a K-basis of R consisting of Γ-homogeneous elements, and $\prec$ is a monomial ordering on $\mathcal{B}$. For instance, R is a commutative polynomial algebra, or a free algebra, or a path (quiver) algebra. Let $\mathcal{G}$ be a Gröbner basis for the ideal $I = \langle\mathcal{G}\rangle$ in R, and $A = R/I$ the corresponding quotient algebra of R. In both the commutative and noncommutative settings concerning the structure theory of A, the most well-known fact is that the effectiveness of using $\mathcal{G}$ is often realized through a combinatorial-algorithmic study of the monomial ideal $\langle\mathbf{LM}(\mathcal{G})\rangle$ generated by the set $\mathbf{LM}(\mathcal{G})$ of $\prec$-leading monomials of $\mathcal{G}$, e.g., in working out a K-basis for A, the Krull (Gelfand-Kirillov) dimension of A, the Hilbert series of A (if A is $\mathbb{N}$-graded), and the syzygy modules, etc. Considering the associated Γ-graded algebra $G^\Gamma(A)$ of A with respect to the Γ-filtration FA induced by the natural Γ-grading filtration of R, however, as soon as the Γ-graded algebra isomorphism $G^\Gamma(A) \cong R/\langle\mathbf{LH}(I)\rangle$ is established, where $\langle\mathbf{LH}(I)\rangle$ is the graded ideal generated by the set $\mathbf{LH}(I)$ of Γ-leading homogeneous elements of I, the graded ideal $\langle\mathbf{LH}(\mathcal{G})\rangle$ generated by the set $\mathbf{LH}(\mathcal{G})$ of Γ-leading homogeneous elements of $\mathcal{G}$ comes immediately into play. Incorporating the filtered-graded structure theory of A, the excellent behavior of $\langle\mathbf{LH}(\mathcal{G})\rangle$ has led to a quite general constructive PBW theory for A, which, in turn, has provided us with an innovative way in recognizing and determining many more structural properties of A. Note, in such a constructive PBW theory, that the power of using $\langle\mathbf{LM}(\mathcal{G})\rangle$ is naturally strengthened. This is because, with the $\mathcal{B}$-graded structure $R = \oplus_{u\in\mathcal{B}}R_u$, if the natural $\mathcal{B}$-grading filtration exists for R (for instance, R is a commutative polynomial algebra or a free algebra), then $G^{\mathcal{B}}(A) \cong R/\langle \mathbf{LH}(I)\rangle$ but with $\mathbf{LH}(I) = \mathbf{LM}(I)$ and $\mathbf{LH}(\mathcal{G}) = \mathbf{LM}(\mathcal{G})$. More concisely, the main story may be pictured into the following diagram:

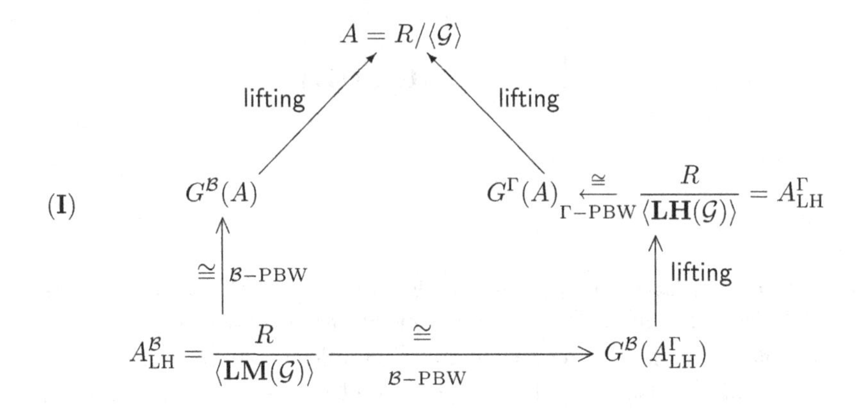

In this monograph, we attempt to clarify the mathematical fundamentals that make the strategy presented above work effectively, not only for A and $G^{\Gamma}(A)$ but also for the Rees algebra $\widetilde{A}$ of A (provided A is $\mathbb{N}$-filtered). This is done through the chapters outlined below.

To make the book self-contained, Ch.1 includes necessary preliminaries, such as presenting algebras by relations, some basics on algebras graded by semigroups and algebras filtered by ordered semigroups.

Let $R = \oplus_{\gamma\in\Gamma}R_{\gamma}$ be a Γ-graded algebra over a field K, where Γ is a well-ordered semigroup, and let I be an arbitrary proper ideal of R. Put $A = R/I$. In the first two sections of Ch.2, we show that many structural properties may be lifted from $G^{\Gamma}(A)$ to A in a classical way, where $G^{\Gamma}(A)$ denotes the associated Γ-graded algebra of A with respect to the Γ-filtration FA induced by the Γ-grading filtration of R. In Section 3 we introduce the Γ-leading homogeneous algebra $A^{\Gamma}_{\rm LH} = R/\langle\mathbf{LH}(I)\rangle$ of A, where $\mathbf{LH}(I)$ is the set of Γ-leading homogeneous elements of I, and we derive the Γ-graded K-algebra isomorphism

$$(*) \qquad A^{\Gamma}_{\rm LH} \xrightarrow{\cong} G^{\Gamma}(A).$$

With the isomorphism $(*)$ mentioned above, in Section 4 we translate the lifting results of "from $G^{\Gamma}(A)$ to A" given in Sections 1, 2 into the lifting results of "from $A^{\Gamma}_{\rm LH}$ to A" via the commutative diagram

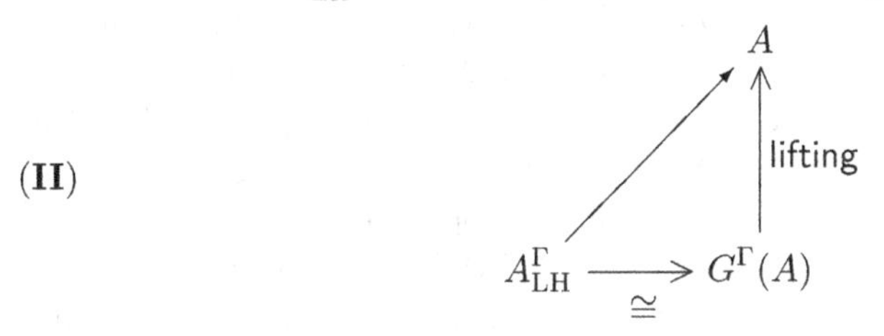

Ch.3 is devoted to a comprehensive introduction to the conception and construction of Gröbner bases in (noncommutative) associative algebras. To suit the context of this book, after introducing the basics of a general (one-sided, two-sided) Gröbner basis theory in the first three sections, starting from Section 4 the Gröbner basis theory focuses on algebras with a skew multiplicative K-basis. In Sections 5 – 7, after covering the classical algorithmic principle of constructing Gröbner bases, we discuss in detail the relation between non-homogeneous Gröbner bases and homogeneous Gröbner bases by employing the (de)homogenization techniques stemming from algebraic geometry. Due to the fact that most well-developed computer algebra systems do a better job of producing homogeneous Gröbner bases (which, in considering ideals of a free algebra, can even be done effectively by passing to the commutative case by the programm LETTERPLACE proposed in [LL]), the results of Section 6 provide a general structure foundation of "non-homogeneous Gröbner bases vs homogeneous Gröbner bases" for further development of calculating Gröbner bases. Moreover, the dh-closed homogeneous Gröbner bases introduced in Section 7 are closely related to the study of Rees algebras in Ch.7.

In Ch.4, we demonstrate, in a quite extensive context, that the Gröbner basis theory is essentially an algorithmic realization of a constructive PBW theory. More precisely, in Sections 1, 2 we first realize, by means of Gröbner bases, the Γ-PBW isomorphism (motivated by the isomorphism $(*)$ mentioned above) with respect to different Γ-filtered-graded structures. This builds the PBW bridges as pictured in the above diagram (I). Moreover, in the context of considering quotient algebras of free algebras, we show that certain specific Gröbner bases and the classical PBW K-bases can be derived from each other. Also we show that quadric Gröbner bases which may yield classical PBW bases and nice associated graded algebras can be recognized and constructed more easily under checkable conditions. Finally in Section 4, by using the results of Section 3 we strengthen some results of ([Li1], Ch.3) concerning solvable polynomial algebras (in the sense of

[K-RW]).

Let the free K-algebra $K\langle X\rangle = K\langle X_1, ..., X_n\rangle$ be equipped with a positive weight $\mathbb{N}$-gradation by assigning to each X_i a positive degree n_i, and $\mathcal{B}$ the standard K-basis of $K\langle X\rangle$. Consider an arbitrary proper ideal I of $K\langle X\rangle$ and put $A = K\langle X\rangle/I$. Along the routes of the foregoing diagram (I), Ch.5 – Ch.7 deal with several practical topics of studying A, as well as the associated graded algebra $G^{\mathbb{N}}(A)$ and the Rees algebra $\widetilde{A}$ of A, by using a Gröbner basis $\mathcal{G}$ of I via the algebras $A^{\mathcal{B}}_{\rm LH} = K\langle X\rangle/\langle \mathbf{LM}(\mathcal{G})\rangle$ and $A^{\mathbb{N}}_{\rm LH} = K\langle X\rangle/\langle \mathbf{LH}(\mathcal{G})\rangle$.

In Ch.5, Section 1 clarifies the working strategy of "using $A^{\mathcal{B}}_{\rm LH}$ in terms of Gröbner bases". Following such a philosophy and combining several significant combinatorial-algorithmic results concerning monomial algebras from the literature, through Sections 2 – 8 we show how to computationally determine or recognize certain structural properties of A and $G^{\mathbb{N}}(A)$, such as the computation of GK dimension and Hilbert series, the recognition of Noetherianity, (semi-)primeness, PI-property, and the finiteness of global homological dimension.

Ch.6 concentrates on a computational determination of both the homogeneous and non-homogeneous p-Koszulity (in the sense of [Pr], [Pos], [PP], [Ber2], [BG1], and [BG2]) for algebras defined by relations. The results obtained in this chapter help us to recognize and construct homogeneous and non-homogeneous p-Koszul algebras.

Ch.7 is for providing an effective study of the Rees algebra $\widetilde{A}$ associated with an $\mathbb{N}$-filtered algebra A (in algebraic geometry, $\widetilde{A}$ corresponds to the coordinate ring of the projective closure of the affine variety defined by the defining relations of A). We show how to obtain the Gröbner defining relations of $\widetilde{A}$ by using a Gröbner defining set of A. This enables us to show that many structural properties of $\widetilde{A}$ may be obtained in a computational way; it also enables us to establish the relation between Rees algebras and several popularly studied topics, such as homogenized algebras, regular central extensions, and PBW-deformations. Finally in this chapter we show that every algebra defined by a dh-closed homogeneous Gröbner basis $\mathscr{G}$ (in the sense of Ch.3, Section 7) may be realized as the Rees algebra $\widetilde{A}$ of the filtered algebra A which has the Gröbner defining set given by the dehomogenization of $\mathscr{G}$.

Due to the paramount importance of having a "nontrivial" (finite) Gröbner basis for an ideal I in various applications, the final Ch.8 looks for more Gröbner bases in certain interesting cases. In Sections 1, 2, we show how to obtain (finite or infinite) Gröbner bases in the free algebra

$K\langle X\rangle = K\langle X_1, ..., X_n\rangle$ by lifting Gröbner bases from a subclass of solvable polynomial algebras (in the sense of [K-RW]) that have a finite Gröbner basis theory. Attracted by the wonderful fact that numerous quantum binomial algebras and their Koszul dual ([G-IV], [Laf], [G-I5]) yield the first model of a proper left Gröbner basis theory, respectively a proper right Gröbner basis theory, through Sections 3 – 5 we introduce (almost) skew 2-nomial algebras and explore the existence of a (left, right, or two-sided) Gröbner basis theory for such algebras. Since our study on this topic is far from reaching a complete solution, the book ends with several related open problems.

Finally, note that if an algebra R has a skew multiplicative K-basis $\mathcal{B}$ such that R has a Gröbner basis theory, then, as indicated in Ch.3, any subset $\Omega \subset \mathcal{B}$ itself is a Gröbner basis, and moreover, Ω represents a family of Gröbner bases with the same set of leading monomials. Inspired by the practical examples presented through Ch.3 – Ch.8, we hope that the principles and methods we have introduced in this book may simultaneously motivate new insights and perspectives of further research on the effective application of Gröbner bases in the structure theory of rings, algebras and their modules (representations).

Conventions used throughout this book

As usual, we use $\mathbb{N}$, $\mathbb{Z}$, $\mathbb{Q}$, $\mathbb{R}$, and $\mathbb{C}$ to denote the sets of *non-negative integers*, *integers*, *rational numbers*, *real numbers*, and *complex numbers*, respectively.

Unless otherwise stated, K always denotes a (commutative) field of arbitrary characteristic, and $K^* = K - \{0\}$.

All rings considered are associative and have identity 1. If A_1 and A_2 are two rings, then $A_1 \subset A_2$ means that A_1 is a subring of A_2 with the same 1. If φ: $A_1 \to A_2$ is a ring homomorphism, then we write Kerφ and Imφ for the kernel and image of φ respectively; if furthermore $\varphi \neq 0$, then we insist that $\varphi(1) = 1$.

All (left or right) modules considered are unitary modules.

Let A be a ring and F a nonempty subset of A. Then we use $\langle F\rangle$ to denote the two-sided ideal of A generated by F, and similarly we write $\langle F]$, respectively $[F\rangle$, for the left ideal, respectively the right ideal of A generated by F.

If we say that Γ is an ordered semigroup (monoid), it is meant that Γ is a semigroup (monoid) equipped with a (partial) ordering $\leq$ which is compatible with the binary operation of Γ, i.e., if $u, v, w \in \Gamma$ and $u < v$, then $wu < wv$ and $uw < vw$. If $\leq$ is a total ordering (well-ordering), then Γ is called a totally ordered (well-ordered) semigroup (monoid).

Chapter 1

Preliminaries

In this preparatory chapter, we first introduce the principle of presenting algebras by relations, which is illustrated through several classical examples. Then we recall some elements on S-graded algebras and modules, where S is a semigroup. We also recall some basics on Γ-filtered algebras and modules, where Γ is a totally ordered semigroup.

1.1 Presenting Algebras by Relations

K-algebra

Let K be a field. A K-*algebra* is a ring A together with a *nonzero* ring homomorphism $\eta_A\colon K \to A$ satisfying

(A1) the image of η_A, denoted $\mathrm{Im}\eta_A$, is contained in the center of A, and

(A2) the map

$$\begin{array}{rcl} K \times A & \longrightarrow & A \\ (\lambda, a) & \mapsto & \eta_A(\lambda)a \end{array}$$

equips A with a vector space structure over K and the multiplicative map $\mu_A\colon A \times A \to A$ is bilinear:

$$\begin{array}{l} \mu_A(a+b,c) = \mu_A(a,c) + \mu_A(b,c),\ a,b,c \in A, \\ \mu_A(a,b+c) = \mu_A(a,b) + \mu_A(a,c),\ a,b,c \in A, \\ \mu_A(\lambda a,b) = \mu_A(a,\lambda b) = \lambda\mu(a,b),\ \ \lambda \in K,\ \ a,b \in A. \end{array}$$

If A is a commutative ring, we say that A is a *commutative* K-*algebra*.

Identifying K with $\eta_A(K)$ in A, it is clear that any A-module M also has a K-vector space structure, and the action of A on M is K-linear in the sense that

$$a(\lambda m_1 + \mu m_2) = \lambda(am_1) + \mu(am_2)$$

for all $a \in A$, $m_1, m_2 \in M$, $\lambda, \mu \in K$, that is, each $a \in A$ defines a linear transformation of the K-vector space M. In general, if V is any K-vector space, then the set of all linear operators on V, denoted $\mathrm{End}_K V$, forms a K-algebra with respect to the operator composition and the scalar multiplication of operators, which is known as the *K-algebra of linear operators* of V (if V is of finite dimension n, then $\mathrm{End}_K V$ is nothing but the matrix algebra consisting of all $n \times n$ matrices over K), and V forms an $\mathrm{End}_K V$-module in a natural way.

Let A and B be K-algebras. A *K-algebra homomorphism* from A to B is a ring homomorphism α: $A \to B$ such that $\alpha \circ \eta_A = \eta_B$, i.e., we have the following commutative diagram:

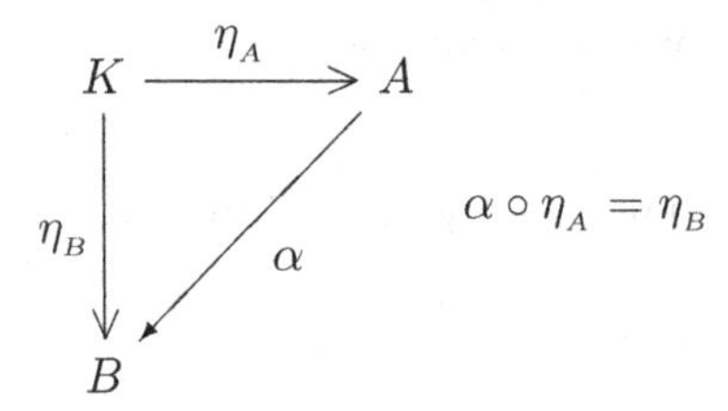

Obviously, α is also a linear transformation from the K-vector space A to the K-vector space B.

If a subring S of a K-algebra A is also a K-algebra with respect to the restriction $\eta_A|_S$: $K \to S$, then S is called a *subalgebra* of A, and in this case, the inclusion map $S \hookrightarrow A$ is a K-algebra homomorphism. Thus, if ℓ: $A \to B$ is an injective algebra homomorphism, we may identify A with $\mathrm{Im}\ell \subseteq B$ and view A as a subalgebra of B.

It follows from the definitions given above that the following facts are clear:

- Since η_A is nonzero, η_A is injective, $K \cong \mathrm{Im}\eta_A$ and $\eta_A(1) = 1$.
- K may be viewed as a subalgebra of A, and we may write λa for $\eta_A(\lambda)a$ (the scalar multiplication of the K-vector space A).
- α preserves units via $\alpha(1) = 1$.

Note that here we have used 1 to denote the identity element in the respective algebra involved.

Let A be a K-algebra and I an ideal of A. Then the K-algebra structure of A induces naturally on the quotient ring A/I a K-algebra structure. In this case, A/I is called the *quotient algebra* of A modulo I. Thus, for any ideal I of a K-algebra A, the natural map $A \to A/I$ is clearly a K-algebra homomorphism.

The tensor (free) algebra

Let V be a K-vector space. For non-negative integers $i \in \mathbb{N}$, define

$$\begin{aligned} &T^0V = K, \text{ the ground field,} \\ &T^1V = V, \\ &T^iV = V^{\otimes i} = \underbrace{V \otimes_K V \otimes_K \cdots \otimes_K V}_{i}, \quad i > 1, \end{aligned}$$

and call T^iV the i-th tensor power of V. Then the *tensor algebra* defined by V over K is the direct sum

$$T(V) = \bigoplus_{i=0}^{\infty} T^iV = K \oplus V \oplus (V \otimes_K V) \oplus (V \otimes_K V \otimes_K V) \oplus \cdots,$$

in which multiplication $T(V) \times T(V) \to T(V)$ is determined by extending linearly the canonical isomorphism

$$T^iV \otimes_K T^jV \xrightarrow{\cong} T^{i+j}V$$

to all of $T(V)$, while the ring homomorphism $\eta_{T(V)}$ is given by the inclusion map ι: $K \hookrightarrow T(V)$. If $X = \{X_\ell\}_{\ell \in J}$ is a K-basis for V, then $T(V)$ has the K-basis

$$\mathcal{B} = \left\{1,\ X_{\ell_1} \otimes X_{\ell_2} \otimes \cdots \otimes X_{\ell_i} \;\middle|\; X_{\ell_k} \in X,\ i \geq 1\right\}.$$

The tensor algebra $T(V)$ defined by V is the "most general" algebra containing V in the sense that $T(V)$ has the following *universal property.*

Theorem 1.1. *Any linear transformation f: $V \to A$ from V to a K-algebra A can be extended uniquely to a K-algebra homomorphism from $T(V)$ to A as illustrated by the following commutative diagram:*

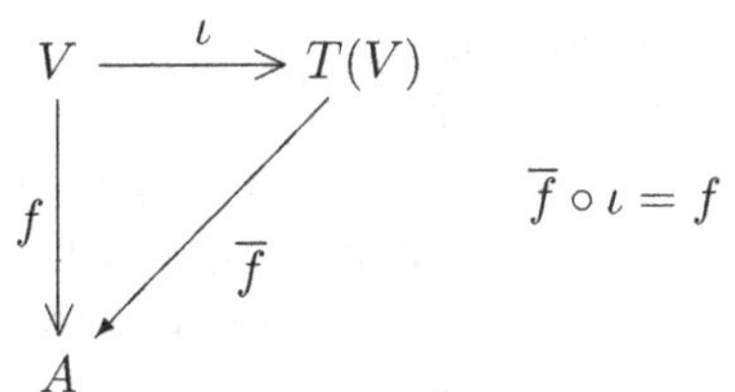

Proof. For basis elements in $\mathcal{B}$, by the linearity of f, the K-algebra homomorphism $\overline{f}$ is defined as

$$\overline{f}(1) = 1_A, \quad \overline{f}(X_{\ell_1} \otimes X_{\ell_2} \otimes \cdots \otimes X_{\ell_i}) = f(X_{\ell_1})f(X_{\ell_2}) \cdots f(X_{\ell_i}).$$

The uniqueness of $\overline{f}$ is clear by looking at the action on V or equivalently on $\mathcal{B}$. □

In fact, one may define the tensor algebra $T(V)$ as the unique K-algebra satisfying the universal property mentioned above, in particular, it is unique up to a unique isomorphism. Due to its construction and the universal property, $T(V)$ is also called the *free algebra* on the vector space V.

Another way of looking at the tensor algebra or free algebra is as the *algebra of polynomials over K in noncommuting variables*, that is, the basis vectors $X_\ell \in X$ for V may be viewed as noncommuting variables (or indeterminants) in $T(V)$, subject to no constraints (beyond associativity, the distributive law and K-linearity). For the convenience of later use, we give below the clear symbolic construction of this structure.

Let $X = \{X_i\}_{i\in J}$ be a set of symbols and $\mathcal{B} = \langle X\rangle$ the (multiplicative) free monoid on X. Then $\mathcal{B}$ consists of all words in the alphabet X plus the empty word $\emptyset$, i.e.,

$$\mathcal{B} = \{X_{i_1}X_{i_2}\cdots X_{i_p} \mid X_{i_j} \in X,\ p \geq 1\} \cup \{\emptyset\},$$

and the multiplication on $\mathcal{B}$ is defined as the *concatenation* of words, that is, if $u = X_{i_1}\cdots X_{i_p}$ and $w = X_{i_{p+1}}\cdots X_{i_n}$ then

$$(*) \qquad \begin{aligned} \emptyset u &= u\emptyset = u, \text{ and} \\ uw &= X_{i_1}\cdots X_{i_p}X_{i_{p+1}}\cdots X_{i_n}. \end{aligned}$$

Consider the K-vector space $K\langle X\rangle$ with basis the set $\mathcal{B}$, and write 1 for $\emptyset$. Then the formula $(*)$ above equips $K\langle X\rangle$ with a K-algebra structure, where the ring homomorphism $\eta_{K\langle X\rangle}$ is given by the map $K \to K1$. If $X = \{X_1, \ldots, X_n\}$ is finite, then $K\langle X\rangle$ is also written as $k\langle X_1, \ldots, X_n\rangle$.

As with the tensor algebra $T(V)$, a similar argumentation shows that the K-algebra $K\langle X\rangle$ has the universal property: Given a k-algebra A and a set-theoretic map φ: $X \to A$, there exists a unique K-algebra homomorphism $\overline{\varphi}$: $K\langle X\rangle \to A$ such that $\overline{\varphi}(X_i) = \varphi(X_i)$ for all $X_i \in X$, making the following commutative diagram:

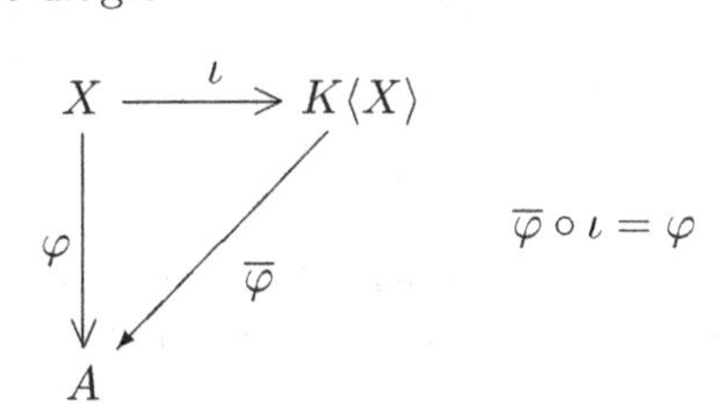

where ι is the inclusion map.

If we write V for the vector space spanned by X over K, then it follows from the universal property that $T(V) \cong K\langle X\rangle$ as K-algebras.

In the case that $X = \{X_1, \ldots, X_n\}$ is a finite set, the free K-algebra $K\langle X\rangle = K\langle X_1, \ldots, X_n\rangle$ is also called a noncommutative polynomial K-algebra in variables $X_1, \ldots, X_n$.

The defining relations of an algebra

Let A be a K-algebra and $T = \{a_i\}_{i\in J} \subset A$ a nonempty subset of A. Write $K[T]$ for the subalgebra of A generated by T over K, i.e.,

$$k[T] = \left\{ \sum_i \lambda_i a_{i_1} a_{i_2} \cdots a_{i_s} \;\middle|\; \lambda_i \in K,\ a_{i_j} \in T,\ s \in \mathbb{N} \right\}.$$

If $K[T] = A$, we say that A *is generated by* T over K and that T is a *generating set* of A (or a *set of generators* of A). If $T = \{a_1, \ldots, a_n\}$ is finite and $A = K[T]$, we say that A is a *finitely generated* K-algebra and write $A = K[a_1, \ldots, a_n]$.

Since every K-algebra A has a generating set T (for instance $T = A$), say $T = \{a_i\}_{i\in J}$, if we take the set of symbols $X = \{X_i\}_{i\in J}$ and consider the free algebra $K\langle X\rangle$ generated by X over K, then, under the set-theoretic map φ: $X \to T$ with $\varphi(X_i) = a_i$, the universal property of $K\langle X\rangle$ yields a unique K-algebra homomorphism $\overline{\varphi}$: $K\langle X\rangle \to A = K[T]$, which is clearly onto, such that

$$(**) \qquad A \cong K\langle X\rangle / I$$

where $I = \mathrm{Ker}\overline{\varphi}$. Thus, we have the following fact.

Proposition 1.2. *Any K-algebra $A = K[T]$ with $T = \{a_i\}_{i\in J}$ is a quotient of the free K-algebra $K\langle X\rangle$ with $X = \{X_i\}_{i\in J}$.* □

Definition 1.3. In the isomorphism $(**)$ above, if the ideal I of $K\langle X\rangle$ is generated by the subset

$$F = \left\{ F_p = \sum_i \lambda_i X_{i1}^{\alpha_1} X_{i2}^{\alpha_2} \cdots X_{is}^{\alpha_s} \;\middle|\; \lambda_i \in K^*,\ X_{ij} \in X,\ \alpha_j \in \mathbb{N},\ s \geq 1 \right\},$$

we say that F is a set of *defining relations* of the K-algebra A. In this case we also say that the k-algebra $A = K[T]$ is generated by $T = \{a_i\}_{i\in J}$ subject to the relations:

$$\widetilde{F_p} = \sum_i \lambda_i a_{i1}^{\alpha_1} a_{i2}^{\alpha_2} \cdots a_{is}^{\alpha_s} = 0,\ F_j \in F.$$

Generally in practice, many important algebras are realized as quotient algebras of free algebras with clear defining relations; yet there are significant algebras appearing as subalgebras generated by chosen elements of interest from known algebras. For such algebras, the next proposition may help to understand and find their defining relations.

Proposition 1.4. *Let $K\langle X\rangle$ be a free K-algebra with $X = \{X_i\}_{i\in J}$ and I an ideal of $K\langle X\rangle$ generated by the set*

$$F = \left\{ F_p = \sum_i \lambda_i X_{i1}^{\alpha_1} X_{i2}^{\alpha_2} \cdots X_{is}^{\alpha_s} \;\middle|\; \lambda_i \in K^*,\ X_{ij} \in X,\ \alpha_j \in \mathbb{N},\ s \geq 1 \right\}.$$

Suppose that A is a K-algebra and there is a set-theoretic map $\varphi\colon X \to A$ such that

$$\widehat{F_p} = \sum_i \lambda_i \varphi(X_{i1})^{\alpha_1} \varphi(X_{i2})^{\alpha_2} \cdots \varphi(X_{is})^{\alpha_s} = 0, \quad F_p \in F,$$

then there is a unique K-algebra homomorphism $\rho\colon K\langle X\rangle/I \to A$ determining the following commutative diagram:

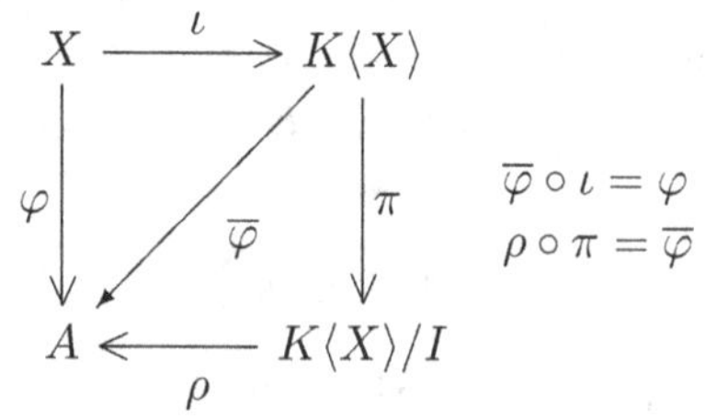

where ι is the inclusion map, π is the canonical epimorphism, and ρ is defined by $\rho(\overline{X}_i) = \varphi(X_i)$. Moreover, if $\overline{\varphi}$ is surjective, then ρ is surjective too.

Proof. This is an easy exercise by applying the universal property of $K\langle X\rangle$ and the fundamental theorem of homomorphism. □

Now, let us review some well-known algebras which are used extensively in scientific studies. The reader is referred to the wonderful online Free Encyclopedia at `http://en.wikipedia.org/wiki/` to find some historical introduction to each algebra we listed below.

The symmetric algebra

This is the K-algebra $S = K[T]$ generated by $T = \{x_i\}_{i\in J}$ subject to the relations

$$x_j x_i = x_i x_j, \quad j,\ i \in J,\ j \neq i,$$

that is, $S \cong K\langle X\rangle/I$, where $K\langle X\rangle$ is the free K-algebra generated by $X = \{X_i\}_{i\in J}$, and I is the ideal of $K\langle X\rangle$ generated by the elements

$$X_j X_i - X_i X_j,\ j, i \in J,\ j \neq i.$$

Clearly, S is a commutative K-algebra. By Proposition 1.4, one can show easily that S has the following universal property:

- Given a commutative algebra B and a set-theoretic map φ: $T \to B$, there exists a unique K-algebra homomorphism $\overline{\varphi}$: $S \to B$ such that $\overline{\varphi}(x_i) = \varphi(x_i)$ for all $x_i \in T$, i.e., we have the following commutative diagram:

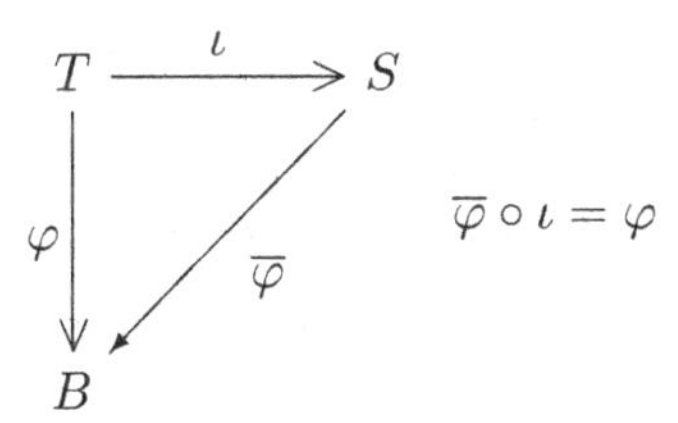

where ι is the inclusion map. Hence, every commutative algebra is a quotient algebra of some symmetric algebra.

The commutative polynomial algebra

Let $R = k[x_1, \ldots, x_n]$ be the commutative polynomial K-algebra in n variables $x_1, \ldots, x_n$, which is also known as the *coordinate ring* of the n-dimensional affine K-space $\mathbb{A}^n(K) = K^n$ in algebraic geometry. Then, by the classical construction of polynomials in commuting variables, R has the standard K-basis consisting of all monomials

$$x_1^{\alpha_1}x_2^{\alpha_2}\cdots x_n^{\alpha_n}, \quad \alpha_i \in \mathbb{N}.$$

Consider the free K-algebra $K\langle X\rangle = K\langle X_1,\dots,X_n\rangle$ generated by $X = \{X_1,\dots,X_n\}$, and let I be the ideal of $K\langle X\rangle$ generated by the elements

$$X_jX_i - X_iX_j, \quad 1 \le i < j \le n.$$

It is easy to see that every element of $K\langle X\rangle/I$ is of the form

$$\sum_\alpha \lambda_\alpha \overline{X_1}^{\alpha_1}\overline{X_2}^{\alpha_2}\cdots\overline{X_n}^{\alpha_n},\ \alpha = (\alpha_1,\dots,\alpha_n),\ \lambda_\alpha \in K^*.$$

where $\overline{X_i}$ is the canonical image of X_i in $K\langle X\rangle/I$, and the map ψ: $R \to K\langle X\rangle/I$ defined by

$$\psi\left(\sum_\alpha \lambda_\alpha x_1^{\alpha_1}x_2^{\alpha_2}\cdots x_n^{\alpha_n}\right) \mapsto \sum_\alpha \lambda_\alpha \overline{X_1}^{\alpha_1}\overline{X_2}^{\alpha_2}\cdots\overline{X_n}^{\alpha_n}$$

is a K-algebra epimorphism. Furthermore, a direct verification shows that ψ determines the inverse of ρ in Proposition 1.4. It follows that $R \cong K\langle X\rangle/I$, that is, the commutative polynomial K-algebra $R = K[x_1,\dots,x_n]$ is nothing but the symmetric K-algebra generated by $x_1,\dots,x_n$ subject to the relations

$$x_jx_i - x_ix_j = 0, \quad 1 \le i < j \le n.$$

Consequently, we recapture the classical result that any finitely generated commutative K-algebra is a quotient of some polynomial algebra $K[x_1,\dots,x_n]$.

The skew polynomial algebra $\mathcal{O}_n(\lambda_{ji})$

For $n \ge 2$, let $\mathcal{O}_n(\lambda_{ji})$ be the K-algebra generated by $x_1,\dots,x_n$ subject to the relations

$$x_jx_i = \lambda_{ji}x_ix_j \text{ with } \lambda_{ji} \in K^*, \qquad 1 \le i < j \le n,$$

that is, $\mathcal{O}_n(\lambda_{ji}) \cong K\langle X\rangle/I$, where $K\langle X\rangle$ is the free K-algebra generated by $X = \{X_1,\dots,X_n\}$, and I is the ideal of $K\langle X\rangle$ generated by the elements

$$X_jX_i - \lambda_{ji}X_iX_j, \quad \lambda_{ji} \in K^*,\ 1 \le i < j \le n.$$

It follows from ([MR], Ch.1) that $\mathcal{O}_n(\lambda_{ji})$ is isomorphic to the iterated skew polynomial algebra $K[z_1;\sigma_1][z_2;\sigma_2]\cdots[z_n;\sigma_n]$ in which σ_1 is the identity automorphism of K, and for $j \geq 2$, the iterated algebra automorphisms σ_j are defined such that $z_j z_i = \sigma_j(z_i) z_j = \lambda_{ji} z_i z_j$, $1 \leq i < j \leq n$. Hence $\mathcal{O}_n(\lambda_{ji})$ is a Noetherian domain with the K-basis

$$\mathcal{B} = \{x_1^{\alpha_1} \cdots x_n^{\alpha_n} \mid \alpha_1, \ldots, \alpha_n \in \mathbb{N}\}.$$

In (Ch.3, Section 5, and Ch.4, Section 4) we will see that these structural properties of $\mathcal{O}_n(\lambda_{ji})$ may be recaptured in a computational way.

Remark (i) Historically in the literature ([Jat], [MP]) the algebra of type $\mathcal{O}_n(\lambda_{ji})$ was referred to as the *multiplicative analogue of the n-th Weyl algebra* ; and if $\lambda_{ji} = q^{-2}$ with $q \in K^*$, $1 \leq i < j \leq n$, then $\mathcal{O}_n(\lambda_{ji})$ was called the *coordinate ring of the n-dimensional quantum affine K-space* [Sm1]. Taking the skew polynomial ring structure into account, we used $\mathcal{O}_n(\lambda_{ji})$ to indicate that this algebra is just a "twisted version" (by the parameters λ_{ji}) of the coordinate ring (i.e., the commutative polynomial ring $K[x_1, \ldots, x_n]$) of the classical affine n-space $\mathbb{A}^n = K^n$.
(ii) In modern computational algebra, $\mathcal{O}_n(\lambda_{ji})$ has been studied as a typical example of noncommutative algebra having a Gröbner basis theory similar to that for a commutative polynomial algebra (e.g., see [K-RW], [G-I1], [Lev], and [Li1]).

Clifford algebra and the exterior algebra

Let $\mathsf{C} = K[T]$ be the K-algbra generated by $T = \{x_i\}_{i \in J}$ subject to the relations

$$\begin{array}{ll} x_i^2 = q_i, & i \in J, \\ x_j x_i + x_i x_j = q_{ji}, & i, j \in J, \; i \neq j, \end{array} \quad q_i, q_{ji} \in K,$$

that is, $\mathsf{C} \cong K\langle X\rangle/I$, where $K\langle X\rangle$ is the free K-algebra generated by $X = \{X_i\}_{i \in J}$, and I is the ideal of $K\langle X\rangle$ generated by the elements

$$\begin{array}{ll} X_i^2 - q_i, & i \in J, \\ X_j X_i + X_i X_j - q_{ji}, & i, j \in J, \; i \neq j. \end{array}$$

Then, C is called a *Clifford algebra* over the field K.

If in the above definition $q_i = q_{ji} = 0$ for all $i, j \in J$, then the obtained algebra is called an *exterior algebra* (or *Grassmann algebra*) over the field K, denoted E, that is, $\mathsf{E} \cong K\langle X\rangle/H$, where H is the ideal of $K\langle X\rangle$ generated by the elements

$$\begin{array}{ll} X_i^2, & i \in J, \\ X_j X_i + X_i X_j, & i, j \in J, \; i \neq j. \end{array}$$

It is well known (or see Example 7 of Ch.3, Section 5) that if $<$ is a well-ordering on J, then both the algebras C and E have the K-basis

$$\mathcal{B}=\left\{1,\ x_{i_1}x_{i_2}\cdots x_{i_s}\ \middle|\ i_1<i_2<\cdots<i_s,\ i_\ell\in J\right\}.$$

In the case that $T=\{x_1,\ldots,x_n\}$, it is a good exercise to show $\mathsf{E}\cong\mathcal{O}_n(\lambda_{ji})/L$, where $\mathcal{O}_n(\lambda_{ji})$ is the skew polynomial algebra defined above, and L is the ideal of $\mathcal{O}_n(\lambda_{ji})$ generated by $\{x_i^2\mid 1\le i\le n\}$ with the $\lambda_{ji}=-1$. (Hint: use the K-basis $\mathcal{B}$ above and Proposition 1.4.)

The enveloping algebra of a finite dimensional Lie algebra

Let g be an n-dimensional vector space over the field K with basis $\{x_1,\ldots,x_n\}$. If there is a binary operation on g, called the bracket product and denoted $[\ ,\ ]$, which is bilinear, i.e., for $a,b,c\in\mathsf{g}$, $\lambda\in K$, the following relations hold:

$$\begin{aligned}[][a+b,c]&=[a,c]+[b,c],\\ [a,b+c]&=[a,b]+[a,c],\\ \lambda[a,b]&=[\lambda a,b]=[a,\lambda b],\end{aligned}$$

and besides, all a, b, $c\in\mathsf{g}$ satisfy:

$$\begin{aligned}&[a,b]=-[b,a],\ [a,a]=0,\\ &[[a,b],c]+[[c,a],b]+[[b,c],a]=0\quad\text{(Jacobi identity)},\end{aligned}$$

then g is called an n-dimensional Lie algebra over K. Note that $[\ ,\ ]$ need not satify the associative law. If $[a,b]=[b,a]$ for every a,b in a Lie algebra g, then g is called an abelian K-Lie algebra.

Example 1. The 3-dimensional special linear K-Lie algebra $\mathsf{sl}(2,K)$, which is a subalgebra of the general linear Lie algebra $\mathsf{gl}(K^2)=\mathrm{End}_K K^2$ with the bracket product given by $[A,B]=AB-BA$. $\mathsf{sl}(2,K)$ has the standard K-basis

$$x=\begin{pmatrix}0&1\\0&0\end{pmatrix},\quad y=\begin{pmatrix}0&0\\1&0\end{pmatrix},\quad z=\begin{pmatrix}1&0\\0&-1\end{pmatrix},$$

satisfying

$$[z,x]=2x,\quad [z,y]=-2y,\quad [x,y]=z.$$

Example 2. The $2n+1$-dimensional Heisenberg K-Lie algebra $\mathsf{h}(n,K)$, which, in connection with Heisenberg groups originally arising in the description of 1-dimensional quantum mechanical systems, has the K-basis

$\{x_i, y_j, z \mid i, j = 1, \ldots, n\}$ and the bracket product is given by

$$\begin{aligned} &[x_i, y_i] = z, && 1 \le i \le n, \\ &[x_i, x_j] = [x_i, y_j] = [y_i, y_j] = 0, && i \ne j, \\ &[z, x_i] = [z, y_i] = 0, && 1 \le i \le n. \end{aligned}$$

Let $\mathbf{g}$ be a Lie algebra with K-basis $\{x_1, \ldots, x_n\}$. Suppose in $\mathbf{g}$ the $[x_j, x_i]$, $1 \le i < j \le n$, is expressed as

$$[x_j, x_i] = \sum_{\ell=1}^{n} \lambda_{ji}^{\ell} x_\ell, \quad \lambda_{ji}^{\ell} \in K.$$

Then the *enveloping algebra* of $\mathbf{g}$, denoted $U(\mathbf{g})$, is defined to be the associative K-algebra generated by $x_1, \ldots, x_n$ subject to the relations

$$x_j x_i - x_i x_j = \sum_{\ell=1}^{n} \lambda_{ji}^{\ell} x_\ell, \quad 1 \le i < j \le n,$$

that is, $U(\mathbf{g}) \cong K\langle X\rangle/I$, where $K\langle X\rangle = K\langle X_1, \ldots, X_n\rangle$ is the free K-algebra generated by $X = \{X_1, \ldots, X_n\}$ and I is the ideal of $K\langle X\rangle$ generated by the elements

$$X_j X_i - X_i X_j - \sum_{\ell=1}^{n} \lambda_{ji}^{\ell} x_\ell, \quad 1 \le i < j \le n.$$

By the celebrated PBW (Poincaré-Birkhoff-Witt) theorem (cf. [Hu]; see also Ch.4, Section 3), $U(\mathbf{g})$ has the K-basis

$$\mathcal{B} = \left\{ x_1^{\alpha_1} \cdots x_n^{\alpha_n} \;\middle|\; \alpha_i \in \mathbb{N} \right\}.$$

The n-th Weyl algebra

This is the well-known algebra $A_n(K)$ that realizes the first quantum algebra stemming from the study of quantum mechanics in 1925. More precisely, the K-algebra $A_n(K)$ is generated by $2n$ generators $x_1, \ldots, x_n, y_1, \ldots, y_n$ subject to the relations

$$\begin{aligned} &x_i x_j = x_j x_i,\ y_i y_j = y_i y_j, && 1 \le i < j \le n, \\ &y_j x_i - x_i y_j = \delta_{ij} \text{ (the Kronecker delta)}, && 1 \le i, j \le n, \end{aligned}$$

that is, $A_n(K) \cong K\langle X, Y\rangle/I$, where $K\langle X, Y\rangle$ is the free algebra generated by $X \cup Y = \{X_1, \ldots, X_n, Y_1, \ldots, Y_n\}$ over K, and I is the ideal of $K\langle X, Y\rangle$ generated by the elements

$$\begin{aligned} &X_i X_j - X_j X_i,\ Y_i Y_j - Y_j Y_i, && 1 \le i < j \le n, \\ &Y_j X_i - X_i Y_j - \delta_{ij} \text{ (the Kronecker delta)}, && 1 \le i, j \le n. \end{aligned}$$

The algebra of partial differential operators

This is the subalgebra $\Delta(K[x_1,\ldots,x_n])$ of the K-algebra $\mathrm{End}_K K[x_1,\ldots,x_n]$ generated by the $2n$ K-linear operators

$$x_1, x_2, \ldots, x_n, \partial_1 = \frac{\partial}{\partial x_1}, \partial_2 = \frac{\partial}{\partial x_2}, \ldots, \partial_n = \frac{\partial}{\partial x_n},$$

where $K[x_1,\ldots,x_n]$ is the commutative polynomial K-algebra in variables $x_1,\ldots,x_n$, each x_i acts on $k[x_1,\ldots,x_n]$ by multiplication on the left-hand side, and each $\frac{\partial}{\partial x_i}$ is the partial differential operator with respect to the i-th coordinate. With respect to the multiplication of operators in $\mathrm{End}_K K[x_1,\ldots,x_n]$, these generators satisfy

$$\begin{aligned} &x_ix_j = x_jx_i,\ \partial_i\partial_j = \partial_j\partial_i, && 1 \le i < j \le n,\\ &\partial_j x_i = x_i\partial_j + \delta_{ij} \text{ (the Kronecker delta)}, && 1 \le i, j \le n.\end{aligned}$$

Thus, every operator $D \in \Delta(K[x_1,\ldots,x_n])$ may be expressed as

$$D = \sum_{\alpha,\beta} \lambda_{\alpha\beta} x^\alpha \partial^\beta,$$

where $\lambda_{\alpha\beta} \in K^*$, x^α stands for $x_1^{\alpha_1}x_2^{\alpha_2}\cdots x_n^{\alpha_n}$ with $\alpha = (\alpha_1,\alpha_2,\ldots,\alpha_n) \in \mathbb{N}^n$, and ∂^β stands for $\partial_1^{\beta_1}\partial_2^{\beta_2}\cdots\partial_n^{\beta_n}$ with $\beta = (\beta_1,\beta_2,\ldots,\beta_n) \in \mathbb{N}^n$. For this reason, elements in $\Delta(K[x_1,\ldots,x_n])$ are called *partial differential operators with polynomial coefficients.*

If $\mathrm{Char}K = 0$, then it is a well-known result that $\Delta(K[x_1,\ldots,x_n])$ is isomorphic to the n-th Weyl algebra constructed before, that is,

$$\Delta(K[x_1,\ldots,x_n]) \cong A_n(K).$$

As a typical example illustrating the principle of Proposition 1.4, we include a proof of this assertion here.

Proof. [Proof of the isomorphism above] We first prove the following conclusion.

- If $\mathrm{Char}K = 0$, then $\mathscr{B} = \{x^\alpha\partial^\beta \mid \alpha,\beta \in \mathbb{N}^n\}$ is a K-basis for the algebra $\Delta(K[x_1,\ldots,x_n])$.

Suppose $\sum \lambda_{\alpha\beta}x^\alpha\partial^\beta = 0$, the zero operator. We want to show that all the $\lambda_{\alpha\beta}$ are 0 in K. To this end, after rewriting $\sum \lambda_{\alpha\beta}x^\alpha\partial^\beta$ in the power of ∂, we have

$$0 = \sum \lambda_{\alpha\beta}x^\alpha\partial^\beta = \sum_{i=1}^{t} f_{\gamma(i)}\partial^{\gamma(i)},$$

where $\gamma(i) = (\gamma_{1i}, \gamma_{2i}, \ldots, \gamma_{ni}) \in \mathbb{N}^n$ and $f_{\gamma(i)} = \sum_\delta \lambda_{\delta\gamma(i)} x^\delta$ in which all x^δ are distinct. Recall that the *lexicographic ordering* for $\alpha = (\alpha_1, \alpha_2, \ldots, \alpha_n)$, $\beta = (\beta_1, \beta_2, \ldots, \beta_n) \in \mathbb{N}^n$ is defined as

$$\alpha \prec \beta \Leftrightarrow \alpha_1 = \beta_1, \alpha_2 = \beta_2, \ldots, \alpha_{i-1} = \beta_{i-1} \text{ and } \alpha_i < \beta_i \text{ for some } i.$$

(In next chapter, this ordering will be discussed in more detail.) Since $\prec$ is a well-ordering, we may assume $\gamma(1) \prec \gamma(2) \prec \cdots \prec \gamma(t)$. Taking the action of $\sum \lambda_{\alpha\beta} x^\alpha \partial^\beta$ on the monomial $x^{\gamma(1)} = x_1^{\gamma_{11}} x_2^{\gamma_{21}} \cdots x_n^{\gamma_{n1}} \in K[x_1, \ldots, x_n]$ into account, it turns out that

$$0 = \left(\sum_{i=1}^{t} f_{\gamma(i)} \partial^{\gamma(i)} \right) (x^{\gamma(1)}) = \gamma_{11}! \gamma_{21}! \cdots \gamma_{n1}! f_{\gamma(1)}.$$

As $\mathrm{Char}K = 0$, we have $f_{\gamma(1)} = \sum_\delta \lambda_{\delta\gamma(1)} x^\delta = 0$. Similarly, $f_{\gamma(2)} = f_{\gamma(3)} = \cdots = f_{\gamma(t)} = 0$. Thus, in view of the rule for rewriting $\sum \lambda_{\alpha\beta} x^\alpha \partial^\beta$, all coefficients $\lambda_{\alpha\beta}$ must be equal to 0. This establishes the conclusion ($\bullet$).

Now, note that every element of $A_n(K) = K\langle X, Y\rangle / I$ is of the form $\sum \lambda_{\alpha\beta} \overline{X}^\alpha \overline{Y}^\beta$, where $\overline{X}^\alpha \overline{Y}^\beta$ denotes the image of $X_1^{\alpha_1} \cdots X_n^{\alpha_n} Y_1^{\beta_1} \cdots Y_n^{\beta_n}$ in $A_n(K)$. With the K-basis $\mathscr{B}$ of $\Delta(K[x_1, \ldots, x_n])$ in hand, it is easy to see that the map

$$\begin{array}{rccc} \psi : \Delta(K[x_1, \ldots, x_n]) & \longrightarrow & K\langle X,\ Y\rangle / I \cong A_n(K) \\ \sum \lambda_{\alpha\beta} x^\alpha \partial^\beta & \mapsto & \sum \lambda_{\alpha\beta} \overline{X}^\alpha \overline{Y}^\beta \end{array}$$

is a K-algebra epimorphism and thereby determines the inverse of ρ in Proposition 1.4, as desired. □

As shown by the previous isomorphism $\mathsf{E} \cong \mathcal{O}_n(-1)/L$ where L is the ideal generated by $\{x_1^2, \ldots, x_n^2\}$, the isomorphism $\Delta(K[x_1, \ldots, x_n]) \cong A_n(K)$ tells us, again, that two algebras coming from quite different scientific research realms may have exactly the same pure algebraic structure. In the remark below we will indicate further that $A_n(K)$ can be realized by a quotient algebra of the enveloping algebra of the $2n+1$-dimensional Heisenberg Lie algebra.

Remark (i) By the above argument we know that the n-th Weyl algebra $A_n(K) \cong K\langle X, Y\rangle / I$ has the K-basis

$$\left\{ \overline{X_1}^{\alpha_1} \cdots \overline{X_n}^{\alpha_n} \overline{Y_1}^{\beta_1} \cdots \overline{Y_n}^{\beta_n} \;\middle|\; \alpha_i, \beta_j \in \mathbb{N} \right\}.$$

Let $K\langle X_1,\dots,X_n,Y_1,\dots,Y_n,Z\rangle$ be the free K-algebra generated by $\{X_1,\dots, X_n,Y_1,\dots,Y_n,Z\}$. It follows from Proposition 1.4 that the map

$$\varphi:\quad \{X_1,\dots,X_n,\ Y_1,\dots,Y_n,\ Z\}\longrightarrow A_n(K)$$

defined by $\varphi(X_i)=\overline{X_i}$, $\varphi(Y_i)=\overline{Y_i}$, and $\varphi(Z)=\varphi(1)=1$ induces the isomorphism

$$\rho:\quad K\langle X_1,\dots,X_n,Y_1,\dots,Y_n,Z\rangle/J\xrightarrow{\cong}A_n(K),$$

where the ideal J is generated by the elements

$$\begin{array}{ll} ZX_i-X_iZ,\ ZY_i-Y_iZ, & 1\le i\le n,\\ X_iY_i-Y_iX_i-Z, & 1\le i\le n,\\ X_iY_j-Y_jX_i, & 1\le i\ne j\le n,\\ Z-1, & \end{array}$$

that is, $A_n(K)\cong U(\mathsf{h})/\langle z-1\rangle$, where $U(\mathsf{h})$ is the enveloping algebra of the $2n+1$-dimensional Heisenberg Lie algebra h.

(ii) If $\mathrm{Char}K=p$ for some prime number p, then since $\partial_i^p(x_i^p)=p!=0$, we see that each ∂_i^p acts as the zero operator for $i=1,2,\dots,n$. Noticing this fact, a similar argument as in the case where $\mathrm{Char}K=0$ proves that

$$\mathscr{B}=\left\{x^\alpha\partial^\beta\ \middle|\ \alpha=(\alpha_1,\dots,\alpha_n),\beta=(\beta_1,\dots,\beta_n)\in\mathbb{N}^n,\ \text{with } \beta_i\le p-1\right\}$$

is a K-basis for $\Delta(K[x_1,\dots,x_n])$. But we cannot have a K-algebra isomorphism between $\Delta(K[x_1,\dots,x_n])$ and $K\langle X,\ Y\rangle/I$, for, the latter does not have divisors of zero and has the K-basis $\{\overline{X}^\alpha\overline{Y}^\beta\mid\alpha,\beta\in\mathbb{N}^n\}$ (detailed proof of such nontrivial properties of $K\langle X,\ Y\rangle/I$ may be found in e.g. [Bj], or in (Ch.4, Sections 3, 4).

As an interesting exercise, one may use Proposition 1.4 to prove that in the case where $\mathrm{Char}K=p$, $\Delta(K[x_1,\dots,x_n])\cong K\langle X,\ Y\rangle/J$, where the ideal J is generated by the elements

$$\begin{array}{ll} X_iX_j-X_jX_i,\ Y_iY_j-Y_iY_j, & 1\le i<j\le n,\\ Y_jX_i-X_iY_j-\delta_{ij}, & 1\le i,j\le n,\\ Y_i^p & 1\le i\le n, \end{array}$$

that is, $\Delta(K[x_1,\dots,x_n])\cong A_n(K)/\langle y_1^p,y_2^p,\dots,y_n^p\rangle$.

The path algebra

This kind of algebra is also called *quiver algebra* in the representation theory of K-algebras.

Let Q be a finite directed graph with the set of vertices $Q_0 = \{e_1, \ldots, e_s\}$ and the set of arrows $Q_1 = \{a_1, \ldots, a_m\}$. Put

$$\mathcal{B} = \{\text{finite directed paths in } Q, \text{ including the vertices}\},$$

where every vertex $e_i \in Q_0$ is viewed as a path of length 0. The *path algebra* defined by Q over K, denoted $R = KQ$, has the K-vector space with basis $\mathcal{B}$ and the multiplication subject to the rule: for $u, v \in \mathcal{B}$,

$$u \cdot v = \begin{cases} 0, & \text{if terminus of } u \neq \text{ origin of } v, \\ uv \in \mathcal{B}, & \text{if terminus of } u = \text{ origin of } v, \end{cases}$$

that is, multiplication of paths in R is given by concatenation of paths (when this makes sense). It is easy to see that R has identity element $1 = \sum_{e_i \in Q_0} e_i$.

Let $K\langle X,\ Y\rangle$ be the free K-algebra generated by $X = \{X_1, \ldots, X_s\}$ and $Y = \{Y_1, \ldots, Y_m\}$ over K. Then the map

$$\begin{aligned} X_i &\mapsto e_i,\ i = 1, \ldots, s, \\ Y_j &\mapsto a_j,\ j = 1, \ldots, m, \end{aligned}$$

induces a K-algebra epimorphism $K\langle X,\ Y\rangle \to KQ$, that is, KQ is a homomorphic image of the free K-algebra $K\langle X,\ Y\rangle$.

On the other hand, if Q has only one vertex v and n loops, then one checks that KQ is isomorphic to the free K-algebra $K\langle X_1, \ldots, X_n\rangle$.

The reader is referred to the lecture notes of E. L. Green entitled *Noncommutative Gröbner Bases, A Computational and Theoretical Tool* at `http://www.math.unl.edu/~shermiller2/hs/green2.ps` for a nice introduction to "Why path algebras?".

Remark If K is a *commutative ring*, then a K-algebra may be defined in a similar way by using module structure in place of vector space, and the construction of a tensor (free) algebra over a field generalizes in straightforward manner to the tensor (free) algebra of any K-module.

1.2 S-Graded Algebras and Modules

Throughout this section, K denotes a field and S denotes a semigroup.

S-*graded algebra*

Let A be a K-algebra. If there is a family $\{A_g\}_{g\in S}$ consisting of K-subspaces A_g of A such that

(Gr1) $A = \oplus_{g\in S} A_g$, and
(Gr2) $A_gA_h \subseteq A_{gh}$, $\quad g,h\in G$,

then A is said to be an S-*graded K-algebra* with the S-*gradation* $\{A_g\}_{g\in S}$, in which A_g is called the *degree-g homogeneous part* of A, and an element $a\in A_g$ is called an S-*homogeneous element of degree g* (we use the phrase "S-homogeneous element" just for emphasizing the S-gradation, for, in the subsequent chapters a K-algebra will be equipped with different gradations simultaneously). If $a\in A_g$, then we write $d(a)=g$ for the degree of a.

In the case that $S=\mathbb{Z}$ and $A=\oplus_{n\in\mathbb{Z}}A_n$ is a $\mathbb{Z}$-graded K-algebra, if $A_n = 0$ for all $n<0$, i.e., $A=\oplus_{n\in\mathbb{N}}A_n$, then we say that A is $\mathbb{N}$-*graded* or *positively graded.*

Let B be a subalgebra of an S-graded K-algebra $A=\oplus_{g\in S}A_g$. If $B=\oplus_{g\in S}B_g$ is S-graded and $B_g\subseteq A_g$ for all $g\in S$, then B is called an S-*graded subalgebra* of A.

Clearly, if B is an S-graded subalgebra of the S-graded K-algebra $A=\oplus_{g\in S}A_g$, then $B_g = B\cap A_g$ for all $g\in S$, that is, $B=\oplus_{g\in S}(B\cap A_g)$.

Weight $\mathbb{N}$-*gradation*

Let $S=\mathbb{N}$ and let $A=K[T]$ be a K-algebra generated by $T=\{a_i\}_{i\in J}$. Then each element $a\in A$ can be written as a finite sum of the form

$$a=\sum_i \lambda_i a_{i_1}^{\alpha_1}\cdots a_{i_s}^{\alpha_s},$$

where $\lambda_i\in K$, $a_{ij}\in T$, $\alpha_j, s\in\mathbb{N}$, $s\ge 1$. By an abuse of language, an element of the form $a_{i_1}^{\alpha_1}\cdots a_{i_s}^{\alpha_s}\in A$ is called a *monomial* of A. Let $\mathcal{B}$ be a K-basis of A consisting of monomials. Unless otherwise stated, from now on, we use lower-case s, t, u, v, w, ... to denote monomials in $\mathcal{B}$.

If $w\in\mathcal{B}$, $w=a_{i_1}^{\alpha_1}\cdots a_{i_s}^{\alpha_s}$, then we write $l(w)$ for the *length* of w, that is,

$$l(w) = \alpha_1 + \cdots + \alpha_s,$$

denoted $l(w) = |\alpha|$. For instance, $l(a_i) = 1$, and if $a_i^p \neq 0$ then $l(a_i^p) = p$ for every $a_i \in T$ and $p \in \mathbb{N}$.

Now, for each $p \in \mathbb{N}$, let A_p denote the K-subspace of A spanned by all monomials of length p in B, that is

$$A_p = K\text{-span}\left\{w \in \mathcal{B} \;\middle|\; l(w) = p\right\}.$$

Clearly, $A = \oplus_{p\in\mathbb{N}} A_p$. If furthermore the condition

$$A_p A_q \subseteq A_{p+q}, \quad p, q \in \mathbb{N},$$

is satisfied, then A is an $\mathbb{N}$-graded K-algebra. In this case we call $\{A_p\}_{p\in\mathbb{N}}$ the *natural* $\mathbb{N}$*-gradation* of A defined by lengths of monomials in B.

Another way to construct an $\mathbb{N}$-gradation for A by using the fixed K-basis $\mathcal{B}$ is to assign to each a_i a *weight* $n_i \in \mathbb{N}$, $i \in J$, and for each $w = a_{i_1}^{\alpha_1} \cdots a_{i_s}^{\alpha_s}$ let

$$|w| = n_{i_1}\alpha_1 + \cdots + n_{i_s}\alpha_s.$$

Then $A = A = \oplus_{p\in\mathbb{N}} A_p$ with

$$A_p = K\text{-span}\left\{w \in \mathcal{B} \;\middle|\; |w| = p\right\}.$$

If, furthermore, the condition

$$A_p A_q \subseteq A_{p+q}, \quad p, q \in \mathbb{N},$$

is satisfied, then A becomes an $\mathbb{N}$-graded K-algebra. In this case we call $\{A_p\}_{p\in\mathbb{N}}$ a *weight* $\mathbb{N}$*-gradation* with the weight $\{n_i\}_{i\in J}$, and if furthermore $n_i > 0$, $i \in J$, then $\{n_i\}_{i\in J}$ is called a *positive weight* for the gradation $\{A_p\}_{p\in\mathbb{N}}$. Clearly, the natural $\mathbb{N}$-gradation is the weight $\mathbb{N}$-gradation with weight $\{n_i = 1\}_{i\in J}$.

Example 1. Let $K\langle X\rangle = K\langle X_1, \ldots, X_n\rangle$ be the free K-algebra generated by $X = \{X_1, \ldots, X_n\}$, and consider the standard K-basis $\mathcal{B} = \{w = X_{i_1} \cdots X_{i_p} \mid X_{i_j} \in X,\ p \geq 1\} \cup \{1\}$ of $K\langle X\rangle$. Put

$$K\langle X\rangle_p = K\text{-span}\left\{w \in \mathcal{B} \;\middle|\; l(w) = p\right\}, \quad p \in \mathbb{N}.$$

Then, by the multiplication of $K\langle X\rangle$, $K\langle X\rangle_p K\langle X\rangle_q = K\langle X\rangle_{p+q}$ holds for all $p \in \mathbb{N}$. Hence, $K\langle X\rangle$ has the natural $\mathbb{N}$-gradation $\{K\langle X\rangle_p\}_{p\in\mathbb{N}}$ such that $K\langle X\rangle = \oplus_{p\in\mathbb{N}} K\langle X\rangle_p$.

On the other hand, since $\mathcal{B}$ is a monoid under the multiplication of $K\langle X\rangle$, $K\langle X\rangle$ is a $\mathcal{B}$-graded K-algebra as well, that is, $K\langle X\rangle = \oplus_{u\in\mathcal{B}}K\langle X\rangle_u$ with $K\langle X\rangle_u = Ku$.

Example 2. Let $K[\mathbf{x}] = K[x_1, \ldots, x_n]$ be the commutative polynomial K-algebra in variables $x_1, \ldots, x_n$, and consider the standard K-basis $\mathcal{B} = \{x^\alpha = x_1^{\alpha_1}\cdots x_n^{\alpha_n} \mid \alpha = (\alpha_1, \ldots, \alpha_n) \in \mathbb{N}^n\}$ of $K[\mathbf{x}]$. Put

$$K[\mathbf{x}]_p = K\text{-span}\left\{x^\alpha \in \mathcal{B} \;\middle|\; l(x^\alpha) = p\right\} \quad p \in \mathbb{N}.$$

Then, the multiplication of $K[\mathbf{x}]$ guarantees that $K[\mathbf{x}]_pK[\mathbf{x}]_q = K[\mathbf{x}]_{p+q}$ for all $p, q \in \mathbb{N}$. Hence, $K[\mathbf{x}]$ has the natural $\mathbb{N}$-gradation $\{K[\mathbf{x}]_p\}_{p\in\mathbb{N}}$ such that $K[\mathbf{x}] = \oplus_{p\in\mathbb{N}}K[\mathbf{x}]_p$.

Moreover, since $\mathcal{B}$ is a monoid under the multiplication of $K[\mathbf{x}]$, $K[\mathbf{x}]$ is a $\mathcal{B}$-graded K-algebra as well, that is, $K[\mathbf{x}] = \oplus_{u\in\mathcal{B}}K[\mathbf{x}]_u$ with $K[\mathbf{x}]_u = Ku$.

Example 3. Let Q be a finite directed graph with the set of vertices $Q_0 = \{v_1, \ldots, v_s\}$ and the set of arrows $Q_1 = \{a_1, \ldots, a_m\}$, and let $R = KQ$ be the path algebra defined by Q over K. From Section 1 we know that R has the K-basis $\mathcal{B}$ consisting of finite directed paths of Q, including the vertices, which are viewed as paths of length 0. Viewing $Q_0 \cup Q_1$ as the set of generators of R, we may adopt our unified notation for elements presented by monomials in an algebra, that is, if $w \in \mathcal{B}$ is of length p, then we write $l(w) = p$. Bearing this convention in mind, we define

$$R_p = K\text{-span}\left\{w \in \mathcal{B} \;\middle|\; l(w) = p\right\}, \quad p \in \mathbb{N}.$$

Then, the multiplication of R guarantees that $R_pR_q \subseteq R_{p+q}$ for all $p, q \in \mathbb{N}$. Hence, R has the natural $\mathbb{N}$-gradation $\{R_p\}_{p\in\mathbb{N}}$ such that $R = \oplus_{p\in\mathbb{N}}R_p$.

Noticing the definition of the multiplication of a path algebra $R = KQ$, R is also graded by its standard K-basis $\mathcal{B}$, that is, $R = \oplus_{u\in\mathcal{B}}R_u$ with $R_u = Ku$.

S-*graded module*

In what follows, modules considered are *left* modules.

Let $A = \oplus_{g\in\mathsf{S}}A_g$ be an S-graded K-algebra. An A-module M is said to be an S-*graded A-module* if $M = \oplus_{g\in\mathsf{S}}M_g$, where the M_g are K-subspaces of the K-vector space M, such that

$$A_hM_g \subset M_{hg} \quad \text{for all } h, g \in \mathsf{S}.$$

For $g \in \mathsf{S}$, M_g is called the *degree-g homogeneous part* of M, and an element $x \in M_g$ is called an S-*homogeneous element of degree* g. If $x \in M_g$, then we write $d(x) = g$ for the degree of x.

Since any element m of an S-graded A-module M can be written uniquely as a sum of finitely many S-homogeneous elements:

$$m = m_{g_1} + m_{g_2} + \cdots + m_{g_s}, \quad m_{g_i} \in M_{g_i},$$

the following proposition may be verified directly.

Proposition 2.1. *Let $M = \oplus_{g \in \mathsf{S}} M_g$ be an S-graded A-module and N a submodule of M. Then the following statements are equivalent.*
(i) $N = \oplus_{g \in \mathsf{S}} (M_g \cap N)$,
(ii) *If $u \in N$ and $u = m_{g_1} + m_{g_2} + \cdots + m_{g_s}$ with $m_{g_i} \in M_{g_i}$, then $m_{g_i} \in N$.*
(iii) *As an A-module, N is generated by S-homogeneous elements of M.*
(iv) *The quotient module M/N is an S-graded A-module with the degree-g homogeneous part $(M/N)_g = (M_g + N)/N$ for each $g \in \mathsf{S}$, that is*

$$M/N = \sum_{g \in \mathsf{S}} (M_g + N)/N = \oplus_{g \in \mathsf{S}} (M_g + N)/N.$$

□

Let A be an S-graded K-algebra and M an S-graded A-module. A submodule N of M satisfying one of the equivalent conditions in Proposition 2.1 is called an S-*graded submodule* of M.

If submodules are replaced by left (two-sided) ideals of A in Proposition 2.1, we obtain the definition of an S-*graded left ideal* (S-*graded ideal*). Thus, If $A = \oplus_{g \in \mathsf{S}} A_g$ is an S-graded K-algebra and I is an S-graded ideal of A, then the quotient K-algebra

$$A/I = \bigoplus_{g \in \mathsf{S}} (A/I)_g \text{ with } (A/I)_g = (A_g + I)/I$$

is not only an S-graded A-module but also an S-graded K-algebra. For instance, by previous Example 1, the symmetric algebra, the exterior algebra and the multiplicative analogues of Weyl algebras constructed in Section 1 are $\mathbb{N}$-graded algebras.

S-*graded homomorphism*

Let $A = \oplus_{g\in\mathsf{S}} A_g$ and $B = \oplus_{g\in\mathsf{S}} B_g$ be S-graded K-algebras. A K-algebra homomorphism φ: $A \to B$ is called an S-*graded K-algebra homomorphism* if $\varphi(A_g) \subset B_g$ for all $g \in \mathsf{S}$.

Let $M = \oplus_{g\in\mathsf{S}} M_g$ and $N = \oplus_{g\in\mathsf{S}} N_g$ be S-graded modules over an S-graded K-algebra $A = \oplus_{g\in\mathsf{S}} A_g$. An A-module homomorphism ψ: $M \to N$ is called an S-*graded A-homomorphism* if $\psi(M_g) \subset N_g$ for all $g \in \mathsf{S}$.

By Proposition 2.1 and the definitions given above, the statements mentioned in the next proposition are easily verified.

Proposition 2.2. (i) *Let $A = \oplus_{g\in\mathsf{S}} A_g$ and $B = \oplus_{g\in\mathsf{S}} B_g$ be S-graded K-algebras. If φ: $A \to B$ is an S-graded K-algebra homomorphism, then $\mathrm{Im}\varphi$ is an S-graded subalgebra of B, and $\mathrm{Ker}\varphi$ is an S-graded ideal of A.*
(ii) *Let A be an S-graded K-algebra, and let $M = \oplus_{g\in\mathsf{S}} M_g$, $N = \oplus_{g\in\mathsf{S}} N_g$ be S-graded A-modules. If ψ: $M \to N$ is an S-graded A-homomorphism, then $\mathrm{Im}\psi$ is an S-graded submodule of N and $\mathrm{Ker}\psi$ is an S-graded submodule of M.*

□

Remark If we replace the field K by $\mathbb{Z}$, then the text of this section becomes that for Γ-graded rings and Γ-graded modules over the rings considered without any modification.

1.3 Γ-Filtered Algebras and Modules

In this section, K denotes a field, and Γ denotes a *totally ordered semigroup*, that is, Γ is a semigroup ordered by a total ordering $\prec$ which is compatible with the binary operation of Γ, i.e., for $\gamma_1, \gamma_2, \gamma \in \Gamma$,

$$\gamma_1 \prec \gamma_2 \text{ implies } \gamma\gamma_1 \prec \gamma\gamma_2 \text{ and } \gamma_1\gamma \prec \gamma_2\gamma.$$

Γ-*filtered Algebra*

A K-algebra A is said to be a Γ-*filtered algebra* if there is a family $FA = \{F_\gamma A\}_{\gamma\in\Gamma}$ consisting of K-subspaces $F_\gamma A$ of A, such that

(F1) $A = \cup_{\gamma\in\Gamma} F_\gamma A$,
(F2) $F_{\gamma_1} A \subseteq F_{\gamma_2} A$ if $\gamma_1 \preceq \gamma_2$,

(F3) $F_{\gamma_1}AF_{\gamma_2}A \subseteq F_{\gamma_1\gamma_2}A$, $\gamma_1, \gamma_2 \in \Gamma$.

If Γ has a smallest element γ_0 and 1 is the identity element of A, we also require that $1 \in F_{\gamma_0}A$.

In the case that $\Gamma = \mathbb{Z}$ and A is a $\mathbb{Z}$-filtered K-algebra, if $F_nA = \{0\}$ for all $n < 0$, then A becomes an $\mathbb{N}$-filtered K-algebra and is also called a *positively filtered* K-algebra.

In the definition given above, the family $FA = \{F_\gamma A\}_{\gamma\in\Gamma}$ of K-subspaces is called a Γ-*filtration* of A.

Let A be a Γ-filtered K-algebra with Γ-filtration $FA = \{F_\gamma A\}_{\gamma\in\Gamma}$. If S is a subalgebra of A and S has a Γ-filtration $FS = \{F_\gamma S\}_{\gamma\in\Gamma}$ satisfying $F_\gamma S \subseteq F_\gamma A$ for all $\gamma \in \Gamma$, then we call S a Γ-*filtered subalgebra* of A.

Indeed, any subalgebra S of A may be made into a Γ-filtered subalgebra by takeing the Γ-filtration $FS = \{F_\gamma S = S \cap F_\gamma A\}_{\gamma\in\Gamma}$ induced by FA.

Natural $\mathbb{N}$*-filtration*

Let $A = K[T]$ be a K-algebra with $T = \{a_i\}_{i\in J}$. As in Section 2, let Ω be the set of all monomials in A, that is,

$$\Omega = \left\{ w = a_{i_1}^{\alpha_1} \cdots a_{i_s}^{\alpha_s} \;\middle|\; a_{i_j} \in T,\ \alpha_i, s \in \mathbb{N},\ s \geq 1 \right\},$$

and for each $w = a_{i_1}^{\alpha_1} \cdots a_{i_s}^{\alpha_s}$, we write $l(w) = \alpha_1 + \cdots + \alpha_s$ for the *length* of w. Thus, for each $p \in \mathbb{N}$, if F_pA denote the K-subspace of A spanned by all monomials of length less than or equal to p, that is

$$F_pA = K\text{-span}\left\{ w \in \Omega \;\middle|\; l(w) \leq p \right\},$$

then it is easy to see that the family $FA = \{F_pA\}_{p\in\mathbb{N}}$ satisfies the foregoing conditions (F1)–(F3). This $\mathbb{N}$-filtration is called the *natural* $\mathbb{N}$*-filtration* of A.

Γ*-grading filtration*

Let Γ be a totally ordered semigroup and $R = \oplus_{\gamma\in\Gamma}R_\gamma$ a Γ-graded K-algebra. Put

$$F_\gamma R = \bigoplus_{\gamma' \preceq \gamma} R_{\gamma'}, \quad \gamma \in \Gamma.$$

Obviously, the family $FR = \{F_\gamma R\}_{\gamma\in\Gamma}$ satisfies the foregoing conditions (F1)–(F2). If furthermore FR satisfies the condition (F3), then the obtained Γ-filtration is called the Γ-*grading filtration* of R.

Example 1. Let $K\langle X\rangle$ be the free K-algebra generated by $X = \{X_1, \ldots, X_n\}$, and consider the standard K-basis $\mathcal{B} = \{w = X_{i_1}\cdots X_{i_p} \mid X_{i_j} \in X,\ p \geq 1\} \cup \{1\}$ of $K\langle X\rangle$. Then it is easy to see that the natural $\mathbb{N}$-filtration of $K\langle X\rangle$ defined by lengths of monomials (words) coincides with the $\mathbb{N}$-grading filtration determined by the natural $\mathbb{N}$-gradation of $K\langle X\rangle$.

Since $K\langle X\rangle$ is also $\mathcal{B}$-graded, in Ch.3 we will see that the $\mathcal{B}$-gradation of $K\langle X\rangle$ gives rise to the $\mathcal{B}$-grading filtration of $K\langle X\rangle$ once $\mathcal{B}$ is endowed with a monomial ordering.

Example 2. Let $K[\mathbf{x}] = K[x_1, \ldots, x_n]$ be the commutative polynomial K-algebra in variables $x_1, \ldots, x_n$, and consider the standard K-basis $\mathcal{B} = \{x^\alpha = x_1^{\alpha_1}\cdots x_n^{\alpha_n} \mid \alpha = (\alpha_1, \ldots, \alpha_n) \in \mathbb{N}^n\}$ of $K[\mathbf{x}]$. Then it is straightforward that the natural $\mathbb{N}$-filtration of $K[\mathbf{x}]$ defined by lengths of monomials coincides with the $\mathbb{N}$-grading filtration determined by the natural $\mathbb{N}$-gradation of $K[\mathbf{x}]$.

Since $K[\mathbf{x}]$ is also $\mathcal{B}$-graded, in Ch.3 we will see that the $\mathcal{B}$-gradation of $K[\mathbf{x}]$ yields the $\mathcal{B}$-grading filtration of $K[\mathbf{x}]$ once $\mathcal{B}$ is endowed with a monomial ordering.

Example 3. Let $R = KQ$ be the path algebra defined by a finite directed graph Q over K. Viewing R as a K-algebra with generators consisting of vertices and arrows, the natural $\mathbb{N}$-filtration of R defined by lengths of paths coincides with the $\mathbb{N}$-grading filtration determined by the natural $\mathbb{N}$-gradation of R.

Though R is also $\mathcal{B}$-graded by its standard K-basis $\mathcal{B}$ consisting of paths, unfortunately the multiplication of R tells us that the $\mathcal{B}$-gradation of R cannot yield the $\mathcal{B}$-grading filtration for R whenever $\mathcal{B}$ is endowed with a monomial ordering (see Ch.3).

Example 4. More generally, let $A = K[T]$ be a K-algebra generated by $T = \{a_i\}_{i\in J}$, and suppose, with respect to some K-basis $\mathcal{B}$ consisting of monomials of the form $a_{i_1}^{\alpha_1}\cdots a_{i_s}^{\alpha_s}$, A has a weight $\mathbb{N}$-gradation $\{A_p\}_{p\in\mathbb{N}}$ with weight $\{n_i\}_{i\in J}$. Then A has the corresponding *weight* $\mathbb{N}$-*grading filtration* $FA = \{F_pA\}_{p\in\mathbb{N}}$ with $F_pA = \oplus_{i\leq p}A_i$.

With the aid of Gröbner bases, Ch.3 and Ch.4 will provide us with more examples of $\mathcal{B}$-$\mathbb{N}$ doubly graded algebras, for instance, the skew polynomial algebras of type $\mathcal{O}_n(\lambda_{ji})$ and the quantum exterior algebras.

The associated Γ*-graded algebra*

Let A be a Γ-filtered K-algebra with Γ-filtration $FA = \{F_\gamma A\}_{\gamma\in\Gamma}$. Put

$$F_\gamma^* A = \bigcup_{\gamma'\prec\gamma} F_{\gamma'}A, \quad \gamma \in \Gamma,$$

where $F_{\gamma_0}^* A = \{0\}$ if Γ has a smallest element γ_0. The *associated* Γ*-graded* K*-algebra* of A, denoted $G^\Gamma(A)$, is defined as

$$G^\Gamma(A) = \bigoplus_{\gamma\in\Gamma} G^\Gamma(A)_\gamma \text{ with } G^\Gamma(A)_\gamma = F_\gamma A/F_\gamma^* A,$$

where the multiplication is obtained by extending the maps

$$\begin{array}{ccc} G^\Gamma(A)_{\gamma_1} \times G^\Gamma(A)_{\gamma_2} & \longrightarrow & G^\Gamma(A)_{\gamma_1\gamma_2} \\ (\overline{a}_{\gamma_1}, \overline{a}_{\gamma_2}) & \mapsto & \overline{a_{\gamma_1}a_{\gamma_2}} \end{array}$$

to a map $G^\Gamma(A) \times G^\Gamma(A) \to G^\Gamma(A)$, in which $a_{\gamma_1} \in F_{\gamma_1}A$, $a_{\gamma_2} \in F_{\gamma_2}A$, and $\overline{a}_{\gamma_1}$ is the canonical image of a_{γ_1} in $G^\Gamma(A)_{\gamma_1}$, $\overline{a}_{\gamma_2}$ is the canonical image of a_{γ_2} in $G^\Gamma(A)_{\gamma_2}$, $\overline{a_{\gamma_1}a_{\gamma_2}}$ is the canonical image of $a_{\gamma_1}a_{\gamma_2}$ in $G^\Gamma(A)_{\gamma_1\gamma_2}$, respectively.

Henceforth, modules considered are *left* modules.

Γ*-filtered module*

Let A be a Γ-filtered K-algebra with Γ-filtration $FA = \{F_\gamma A\}_{\gamma\in\Gamma}$ and M an A-module. We say that M is a Γ*-filtered* A*-module* if there is a family $FM = \{F_\gamma M\}_{\gamma\in\Gamma}$ consisting of K-subspaces $F_\gamma M$ of M, such that

(FM1) $M = \cup_{\gamma\in\Gamma} F_\gamma M$,
(FM2) $F_{\gamma_1}M \subseteq F_{\gamma_2}M$ if $\gamma_1 \preceq \gamma_2$,
(FM3) $F_{\gamma_1}AF_{\gamma_2}M \subseteq F_{\gamma_1\gamma_2}M$, $\gamma_1, \gamma_2 \in \Gamma$.

In the definition given above, the family $FM = \{F_\gamma M\}_{\gamma\in\Gamma}$ is called a Γ*-filtration* of M.

Example 5. Given a Γ-filtered K-algebra A with Γ-filtration FA, if Γ has a smallest element γ_0 (for instance $\Gamma = \mathbb{N}$), then by the convention we made for FA, the identity element 1 of A is contained in $F_{\gamma_0}A$. In this case, any A-module M has a Γ-filtration FM. To see this, let $\{\xi_i\}_{i\in J}$ be a generating set of M, i.e., $M = \sum_{i\in J} A\xi_i$. Put $V = \sum_{i\in J} F_{\gamma_0}A\xi_i$. Then it may be verified directly that the family

$$FM = \{F_\gamma M = F_\gamma AV\}_{\gamma\in\Gamma}$$

forms a Γ-filtration of M.

The associated Γ*-graded module*

Let A be a Γ-filtered K-algebra with Γ-filtration $FA = \{F_\gamma A\}_{\gamma\in\Gamma}$ and $G^\Gamma(A)$ the associated Γ-graded K-algebra of A. For a Γ-filtered A-module M with Γ-filtration $FM = \{F_\gamma M\}_{\gamma\in\Gamma}$, put

$$F_\gamma^* M = \bigcup_{\gamma'\prec\gamma} F_{\gamma'}M, \quad \gamma \in \Gamma,$$

where $F_{\gamma_0}^* M = \{0\}$ if Γ has a smallest element γ_0. The *associated* Γ*-graded module* of M, denoted $G^\Gamma(M)$, is the Γ-graded $G^\Gamma(A)$-module defined as

$$G^\Gamma(M) = \bigoplus_{\gamma\in\Gamma} G^\Gamma(M)_\gamma \text{ with } G^\Gamma(M)_\gamma = F_\gamma M/F_\gamma^* M,$$

where the module action is given by extending the maps

$$\begin{array}{ccc} G^\Gamma(A)_{\gamma_1} \times G^\Gamma(M)_{\gamma_2} & \longrightarrow & G^\Gamma(M)_{\gamma_1\gamma_2} \\ (\overline{a}_{\gamma_1}, \overline{m}_{\gamma_2}) & \mapsto & \overline{a_{\gamma_1}m_{\gamma_2}} \end{array}$$

to a map $G^\Gamma(A) \times G^\Gamma(M) \to G^\Gamma(M)$, in which $a_{\gamma_1} \in F_{\gamma_1}A$, $m_{\gamma_2} \in F_{\gamma_2}M$, and $\overline{a}_{\gamma_1}$ is the canonical image of a_{γ_1} in $G^\Gamma(A)_{\gamma_1}$, $\overline{m}_{\gamma_2}$ is the canonical image of m_{γ_2} in $G^\Gamma(M)_{\gamma_2}$, $\overline{a_{\gamma_1}m_{\gamma_2}}$ is the canonical image of $a_{\gamma_1}m_{\gamma_2}$ in $G^\Gamma(M)_{\gamma_1\gamma_2}$, respectively.

Γ*-filtered submodule and the induced* Γ*-filtration*

Let A be a Γ-filtered K-algebra with Γ-filtration $FA = \{F_\gamma A\}_{\gamma\in\Gamma}$ and M a Γ-filtered A-module with Γ-filtration $FM = \{F_\gamma M\}_{\gamma\in\Gamma}$. If N is an A-submodule of M and N has a Γ-filtration $FN = \{F_\gamma N\}_{\gamma\in\Gamma}$ satisfying $F_\gamma N \subseteq F_\gamma M$ for all $\gamma \in \Gamma$, then we call N a Γ*-filtered A-submodule* of M.

Indeed, any A-submodule N of M can be made into a Γ-filtered A-submodule by using the *induced* Γ*-filtration* FN consisting of

$$F_\gamma N = N \cap F_\gamma M, \quad \gamma \in \Gamma.$$

Furthermore, for an A-submodule N of M, the quotient A-module M/N has the *induced* Γ-filtration $F(M/N)$ consisting of

$$F_\gamma(M/N) = (F_\gamma M + N)/N, \quad \gamma \in \Gamma.$$

Γ-*filtered homomorphism*

Let A, B be Γ-filtered K-algebras with $FA = \{F_\gamma A\}_{\gamma\in\Gamma}$, $FB = \{F_\gamma B\}_{\gamma\in\Gamma}$, respectively. A K-algebra homomorphism φ: $A \to B$ is called a Γ-*filtered K-algebra homomorphism* if $\varphi(F_\gamma A) \subseteq F_\gamma B$ for all $\gamma \in \Gamma$.

Given a Γ-filtered K-algebra A and two Γ-filtered A-modules M, N with $FM = \{F_\gamma M\}_{\gamma\in\Gamma}$, $FN = \{F_\gamma N\}_{\gamma\in\Gamma}$, respectively, a Γ-*filtered A-homomorphism* from M to N is an A-module homomorphism ψ: $M \to N$ such that $\psi(F_\gamma M) \subseteq F_\gamma N$ for all $\gamma \in \Gamma$.

A Γ-filtered A-homomorphism ψ: $M \to N$ is said to be *strict* if it satisfies

$$\psi(F_\gamma M) = \psi(M) \cap F_\gamma N, \quad \gamma \in \Gamma.$$

Let M be a Γ-filtered A-module with Γ-filtration FM and N a submodule of M, Then, with respect to the Γ-filtration $FN = \{F_\gamma N = N \cap F_\gamma M\}_{\gamma\in\Gamma}$ of N and the Γ-filtration $F(M/N) = \{(F_\gamma M + N)/N\}_{\gamma\in\Gamma}$ of the quotient module M/N, induced by FM, the inclusion map $N \hookrightarrow M$ and the canonical epimorphism $M \to M/N$ are obviously strict Γ-filtered A-homomorphisms.

Verification of the next proposition is an easy exercise.

Proposition 3.1. *If ψ: $M \to N$ is a Γ-filtered A-homomorphism, then $V = Im\psi$ is a Γ-filtered A-submodule of N with the Γ-filtration $FV = \{F_\gamma V = \psi(F_\gamma M)\}_{\gamma\in\Gamma}$, and $W = Ker\psi$ is a Γ-filtered A-submodule of M with the induced filtration $FW = \{F_\gamma W = W \cap F_\gamma M\}_{\gamma\in\Gamma}$.*

□

The associated Γ-*graded* $G^\Gamma(A)$-*homomorphism*

If φ: $M \to N$ is a Γ-filtered A-homomorphism, then φ induces naturally a Γ-graded $G^\Gamma(A)$-homomorphism:

$$\begin{array}{rcl} G^\Gamma(\varphi) : G^\Gamma(M) = \bigoplus\limits_{\gamma\in\Gamma} G^\Gamma(M)_\gamma & \longrightarrow & \bigoplus\limits_{\gamma\in\Gamma} G^\Gamma(N)_\gamma = G^\Gamma(N) \\ \sum \overline{m}_\gamma & \mapsto & \sum \overline{\varphi(m_\gamma)} \end{array}$$

where $m_\gamma \in F_\gamma M$, and $\overline{m}_\gamma$, respectively $\overline{\varphi(m_\gamma)}$, is the canonical image of m_γ in $G^\Gamma(M)_\gamma$, respectively the canonical image of $\varphi(m_\gamma)$ in $G^\Gamma(N)_\gamma$.

Remark If we replace the field K by $\mathbb{Z}$, then the text of this section concerning K-algebras and their modules becomes that for Γ-filtered rings and Γ-filtered modules without any modification.

Chapter 2

The Γ-Leading Homogeneous Algebra $A_{\rm LH}^{\Gamma}$

Let $R = \oplus_{\gamma\in\Gamma}R_{\gamma}$ be a Γ-graded algebra over a field K, where Γ is a well-ordered monoid, and let I be an arbitrary proper ideal of R. The aim of this chapter is for establishing a solid foundation of studying the quotient algebra $A = R/I$ via its Γ-leading homogeneous algebra $A_{\rm LH}^{\Gamma} = R/\langle \mathbf{LH}(I)\rangle$ (Section 4). The realization of this idea is based on the comparison principle of the Γ-filtered K-algebra A with its associated Γ-graded K-algebra $G^{\Gamma}(A)$ (Sections 1, 2), which is an analogue of the classical comparison principle of an $\mathbb{N}$-filtered K-algebra A with its associated $\mathbb{N}$-graded K-algebra $G^{\mathbb{N}}(A)$ (e.g. [Sj], [MR], [LVO4]), while the key bridge connecting A and $A_{\rm LH}^{\Gamma}$ is the Γ-graded K-algebra isomorphism $A_{\rm LH}^{\Gamma} \cong G^{\Gamma}(A)$ (Section 3).

All notations used in Ch.1 are maintained.

2.1 Recognizing A via $G^{\Gamma}(A)$: Part 1

In this and the next section we let Γ be a *well-ordered monoid*, that is, Γ is a monoid ordered by a well-ordering $\prec$, which is a total ordering with the property that every nonempty subset of Γ has a minimal element, or equivalently, for every descending chain of elements, $\gamma_1 \succeq \gamma_2 \succeq \cdots \succeq \gamma_n \succeq \cdots$, there exists some $N > 0$ such that $\gamma_N = \gamma_{N+1} = \gamma_{N+2} = \cdots$. If γ_0 is the identity element of Γ, we assume that γ_0 is the *smallest* element in Γ.

Let A be a Γ-filtered K-algebra with Γ-filtration FA, and let $G^{\Gamma}(A)$ be the associated Γ-graded K-algebra of A. Then the identity element 1 of A is contained in $F_{\gamma_0}A$ by the assumption on Γ made above and the convention fixed in the definition of FA (Ch.1, Section 3). Let M be a Γ-filtered A-module with Γ-filtration $FM = \{F_{\gamma}M\}_{\gamma\in\Gamma}$, and let $G^{\Gamma}(M)$ be the associated Γ-graded $G^{\Gamma}(A)$-module of M, in the sense of (Ch.1, Section 3). Since $\prec$ is a well-ordering on Γ, for each element $m \in M$, we define the *degree* of m, denoted $d(m)$, as

$$d(m) = \min\left\{\gamma \in \Gamma \;\middle|\; m \in F_\gamma M\right\}.$$

If $m \neq 0$ and $d(m) = \gamma$, then we write $\sigma(m)$ for the corresponding *nonzero* Γ-homogeneous element of degree γ in $G^\Gamma(M)_\gamma = F_\gamma M/F^*_\gamma M$.

Concerning the σ-element defined above, we note an easily verified but useful property that may help to understand the subsequent argument:

(σ) $\quad\forall\, a \in A,\ m \in M,$ either $\sigma(a)\sigma(m) = 0$ or $\sigma(a)\sigma(m) = \sigma(am)$.

We deal first with some very basic structural properties that may be lifted from $G^\Gamma(A)$ to A.

Recall that a ring S is a domain if S does not have divisors of zero; and S is called a prime (semi-prime) ring if $s_1Ss_2 \neq 0$ for any nonzero $s_1, s_2 \in S$ (if $sSs \neq 0$ for any nonzero $s \in S$).

Theorem 1.1. *Let A be a Γ-filtered K-algebra with Γ-filtration FA, and let $G^\Gamma(A)$ be the associated Γ-graded K-algebra of A. The following statements hold.*
(i) *Suppose that $\{a_i\}_{i\in J}$ is a subset of A such that $\{\sigma(a_i)\}_{i\in J}$ forms a K-basis for $G^\Gamma(A)$, then $\{a_i\}_{i\in J}$ is a K-basis of A. Hence, if $G^\Gamma(A)$ is finite dimensional over K, then A has the same property.*
(ii) *If $G^\Gamma(A)$ is a domain, then so too is A.*
(iii) *If $G^\Gamma(A)$ is a prime (semi-prime) ring, then so too is A.*

Proof. (i). First, we show that $\{a_i\}_{i\in J}$ is K-linearly independent, namely, if $a = \sum_{j=1}^{s}\lambda_{i_j}a_{i_j} = 0$, where $\lambda_i \in K$ and $a_{i_j} \in \{a_i\}_{i\in J}$, then all coefficients $\lambda_{i_j} = 0$. To this end, assume that $a_{i_1}, a_{i_2}, \ldots, a_{i_t}$, $t \leq s$, all have the same degree γ that is the highest degree among all terms in the linear expression of a. Then since $K \subseteq F_{\gamma_0}A$, taking the image of a in $G^\Gamma(A)_\gamma = F_\gamma A/F^*_\gamma A$ into account, we have

$$\lambda_{i_1}\sigma(a_{i_1}) + \lambda_{i_2}\sigma(a_{i_2}) + \cdots + \lambda_{i_t}\sigma(a_{i_t}) = 0$$

in $G^\Gamma(A)_\gamma$. By the K-linear independence of $\{\sigma(a_i)\}_{i\in J}$, $\lambda_{i_1} = \lambda_{i_2} = \cdots = \lambda_{i_t} = 0$. Similarly we assert that all other coefficients in the linear expression of a are equal to 0. This proves the K-linear independence of $\{a_i\}_{i\in J}$.

Next, we conclude that A, as a K-space, is spanned by $\{a_i\}_{i\in J}$. To see this, let $a \in F_\gamma A - F^*_\gamma A$, i.e., $d(a) = \gamma$. Then, by the assumption, $\sigma(a)$ can be written uniquely as a linear combination of $\sigma(a_i)$'s, say

$$\sigma(a) = \sum_{j=1}^{s} \lambda_{i_j}\sigma(a_{i_j}), \quad \lambda_{i_j} \in K,\ a_{i_j} \in \{a_i\}_{i\in J}.$$

As $\sigma(a)$ is a Γ-homogeneous element of degree γ in $G^\Gamma(A)$ and $K \subseteq G^\Gamma(A)_{\gamma_0} = F_{\gamma_0}A$, it follows that all homogeneous elements $\sigma(a_{i_j})$ in the linear expression of $\sigma(a)$ have the same degree γ. Thus, by the definition of a σ-element, we have

$$a' = a - \sum_{j=1}^{s} \lambda_{i_j}a_{i_j} \in F^*_\gamma A.$$

Suppose $a' \in F_{\gamma'}A - F^*_{\gamma'}A$, i.e., $d(a') = \gamma' \prec \gamma$. By a similar argument we may get $a'' = a' - \sum_{\ell=1}^{m} \lambda_{i_\ell}a_{i_\ell} \in F^*_{\gamma'}A$ with $\lambda_{i_\ell} \in K$ and $a_{i_\ell} \in \{a_i\}_{i\in J}$. If $d(a'') = \gamma''$, then $\gamma \succ \gamma' \succ \gamma''$. Since $\prec$ is a well-ordering, the procedure of reducing degrees must stop after a finite number of steps, and consequently we obtain $a \in K\text{-span}\{a_i\}_{i\in J}$, as desired.

(ii). Let $a, b \in A$ be nonzero elements of degree γ_1, γ_2 respectively. Then $\sigma(a)$, $\sigma(b)$ are nonzero Γ-homogeneous elements of $G^\Gamma(A)$ and so $\sigma(a)\sigma(b) = \sigma(ab) \neq 0$. This means $ab \notin F^*_{\gamma_1\gamma_2}A$, and hence $ab \neq 0$, showing that A is a domain.

(iii). If $a, b \in A$ are nonzero, then $\sigma(a)$, $\sigma(b)$ are nonzero Γ-homogeneous elements of $G^\Gamma(A)$ and so $\sigma(a)G^\Gamma(A)\sigma(b) \neq \{0\}$. It follows that there is a homogeneous element $\sigma(c) \in G^\Gamma(A)$, represented by $c \in A$, such that $\sigma(a)\sigma(c)\sigma(b) \neq 0$. But this means $acb \neq 0$. Hence $aAb \neq \{0\}$, i.e., A is prime. A similar argument achieves the proof concerning the semiprimeness of A. □

Next, we focus on modules over a Γ-filtered K-algebra A with Γ-filtration FA.

Proposition 1.2. *Let M be an arbitrary A-module.*
(i) *Suppose that M is a Γ-filtered A-module with Γ-filtration $FM = \{F_\gamma M\}_{\gamma\in\Gamma}$. If $G^\Gamma(M) = \sum_{i\in J} G^\Gamma(A)\sigma(\xi_i)$ with $\xi_i \in M$ and $d(\xi_i) = \gamma_i \in \Gamma$, then $M = \sum_{i\in J} A\xi_i$ with*

$$F_\gamma M = \sum_{i\in J}\left(\sum_{s_i\gamma_i \preceq \gamma} F_{s_i}A\right)\xi_i, \quad \gamma \in \Gamma.$$

In particular, if $G^\Gamma(M)$ is finitely generated, then so too is M.
(ii) *If M is finitely generated over A, then M has a Γ-filtration $FM = \{F_\gamma M\}_{\gamma\in\Gamma}$ such that $G^\Gamma(M)$ is a finitely generated $G^\Gamma(A)$-module. Indeed, if $M = \sum_{i=1}^n A\xi_i$ and $\{\xi_1,\ldots,\xi_n\}$ is a minimal set of generators for M, then the desired Γ-filtration FM consists of*

$$F_\gamma M = \sum_{i=1}^n \left(\sum_{s_i\gamma_i\preceq\gamma} F_{s_i}A\right)\xi_i,\quad \gamma\in\Gamma,$$

where $\gamma_1,\ldots,\gamma_n\in\Gamma$ are chosen arbitrarily.

Proof. (i). Since $G^\Gamma(M) = \sum_{i\in J} G^\Gamma(A)\sigma(\xi_i)$ with $\xi_i\in M$ and $d(\xi_i) = \gamma_i\in\Gamma$, we have

$$G^\Gamma(M)_\gamma = \sum_{\rho_i\gamma_i=\gamma,\ i\in J} G^\Gamma(A)_{\rho_i}\sigma(\xi_i),\quad \gamma\in\Gamma.$$

Hence, for any $m\in F_\gamma M$, $m = \sum a_{\rho_i}\xi_i + m'$, where $a_{\rho_i}\in F_{\rho_i}A$, $\rho_i\gamma_i = \gamma$, and $m'\in F^*_\gamma M$. Assume $d(m') = \gamma'$. Then similarly we have $m' = \sum a_{\mu_i}\xi_i + m''$, where $a_{\mu_i}\in F_{\mu_i}A$, $\mu_i\gamma_i = \gamma'$, and $m''\in F^*_{\gamma'}M$. Suppose $d(m'') = \gamma''$. Then, $\gamma\succ\gamma'\succ\gamma''$. As $\prec$ is a well-ordering, after repeating the procedure of reducing degrees for a finite number of steps, we should arrive at

$$m\in\sum_{i\in J}\left(\sum_{s_i\gamma_i\preceq\gamma} F_{s_i}A\right)\xi_i.$$

Since m is taken arbitrarily, this shows that

$$F_\gamma M = \sum_{i\in J}\left(\sum_{s_i\gamma_i\preceq\gamma} F_{s_i}A\right)\xi_i,$$

and therefore $M = \sum_{i\in J} A\xi_i$.

(ii). Suppose $M = \sum_{i=1}^n A\xi_i$ and $\{\xi_1,\ldots,\xi_n\}$ is a *minimal* set of generators for M. Choose $\gamma_1,\ldots,\gamma_n\in\Gamma$ arbitrarily. Then, since each $m\in M$ has a representation $m = \sum_{i=1}^n a_i\xi_i$ with $a_i\in F_{s_i}A - F^*_{s_i}A$ for some $s_i\in\Gamma$, it is easy to see that the K-subspaces

$$F_\gamma M = \sum_{i=1}^n \left(\sum_{s_i\gamma_i\preceq\gamma} F_{s_i}A\right)\xi_i,\quad \gamma\in\Gamma,$$

form a Γ-filtration $FM = \{F_\gamma M\}_{\gamma\in\Gamma}$ for M. Furthermore, note that $1\in F_{\gamma_0}A$, where γ_0 is the identity element of Γ. It follows from the construction of FM and the minimality of $\{\xi_1,\ldots,\xi_n\}$ (as a set of generators of M) that $\xi_i\in F_{\gamma_i}M-F^*_{\gamma_i}M$, i.e., $d(\xi_i)=\gamma_i$, $i=1,\ldots,n$. Thus, by the foregoing property (σ) of the associated σ-elements, it is not difficult to verify that

$$G^{\Gamma}(M)_{\gamma}=F_{\gamma}M/F^*_{\gamma}M=\sum_{s_i\gamma_i=\gamma,\ 1\le i\le n}G^{\Gamma}(A)_{s_i}\sigma(\xi_i),\quad \gamma\in\Gamma,$$

and hence $G^{\Gamma}(M)=\oplus_{\gamma\in\Gamma}G^{\Gamma}(M)_{\gamma}=\sum_{i=1}^{n}G^{\Gamma}(A)\sigma(\xi_i)$. □

Recall that a sequence

$$\cdots\overset{\varphi_{i-2}}{\longrightarrow}M_{i-1}\overset{\varphi_{i-1}}{\longrightarrow}M_i\overset{\varphi_i}{\longrightarrow}M_{i+1}\overset{\varphi_{i+1}}{\longrightarrow}\cdots$$

of A-modules and A-homomorphisms is said to be *exact* if $\mathrm{Ker}\varphi_i=\mathrm{Im}\varphi_{i-1}$ holds for every i.

The Γ-graded homomorphisms and Γ-filtered homomorphisms considered below are in the sense of (Ch.1, Sections 2, 3).

Proposition 1.3. *Let*

$$(*)\qquad L\overset{\varphi}{\longrightarrow}M\overset{\psi}{\longrightarrow}N$$

be a sequence of Γ-filtered A-modules and Γ-filtered A-homomorphisms satisfying $\psi\circ\varphi=0$. *Then the following two properties are equivalent.*
(i) *The sequence* $(*)$ *is exact and* φ *and* ψ *are strict Γ-filtered homomorphisms in the sense of (Ch.1, Section 3).*
(ii) *The associated sequence of Γ-graded* $G^{\Gamma}(A)$*-modules and Γ-graded* $G^{\Gamma}(A)$*-homomorphisms*

$$G^{\Gamma}(*)\qquad G^{\Gamma}(L)\overset{G^{\Gamma}(\varphi)}{\longrightarrow}G^{\Gamma}(M)\overset{G^{\Gamma}(\psi)}{\longrightarrow}G^{\Gamma}(N)$$

is exact.

Proof. For an element $x\in F_{\gamma}M$, throughout the proof we use $\overline{x}$ to denote the canonical image of x in $F_{\gamma}M/F^*_{\gamma}M=G^{\Gamma}(M)_{\gamma}$. Similar notation is used for elements in L and N.

(i) ⇒ (ii). By the definition of the associated Γ-graded $G^{\Gamma}(A)$-homomorphism of a Γ-filtered A-homomorphism, it is clear that $\mathrm{Im}G^{\Gamma}(\varphi)\subseteq\mathrm{Ker}G^{\Gamma}(\psi)$. To prove the inverse inclusion, for $m\in F_{\gamma}M-F^*_{\gamma}M$, i.e., $d(m)=\gamma$, suppose $0=G^{\Gamma}(\psi)(\overline{m})=\overline{\psi(m)}$. If $\psi(m)=0$, then $m\in\mathrm{Ker}\psi=\mathrm{Im}\varphi$ and there is some $\ell\in L$ such that

$$m=\varphi(\ell)\in\varphi(L)\cap F_{\gamma}M=\varphi(F_{\gamma}L).$$

Obviously we may assume $\ell \in F_\gamma L$, and thus, $\overline{m} = \overline{\varphi(\ell)} = G^\Gamma(\varphi)(\overline{\ell})$, i.e., $\overline{m} \in \mathrm{Im}G^\Gamma(\varphi)$. If $\psi(m) \neq 0$, then since $0 = G^\Gamma(\psi)(\overline{m}) = \overline{\psi(m)} \in G^\Gamma(N)_\gamma$, we have $\psi(m) \in F_{\gamma'}N - F^*_{\gamma'}N$ for some $\gamma' \prec \gamma$, i.e., $\psi(m) \in \psi(M) \cap F_{\gamma'}N = \psi(F_{\gamma'}M)$. This yields $\psi(m) = \psi(m')$ for some $m' \in F_{\gamma'}M$, and hence

$$m - m' \in \ \mathrm{Ker}\psi \cap F_\gamma M = \varphi(L) \cap F_\gamma M = \varphi(F_\gamma L).$$

Let $m - m' = \varphi(\ell')$ for some $\ell' \in F_\gamma L$. Then $\overline{m} = \overline{m - m'} = \overline{\varphi(\ell')} = G^\Gamma(\varphi)(\overline{\ell'})$. this shows that $\overline{m} \in \mathrm{Im}G^\Gamma(\varphi)$. As m is taken arbitrarily, so we have $\mathrm{Ker}G^\Gamma(\psi) \subseteq \mathrm{Im}G^\Gamma(\varphi)$. Therefore, we conclude $\mathrm{Ker}G^\Gamma(\psi) = \mathrm{Im}G^\Gamma(\varphi)$, that is, the sequence $G^\Gamma(*)$ is exact.

(ii) $\Rightarrow$ (i). Suppose that the graded sequence $G^\Gamma(*)$ is exact. Let us show that the sequence $(*)$ is exact first. If $\psi(m) = 0$ with $m \in F_\gamma M - F^*_\gamma M$, i.e., $d(m) = \gamma$, then $G^\Gamma(\psi)(\overline{m}) = 0$ with $\overline{m} \in G^\Gamma(M)_\gamma$. It follows that $\overline{m} = G^\Gamma(\varphi)(\overline{\ell'}) = \overline{\varphi(\ell')}$ for some $\ell' \in F_\gamma L - F^*_\gamma L$. Hence $m - \varphi(\ell') = m'$ for some $m' \in F_{\gamma'}M$ with $\gamma' \prec \gamma$. Thus, $\psi(m') = \psi(m - \varphi(\ell')) = 0$. Similarly, $m' - \varphi(\ell'') = m''$ with $\ell'' \in L$ and $m'' \in F_{\gamma''}M$ with $\gamma'' \prec \gamma'$. As the chain

$$\gamma \succ \gamma' \succ \gamma'' \succ \cdots$$

cannot be infinite, for $\prec$ is a well-ordering, after repeating the reduction procedure for a finite number of steps we arrive at $m = \varphi(\ell)$ for some $\ell \in L$. This shows that $\mathrm{Ker}\psi \subseteq \varphi(L)$. Therefore, $\mathrm{Ker}\psi = \varphi(L)$ and the exactness of the sequence $(*)$ follows.

Concerning the strictness of φ and ψ, we prove it only for ψ because a similar argument is valid for φ. Let $f \in F_\gamma N \cap \psi(M)$ and $f \notin F^*_\gamma N$. Then $f = \psi(m)$ for some $m \in F_w M$. Suppose $\gamma \preceq w$. If $w = \gamma$, then $f = \psi(m) \in \psi(F_\gamma M)$. If $\gamma \prec w$, then since $f \in F_\gamma N$, we have $G^\Gamma(\psi)(\overline{m}) = \overline{\psi(m)} = 0$ in $G^\Gamma(N)$. By the exactness, $\overline{m} = G^\Gamma(\varphi)(\overline{\ell}) = \overline{\varphi(\ell)}$ for some $\ell \in F_w L$. Put $m' = m - \varphi(\ell)$. Then $m' \in F_{w'}M$ with $w' \prec w$, and $\psi(m') = \psi(m - \varphi(\ell)) = \psi(m) = f$. If $\gamma \prec w'$, then similarly we may find $m'' \in F_{w''}M$ with $w'' \prec w'$, such that $\psi(m'') = f$. Note that the chain

$$w \succ w' \succ w'' \succ \cdots \succ \gamma$$

has finite length in Γ. So the above reduction procedure stops after a finite number of steps, and eventually we have $f = \psi(m_\gamma) \in \psi(F_\gamma M)$. This shows that $F_\gamma N \cap \psi(M) \subset \psi(F_\gamma M)$, that is, ψ is strict. □

Corollary 1.4. (i) *Let $\varphi\colon M \to N$ be a Γ-filtered A-homomorphism. Then $G^\Gamma(\varphi)$ is injective, respectively surjective, if and only if φ is injective, respectively surjective, and φ is strict.*

(ii) *Let* N, W *be submodules of a* Γ*-filtered* A*-module* M *with* Γ*-filtration* $FM = \{F_\gamma M\}_{\gamma\in\Gamma}$. *Consider the* Γ*-filtration* $FN = \{F_\gamma N = N \cap F_\gamma M\}_{\gamma\in\Gamma}$ *of* N *and the* Γ*-filtration* $FW = \{F_\gamma W = W \cap F_\gamma M\}_{\gamma\in\Gamma}$ *of* W, *induced by* FM *respectively. If* $N \subseteq W$, *then* $G^{\Gamma}(N) \subseteq G^{\Gamma}(W)$; *and if* $G^{\Gamma}(N) = G^{\Gamma}(W)$ *then* $N = W$.

□

We summarize some immediate applications of previous results into the next theorem. Before mentioning the theorem, let us point out that the notion of Krull dimension used in part (iv) of the theorem is defined for rings and modules in both the commutative and noncommutative case. This dimension is a measure of how close the ring or module is to being Artinian, and it coincides with the classical Krull dimension in commutative case. Roughly speaking, if M is a left A-module over a ring A, then the Krull dimension of M, denoted K.dim$M$ is defined to be the *deviation* of $\mathscr{L}(M)$, the lattice of submodules of M, that is, K.dim$M = -\infty$ if M is trivial; K.dim$M = 0$ if M is nontrivial but satisfied the descending chain condition; and K.dim$M = \alpha$ for some ordinal α provided: (1) K.dim$M \neq \beta < \alpha$, and (2) in any descending chain of elements of $\mathscr{L}(M)$ all but finitely many factors have K.dim $< \alpha$. K.dim$A$ is defined to be the Krull dimension of the left A-module A. The reader is referred to ([MR], Chapter 6) for details on Krull dimension which, in the noncommutative case, was introduced by P. Gabriel and R. Rentschler in [GR].

Theorem 1.5. *Let* A *be a* Γ*-filtered* K*-algebra with* Γ*-filtration* FA, *and let* $G^{\Gamma}(A)$ *be the associated* Γ*-graded* K*-algebra of* A. *The following statements hold.*

(i) *Suppose that* $G^{\Gamma}(A)$ *is* Γ*-graded left Noetherian, that is, every* Γ*-graded left ideal of* $G^{\Gamma}(A)$ *is finitely generated, or equivalently,* $G^{\Gamma}(A)$ *satisfies the ascending chain condition for* Γ*-graded left ideals. Then every finitely generated* A*-module is left Noetherian, in particular,* A *is left Noetherian.*

(ii) *Suppose that* $G^{\Gamma}(A)$ *is* Γ*-graded left Artinian, that is,* $G^{\Gamma}(A)$ *satisfies the descending chain condition for* Γ*-graded left ideals. Then every finitely generated* A*-module is left Artinian, in particular,* A *is left Artinian.*

(iii) *If* $G^{\Gamma}(A)$ *is a* Γ*-graded simple* K*-algebra, that is,* $G^{\Gamma}(A)$ *does not have nontrivial* Γ*-graded ideal, then* A *is a simple* K*-algebra.*

(iv) *Let* M *be a* Γ*-filtered* A*-module with* Γ*-filtration* FM. *If the Krull dimension (in the sense of Gabriel and Rentschler) of* $G^{\Gamma}(M)$ *is well defined,*

then the Krull dimension of M is defined and $K.dimM \leq K.dimG^{\Gamma}(M)$. In particular, this is true for $M = A$ and $G^{\Gamma}(M) = G^{\Gamma}(A)$.
(v) *Let M be a Γ-filtered A-module with Γ-filtration FM. If $G^{\Gamma}(M)$ is a Γ-graded simple $G^{\Gamma}(A)$-module, that is, $G^{\Gamma}(M)$ does not have nontrivial Γ-graded submodule, then M is a simple A-module.*
(vi) *If $G^{\Gamma}(A)$ is semisimple (simple) Artinian, then A is semisimple (simple) Artinian.*

Proof. The assertions of (i) – (v) follow from Corollary 1.4 immediately. It remains to prove the semisimplicity of A in (vi).

Recall (cf. [Lam]) that if A is an Artinian ring, then the Jacobson radical $J(A)$ of A, which is by definition the intersection of all the maximal left (right) ideals of A, is nilpotent, and A is semisimple if and only if $J(A) = 0$. Thus, if $G^{\Gamma}(A)$ is semisimple Artinian, then it does not contain nonzero nilpotent ideal. Now consider the Γ-filtration $FJ(A) = \{F_{\gamma}J(A) = J(A) \cap F_{\gamma}A\}_{\gamma\in\Gamma}$ of $J(A)$ induced by FA. Then $G^{\Gamma}(J(A))$ is a nilpotent ideal of $G^{\Gamma}(A)$ and hence $G^{\Gamma}(J(A)) = \{0\}$. By Corollary 1.4, $J(A) = \{0\}$ as desired. □

We finish this section by lifting the structural property of being a maximal order. Recall from (e.g. [MR]) that a ring Q is called a quotient ring if every regular element (i.e., not a divisor of zero) of Q is a unit. A subring B of a quotient ring Q is called a right order in Q if each $q \in Q$ has the form bs^{-1} for some $b, s \in B$. A left order is defined in a similar way. A left and right order is called an order. Let Q be a quotient ring. An order B of Q is called a *maximal order* if B is maximal within its equivalence class determined by the equivalence relation: two orders $B_1 \sim B_2$ if there are units $a_1, a_2, b_1, b_2 \in Q$ such that $a_1B_1b_1 \subseteq B_2$ and $a_2B_2b_2 \subseteq B_1$. It is known that any Noetherian domain A is an order in its quotient ring $Q = \{bs^{-1} \mid b, s \in A,\ s \neq 0\}$ which is indeed a division ring.

Theorem 1.6. *Let A be a Γ-filtered K-algebra with Γ-filtration FA, and let $G^{\Gamma}(A)$ be the associated Γ-graded K-algebra of A. If $G^{\Gamma}(A)$ is a Noetherian domain which is a maximal order in its quotient ring, then A is a Noetherian domain and a maximal order in its quotient ring.*

Proof. By Theorem 1.1 and Theorem 1.5, A is a Noetherian domain. Hence A has the quotient ring of fractions Q. To prove A is a maximal order in Q, by ([MR], Ch.5 Proposition 1.4) it is sufficient to show that

$$\mathcal{O}_r(I) = \{q \in Q \mid Iq \subseteq I\} = \mathcal{O}_l(I) = \{q \in Q \mid Iq \subseteq I\} = A$$

for any nonzero ideal I of A.

Let $q \in \mathcal{O}_r(I)$, say $q = fg^{-1}$ with $f, g \in A$. Then $If \subseteq Ig$. Suppose that f has the least degree among all f's with this property. Considering the associated Γ-graded ideal $G^{\Gamma}(I)$ of I with respect to the Γ-filtration FI induced by FA, we have $G(I)\sigma(f) \subseteq G(I)\sigma(g)$ (note that $G^{\Gamma}(A)$ is a domain). Since $G^{\Gamma}(I) \neq \{0\}$ and $G^{\Gamma}(A)$ is a maximal order in its quotient ring, it follows that $\sigma(f)\sigma(g)^{-1} \in G^{\Gamma}(A)$ and this turns out that $\sigma(f) \in G^{\Gamma}(A)\sigma(g)$. Thus, $f = hg + f_1$ with $h \in A$, $d(f_1) \prec d(f)$. Note that $If_1 = I(f - hg) \subseteq Ig$. So the hypothesis on the degree of f yields $f_1 = 0$. Hence $fg^{-1} = h \in A$. As $\prec$ is a well-ordering, it is then clear that $\mathcal{O}_r(I) = A$. By symmetry, $\mathcal{O}_l(I) = A$. This completes the proof. □

2.2 Recognizing A via $G^{\Gamma}(A)$: Part 2

In this section we keep the assumption that Γ *is a well-ordered monoid* with the well-ordering $\prec$, and that the identity element γ_0 of Γ is the *smallest* element.

Let A be a Γ-filtered K-algebra with Γ-filtration $FA = \{F_{\gamma}A\}_{\gamma\in\Gamma}$, and let $G^{\Gamma}(A)$ be the associated Γ-graded K-algebra of A. With notation as before, the aim of this section is to lift several homological properties from $G^{\Gamma}(A)$ to A. For a general theory on homological algebra, the reader is referred to (e.g. [Rot]).

Unless otherwise stated, all modules considered are *left* modules.

We begin with some basics on graded free modules and graded projective modules.

Let $R = \oplus_{\gamma\in\Gamma}R_{\gamma}$ be a Γ-graded K-algebra. A Γ-*graded free* R-module is a free R-module $T = \oplus_{i\in J}Re_i$ on the basis $\{e_i\}_{i\in J}$, which is also Γ-graded such that each e_i is a Γ-homogeneous element, that is, if $\mathrm{d}(e_i) = \gamma_i$, $i \in J$, then $T = \oplus_{\gamma\in\Gamma}T_{\gamma}$ with

$$T_{\gamma} = \sum_{w_i\gamma_i=\gamma,\ i\in J} R_{w_i}e_i, \quad \gamma \in \Gamma.$$

By the definition, to construct a Γ-graded free R-module $T = \oplus_{i\in J}Re_i$ with the R-basis $\{e_i\}_{i\in J}$, it is sufficient to assign each e_i to a chosen degree and define the Γ-gradation as described above.

Given any Γ-graded R-module $M = \oplus_{\gamma\in\Gamma} M_\gamma$, M has a generating set $\{m_i\}_{i\in J}$ consisting of Γ-homogeneous elements, i.e., $M = \sum_{i\in J} Rm_i$. Suppose $d(m_i) = \gamma_i$, $\gamma_i \in \Gamma$, $i \in J$. Then it is easy to see that

$$M_\gamma = \sum_{w_i\gamma_i=\gamma,\ i\in J} R_{w_i} m_i, \quad \gamma \in \Gamma.$$

Thus, considering the Γ-graded free R-module $T = \oplus_{i\in J} Re_i = \oplus_{\gamma\in\Gamma} T_\gamma$ with $d(e_i) = \gamma_i$, the map φ: $e_i \mapsto m_i$ defines a Γ-graded R-epimorphism φ: $T \to M$ in the sense of (Ch.1, Section 2).

Let T be a Γ-graded free R-module and P a Γ-graded R-module. If there is another Γ-graded R-module Q such that $T = P \oplus Q$ and

$$T_\gamma = P_\gamma + Q_\gamma, \quad \gamma \in \Gamma,$$

then P is called a *Γ-graded projective* R-module.

Concerning Γ-graded projective R-modules, the following analogue of ([NVO], Corollary I.2.2 with $\Gamma = G$ a group; [MR], Proposition 7.6.6 with $\Gamma = \mathbb{N}$ the additive monoid of all non-negative integers) is fundamental. Before stating and proving this proposition, one is reminded to notice that in the proof of (iii) $\Rightarrow$ (i) below, we require that Γ satisfies the *left cancelation law*: $\gamma\gamma_1 = \gamma\gamma_2$ implies $\gamma_1 = \gamma_2$ for all $\gamma, \gamma_1, \gamma_2 \in \Gamma$. But this property is guaranteed since we are working with a totally ordered monoid Γ.

Proposition 2.1. *For a Γ-graded R-module P, the following statements are equivalent.*
(i) *P is a Γ-graded projective R-module.*
(ii) *Given any exact sequence of Γ-graded R-modules and Γ-graded R-homomorphisms $M\xrightarrow{\psi}N \to 0$, if $P\xrightarrow{\alpha}N$ is any Γ-graded R-homomorphism, then there exists a unique Γ-graded R-homomorphism $P\xrightarrow{\varphi}M$ such that the following diagram commutes:*

$$\begin{array}{ccc} & P & \\ \varphi\swarrow & \downarrow \alpha & \qquad \psi\circ\varphi = \alpha \\ M & \xrightarrow[\psi]{} N \to 0 & \end{array}$$

(iii) *P is projective as an ungraded R-module.*

Proof. By virtue of graded A-homomorphisms, the proof of (i) $\Leftrightarrow$ (ii) is similar to the ungraded case. (i) $\Rightarrow$ (iii) is obvious because, a Γ-graded free R-module is certainly free as an ungraded R-module.

(iii) $\Rightarrow$ (i) Let $\{\xi_i \mid i \in J\}$ be a homogeneous generating set of P, and ϑ: $T \to P$ the Γ-graded R-epimorphism defined by $\vartheta(e_i) = \xi_i$, where $T = \oplus_{i\in J}Re_i = \oplus_{\gamma\in\Gamma}T_\gamma$ is the Γ-graded free R-module with homogeneous free R-basis $\{e_i \mid d(e_i) = d(\xi_i),\ i \in J\}$. Since P is projective as an ungraded R-module, ϑ splits, i.e., there is an R-homomorphism β: $P \to T$ such that $\vartheta\beta = 1_P$. Note that the splitting homomorphism β need not be a Γ-graded R-homomorphism. But if we define φ: $P \to T$ with $\varphi(x_\gamma) = y_\gamma$, where $x_\gamma \in P_\gamma$, $\beta(x_\gamma) = y_\gamma + \sum_j y_{\gamma_j}$ with $y_\gamma \in T_\gamma$, $y_{\gamma_j} \in T_{\gamma_j}$ and $\gamma \neq \gamma_j$, then, with the aid of left cancelation law on Γ, a direct verification shows that φ is a Γ-graded R-homomorphism, in particular $\varphi(r_{\gamma'}x_\gamma) = r_{\gamma'}\varphi(x_\gamma)$ for all $r_{\gamma'} \in R_{\gamma'}$ and $x_\gamma \in P_\gamma$, $\gamma', \gamma \in \Gamma$ (this is the *key* point), such that $\vartheta\varphi = 1_P$. Thus, the Γ-graded R-homomorphism φ splits ϑ in degrees and therefore, (i) is proved. □

Returning to modules over a Γ-filtered K-algebra A with Γ-filtration FA, we first construct a *Γ-filtered free A-module* L with a Γ-filtration FL such that its associated Γ-graded module $G^\Gamma(L)$ is a Γ-graded free $G^\Gamma(A)$-module. To this end, let $L = \oplus_{i\in J}Ae_i$ be a free A-module on the basis $\{e_i\}_{i\in J}$. Then, as we did in Section 1 (see the proof of Proposition 1.2(ii)), a Γ-filtration FL for L can be constructed by using the A-basis $\{e_i\}_{i\in J}$ of L and arbitrarily chosen $\gamma_i \in \Gamma$, $i \in J$, that is,

$$F_\gamma L = \bigoplus_{i\in J}\left(\sum_{s_i\gamma_i\preceq\gamma} F_{s_i}A\right)e_i, \quad \gamma \in \Gamma.$$

Observation 2.2. Note that Γ is a monoid with the identity element γ_0 which is the smallest element in Γ. It is not difficult to see that in the Γ-filtration $FL = \{F_\gamma L\}_{\gamma\in\Gamma}$ constructed above, for each $i \in J$, we have $e_i \in F_{\gamma_i}L - F^*_{\gamma_i}L$, i.e., each e_i is of degree γ_i.

New convention In what follows, if we say that L is a Γ-filtered free A-module, then it means that L is certainly the type constructed above.

Proposition 2.3. *With notation as above, the following statements hold.*
(i) *Let $L = \oplus_{i\in J}Ae_i$ be a Γ-filtered free A-module with Γ-filtration FL. Then the associated Γ-graded $G^\Gamma(A)$-module $G^\Gamma(L)$ of L is a Γ-graded free $G^\Gamma(A)$-module. More precisely, we have $G^\Gamma(L) = \oplus_{i\in J}G^\Gamma(A)\sigma(e_i) =$*

$\oplus_{\gamma\in\Gamma}G^{\Gamma}(L)_{\gamma}$ *with*

$$G^{\Gamma}(L)_{\gamma}=\sum_{s_i\gamma_i=\gamma,\ i\in J}G^{\Gamma}(A)_{s_i}\sigma(e_i),\quad \gamma\in\Gamma.$$

(ii) *If* $L'=\oplus_{i\in J}G^{\Gamma}(A)\eta_i$ *is a* Γ*-graded free* $G^{\Gamma}(A)$*-module with the* $G^{\Gamma}(A)$*-basis* $\{\eta_i\}_{i\in J}$ *consisting of* Γ*-homogeneous elements, then there is some* Γ*-filtered free* A*-module* L *such that* $L'\cong G^{\Gamma}(L)$ *as* Γ*-graded* $G^{\Gamma}(A)$*-modules.*
(iii) *Let* M *be a* Γ*-filtered* A*-module with* Γ*-filtration* $FM=\{F_{\gamma}M\}_{\gamma\in\Gamma}$*. Then there is an exact sequence of* Γ*-filtered* A*-modules and strict* Γ*-filtered* A*-homomorphisms*

$$0\rightarrow N\xrightarrow{\ \iota\ }L\xrightarrow{\ \varphi\ }M\rightarrow 0$$

where L *is a* Γ*-filtered free* A*-module with* Γ*-filtration* FL*,* N *is the kernel of the* Γ*-filtered* A*-epimorphism* φ *that has the* Γ*-filtration* $FN=\{F_{\gamma}N=N\cap F_{\gamma}L\}_{\gamma\in\Gamma}$ *induced by* FL*, and* ι *is the inclusion map.*
(iv) *If* L *is a* Γ*-filtered free* A*-module with* Γ*-filtration* FL*,* N *is a* Γ*-filtered* A*-module with* Γ*-filtration* FN*, and* φ*:* $G^{\Gamma}(L)\rightarrow G^{\Gamma}(N)$ *is a* Γ*-graded* $G^{\Gamma}(A)$*-epimorphism, then* $\varphi=G^{\Gamma}(\psi)$ *for some strict* Γ*-filtered* A*-epimorphism* ψ*:* $L\rightarrow N$*.*

Proof. (i). By the construction of FL, Observation 2.2 and the property (σ) of σ-elements formulated in Section 1, the argument on this assertion is straightforward.

(ii). Suppose $d(\eta_i)=\gamma_i$, $\gamma_i\in\Gamma$, $i\in J$. Then by (i) we see that the Γ-filtered free A-module $L=\oplus_{i\in J}Ae_i$ with $d(e_i)=\gamma_i$ satisfies $G^{\Gamma}(L)\cong L'$.

(iii). By Proposition 1.2(i), if $G^{\Gamma}(M)=\sum_{i\in J}G^{\Gamma}(A)\sigma(\xi_i)$ with $\xi_i\in M$ and $d(\xi_i)=\gamma_i\in\Gamma$, then the Γ-filtered free A-module $L=\oplus_{i\in J}Ae_i$ with $d(e_i)=\gamma_i$, $i\in J$, and the map φ: $e_i\mapsto\xi_i$ together make the desired exact sequence.

(iv). Let $L=\oplus_{i\in J}Ae_i$ be the Γ-filtered free A-module with $d(e_i)=\gamma_i$, $i\in J$. For each $i\in J$, choose $\xi_i\in F_{\gamma_i}N$ such that $\varphi(\sigma(e_i))=\overline{\xi_i}$, where $\overline{\xi_i}$ is the Γ-homogeneous element in $G^{\Gamma}(N)_{\gamma_i}$ represented by ξ_i. Then ψ: $L\rightarrow N$ can be defined by putting

$$\psi\left(\sum a_ie_i\right)=\sum a_i\xi_i,\ \text{where}\ \sum a_ie_i\in L.$$

Clearly, ψ is a Γ-filtered A-homomorphism. Since $G^{\Gamma}(\psi)$ and φ agree on generators, we have $G^{\Gamma}(\psi)=\varphi$. Hence, by Corollary 1.4, ψ is a strict Γ-filtered surjection. □

Proposition 2.4. *Let P be a Γ-filtered A-module with Γ-filtration $FP = \{F_\gamma P\}_{\gamma\in\Gamma}$. The following two statements hold.*
(i) *If $G^\Gamma(P)$ is a projective $G^\Gamma(A)$-module, then P is a projective A-module.*
(ii) *If $G^\Gamma(P)$ is a Γ-graded free $G^\Gamma(A)$-module, then P is a free A-module.*

Proof. (i). By Proposition 2.3(iii), there is an exact sequence of Γ-filtered A-modules and strict Γ-filtered A-homomorphisms

$$0 \longrightarrow N \stackrel{\iota}{\longrightarrow} L \stackrel{\varphi}{\longrightarrow} P \longrightarrow 0$$

where L is a Γ-filtered free A-module with Γ-filtration FL, N is the kernel of the Γ-filtered A-epimorphism φ that has the Γ-filtration $FN = \{F_\gamma N = N \cap F_\gamma L\}_{\gamma\in\Gamma}$ induced by FL, and ι is the inclusion map. It follows from Proposition 1.3 and Corollary 1.4 that the associated Γ-graded sequence

$$0 \longrightarrow G^\Gamma(N) \stackrel{G^\Gamma(\iota)}{\longrightarrow} G^\Gamma(L) \stackrel{G^\Gamma(\psi)}{\longrightarrow} G^\Gamma(P) \longrightarrow 0$$

is exact. Since $G^\Gamma(P)$ is a projective $G^\Gamma(A)$-module, by Proposition 2.1, this sequence splits through Γ-graded $G^\Gamma(A)$-homomorphisms. Consequently, $G^\Gamma(L) = G^\Gamma(P) \oplus G^\Gamma(N)$ with $G^\Gamma(L)_\gamma = G^\Gamma(P)_\gamma \oplus G^\Gamma(N)_\gamma$, $\gamma \in \Gamma$, and the projection of $G^\Gamma(L)$ onto $G^\Gamma(N)$ gives a Γ-graded $G^\Gamma(A)$-epimorphism ψ: $G^\Gamma(L) \to G^\Gamma(N)$ such that $\psi \circ G^\Gamma(\iota) = 1_{G^\Gamma(N)}$. Further, by Proposition 2.3(iv), $\psi = G^\Gamma(\beta)$ for some strict Γ-filtered A-epimorphism β: $L \to N$. Note that $G^\Gamma(\beta) \circ G^\Gamma(\iota) = G^\Gamma(\beta \circ \iota) = 1_{G^\Gamma(N)}$. It follows from Corollary 1.4 that $\beta \circ \iota$ is an automorphism of N. Hence, $L \cong K \oplus P$. This shows that P is projective.

(ii). Suppose $G^\Gamma(P) = \oplus_{i\in J} G^\Gamma(A)\sigma(\xi_i)$ with the free $G^\Gamma(A)$-basis $\{\sigma(\xi_i)\}_{i\in J}$, where each $\xi_i \in F_{\gamma_i}P - F^*_{\gamma_i}P$, i.e., $d(\xi_i) = \gamma_i \in \Gamma$, $i \in J$. Then, by Proposition 1.2 (or its proof), $P = \sum_{i\in J} A\xi_i$ with

$$F_\gamma P = \sum_{i\in J}\left(\sum_{s_i\gamma_i\preceq\gamma} F_{s_i}A\right)\xi_i, \quad \gamma \in \Gamma.$$

By using the freeness of $\{\sigma(\xi_i)\}_{i\in J}$ and the property (σ) of σ-elements given in Section 1, one may directly check that $\{\xi_i\}_{i\in J}$ is a free basis for P over A; or else it may also be seen as follows. Construct as before the Γ-filtered free A-module $L = \oplus_{i\in J} Ae_i$ with Γ-filtration

$$F_\gamma L = \bigoplus_{i\in J}\left(\sum_{s_i\gamma_i\preceq\gamma} F_{s_i}A\right)e_i, \quad \gamma \in \Gamma,$$

such that each e_i has the degree $\gamma_i = d(\xi_i)$. Then we have an exact sequence of Γ-filtered A-modules and strict Γ-filtered A-homomorphisms

$$0 \longrightarrow N \longrightarrow L \xrightarrow{\varphi} P \longrightarrow 0$$

where N has the Γ-filtration $FN = \{F_\gamma N = N \cap F_\gamma L\}_{\gamma\in\Gamma}$ induced by FL. It follows from Proposition 1.3 that this sequence yields an exact sequence of Γ-graded $G^\Gamma(A)$-modules and Γ-graded $G^\Gamma(A)$-homomorphisms

$$0 \longrightarrow G^\Gamma(N) \longrightarrow G^\Gamma(L) \xrightarrow{G^\Gamma(\varphi)} G^\Gamma(P) \longrightarrow 0.$$

However, by Proposition 2.3(i), $G^\Gamma(\varphi)$ is an isomorphism. Hence $G^\Gamma(N) = \{0\}$ and consequently $N = \{0\}$ by Corollary 1.4. This proves that φ is an isomorphism, or in other words, P is free. □

Proposition 2.5. *Let M be a Γ-filtered A-module with Γ-filtration $FM = \{F_\gamma M\}_{\gamma\in\Gamma}$. Given an exact sequence of Γ-graded $G^\Gamma(A)$-modules and Γ-graded $G^\Gamma(A)$-homomorphisms*

$$0 \to N' \to L'_n \to \cdots \to L'_0 \to G^\Gamma(M) \to 0 \tag{1}$$

where the L'_i are Γ-graded free $G^\Gamma(A)$-modules, the following statements hold.

(i) *There exists an exact sequence of Γ-filtered A-modules and strict Γ-filtered A-homomorphisms*

$$0 \to N \to L_n \to \cdots \to L_0 \to M \to 0 \tag{2}$$

in which the L_i are Γ-filtered free A-modules, such that we have the isomorphism of chain complexes

$$\begin{array}{ccccccccc}
0 \to & N' & \to & L'_n & \to \cdots \to & L'_0 & \to G^\Gamma(M) \to 0 \\
& \cong\downarrow & & \cong\downarrow & & \cong\downarrow & \quad =\downarrow \\
0 \to & G^\Gamma(N) & \to & G^\Gamma(L_n) & \to \cdots \to & G^\Gamma(L_0) & \to G^\Gamma(M) \to 0
\end{array}$$

(ii) *If N' is a Γ-graded free $G^\Gamma(A)$-module, then N is a free A-module.*
(iii) *If the Γ-graded $G^\Gamma(A)$-module N' is a projective $G^\Gamma(A)$-module, then N is a projective A-module.*
(iv) *If all modules in the sequence (1) are finitely generated over $G^\Gamma(A)$, then all modules in the sequence (2) are finitely generated over A.*

Proof. (i). By Proposition 2.3, the homomorphism $L'_0 \to G^\Gamma(M)$ in sequence (1) has the form $G^\Gamma(\beta)$ for some strict Γ-filtered surjection β:

$L_0 \rightarrow M$, where L_0 is a Γ-filtered free A-module such that $L'_0 \cong G^{\Gamma}(L_0)$ as Γ-graded $G^{\Gamma}(A)$-modules. Let $N_0 = \mathrm{Ker}\beta$ and consider the Γ-filtration $FN_0 = \{F_{\gamma}N_0 = N_0 \cap F_{\gamma}L_0\}_{\gamma\in\Gamma}$ induced by FL_0. Then we have the diagram of Γ-graded $G^{\Gamma}(A)$-modules and Γ-graded $G^{\Gamma}(A)$-homomorphisms

$$\begin{array}{ccccccccc} \cdots \rightarrow L'_2 \rightarrow & L'_1 & \rightarrow & L'_0 & \rightarrow & G^{\Gamma}(M) & \rightarrow 0 \\ & & & \downarrow\cong & & \downarrow = & \\ 0 \rightarrow & G^{\Gamma}(N_0) & \rightarrow & G^{\Gamma}(L_0) & \rightarrow & G^{\Gamma}(M) & \rightarrow 0 \end{array}$$

which has two exact rows. Note that the directed square involved in the above diagram commutes. It turns out that the homomorphism $L'_1 \rightarrow L'_0$ factors through $G^{\Gamma}(N_0)$, that is, we obtain the diagram

$$\begin{array}{ccccccccc} \cdots \rightarrow L'_2 \rightarrow & L'_1 & \rightarrow & L'_0 & \rightarrow & G^{\Gamma}(M) & \rightarrow 0 \\ & \downarrow & & \downarrow\cong & & \downarrow = & \\ 0 \rightarrow & G^{\Gamma}(N_0) & \rightarrow & G^{\Gamma}(L_0) & \rightarrow & G^{\Gamma}(M) & \rightarrow 0 \\ & \downarrow & & & & & \\ & 0 & & & & & \end{array}$$

in which both the rows are exact and the diagram given by both the directed squares commutes. Starting with $L'_1 \rightarrow G^{\Gamma}(N_0) \rightarrow 0$, the foregoing constructive procedure can be repeated step by step to yield the desired sequence (2).

The assertions of (ii), (iii) and (iv) follow immediately from Proposition 2.4 and Proposition 1.2, respectively. □

Let A be a Γ-filtered K-algebra with Γ-filtration FA. To deal with flat modules over A, we need to define a Γ-filtration, respectively a Γ-gradation, for a tensor product of two Γ-filtered A-modules, respectively for a tensor product of two Γ-graded $G^{\Gamma}(A)$-modules.

Let M be a Γ-filtered left A-module with Γ-filtration FM, and let N be a Γ-filtered right A-module with Γ filtration FN. Viewing $N \otimes_A M$ as a $\mathbb{Z}$-module, we define the Γ-filtration $F(N \otimes_A M)$ of $N \otimes_A M$ as

$$F_{\gamma}(N \otimes_A M) = \mathbb{Z}\text{-span}\left\{x \otimes y \;\middle|\; x \in F_vN,\ y \in F_wM,\ vw \preceq \gamma\right\}, \quad \gamma \in \Gamma.$$

The associated Γ-graded $\mathbb{Z}$-module of $N \otimes_A M$ with respect to $F(N \otimes_A M)$ is then defined as $G^{\Gamma}(N \otimes_A M) = \oplus_{\gamma\in\Gamma} G^{\Gamma}(N \otimes_A M)_{\gamma}$ with

$$G^{\Gamma}(N \otimes_A M)_{\gamma} = F_{\gamma}(N \otimes_A M)/F^*_{\gamma}(N \otimes_A M), \quad \gamma \in \Gamma,$$

where $F^*_\gamma(N\otimes_A M)=\cup_{\gamma'\prec\gamma}F_{\gamma'}(N\otimes M)$.

Let P be a Γ-graded left $G^\Gamma(A)$-module, and let Q be a Γ-graded right $G^\Gamma(A)$-module. Viewing $Q\otimes_{G^\Gamma(A)}P$ as a $\mathbb{Z}$-module, we define the Γ-gradation of $Q\otimes_{G^\Gamma(A)}P$ as

$$(Q\otimes_{G^\Gamma(A)}P_\bullet)_\gamma=\mathbb{Z}\text{-span}\left\{z\otimes t\;\middle|\;z\in Q_v,\ t\in P_w,\ vw=\gamma\right\},\quad \gamma\in\Gamma.$$

Lemma 2.6. *Let M be a Γ-filtered left A-module with Γ-filtration FM, and let N be a Γ-filtered right A-module with Γ filtration FN. With the definition made above, the following statements hold.*
(i) *For $\overline{x}_v\in G^\Gamma(N)_v$ represented by $x\in F_vN$, and $\overline{y}_w\in G^\Gamma(M)_w$ represented by $y\in F_wM$, the mapping defined by*

$$\begin{array}{rcl}\varphi(M,N):G^\Gamma(N)\otimes_{G^\Gamma(A)}G^\Gamma(M) & \longrightarrow & G^\Gamma(N\otimes_A M)\\ \overline{x}_v\otimes\overline{y}_w & \mapsto & (\overline{x\otimes y})_{vw}\end{array}$$

is an epimorphism of Γ-graded $\mathbb{Z}$-modules.
(ii) *The canonical A-isomorphisms*

$$A\otimes_A M\xrightarrow{\cong}M \text{ and } N\otimes_A A\xrightarrow{\cong}N$$

are strict Γ-filtered A-isomorphisms.
(iii) *The strict Γ-filtered A-isomorphisms in* (ii) *induce Γ-graded $G^\Gamma(A)$-isomorphisms*

$$G^\Gamma(A\otimes_A M)\xrightarrow{\cong}G^\Gamma(M) \text{ and } G^\Gamma(N\otimes_A A)\xrightarrow{\cong}G^\Gamma(N).$$

(iv) *The canonical $G^\Gamma(A)$-isomorphisms*

$$G^\Gamma(A)\otimes_{G^\Gamma(A)}G^\Gamma(M)\xrightarrow{\cong}G^\Gamma(M) \text{ and } G^\Gamma(N)\otimes_{G^\Gamma(A)}G^\Gamma(A)\xrightarrow{\cong}G^\Gamma(N)$$

are Γ-graded $G^\Gamma(A)$-isomorphisms.

Proof. Verification is straightforward. □

Proposition 2.7. *Let M be a Γ-filtered left A-module with Γ-filtration FM. If $G^\Gamma(M)$ is a flat Γ-graded $G^\Gamma(A)$-module, then M is a flat A-module.*

Proof. Let J be a right ideal of A and $FJ=\{F_\gamma J=J\cap F_\gamma A\}_{\gamma\in\Gamma}$ the Γ-filtration of J induced by FA. Consider the inclusion map $\iota\colon J\hookrightarrow A$. Then the strict exactness of the Γ-filtered sequence

$$0\longrightarrow J\xrightarrow{\iota}A$$

yields the exact Γ-graded sequence

$$0\longrightarrow G^\Gamma(J)\xrightarrow{G^\Gamma(\iota)}G^\Gamma(A).$$

Furthermore, it follows from the flatness of $G^{\Gamma}(M)$ and Lemma 2.6 that we have the following commutative diagram of Γ-graded $\mathbb{Z}$-modules and Γ-graded $\mathbb{Z}$-homomorphisms:

$$\begin{array}{ccccc}
0 \to G^{\Gamma}(J) \otimes_{G^{\Gamma}(A)} G^{\Gamma}(M) & \overset{G^{\Gamma}(\iota)\otimes 1_{G^{\Gamma}(M)}}{\longrightarrow} & G^{\Gamma}(A) \otimes_{G^{\Gamma}(A)} G^{\Gamma}(M) & & \\
 & & & \cong\searrow & \\
\varphi(M,J)\Big\downarrow & & \varphi(M,A)\Big\downarrow & & G^{\Gamma}(M) \\
 & & & \cong\swarrow & \\
G^{\Gamma}(J \otimes_A M) & \underset{G^{\Gamma}(\iota\otimes 1_M)}{\longrightarrow} & G^{\Gamma}(A \otimes_A M) & &
\end{array}$$

As $\varphi(M, A)$ is an isomorphism and $G^{\Gamma}(\ell) \otimes 1_{G^{\Gamma}(M)}$ is a monomorphism, it turns out that $\varphi(M, J)$ is an isomorphism. So $G^{\Gamma}(\iota \otimes 1_M)$ must be a monomorphism. By previous Corollary 1.4, we conclude that $\iota \otimes 1_M$ is a strict Γ-filtered monomorphism. This proves the flatness of M. □

Let A be a Γ-filtered K-algebra with Γfiltration FA. Noticing that every A-module can be endowed with a Γ-filtration (Example 5 of Section 3 in Ch.1), Proposition 2.5 and Proposition 2.7 enable us to reach the main results of this section. In the text below we write p.dim to denote the projective dimension of a module, gl.dim to denote the (*left*) global homological dimension of a ring, and gl.wdim to denote the global weak dimension of a ring, respectively; and moreover, we write w.dim for the weak dimension of a module.

Theorem 2.8. *Let A be a Γ-filtered K-algebra with Γ-filtration FA, and let $G^{\Gamma}(A)$ be the associated Γ-graded K-algebra of A. The following statements hold.*

(i) *Let M be an A-module with Γ-filtration FM. Then $p.dim_A M \le p.dim_{G^{\Gamma}(A)} G^{\Gamma}(M)$. In particular, if $G^{\Gamma}(M)$ has a (finite or infinite Γ-graded) free resolution, then M has a (finite or infinite) free resolution.*

(ii) *$gl.dim A \le gl.dom G^{\Gamma}(A)$.*

(iii) *If $G^{\Gamma}(A)$ is left hereditary, then A is left hereditary.*

(iv) *Let M be an A-module with Γ-filtration FM. Then $w.dim_A M \le w.dim_{G^{\Gamma}(A)} G^{\Gamma}(M)$.*

(v) *$gl.wdim A \le gl.wdim G^{\Gamma}(A)$.*

(vi) *If $G^{\Gamma}(A)$ is a von Neuman regular ring then so is A.*

Proof. (i) and (ii) are immediate consequences of Proposition 2.4 and Proposition 2.5.

(iii). If $G^\Gamma(A)$ is left hereditary, then every left ideal of $G^\Gamma(A)$ is a projective $G^\Gamma(A)$-module. Let L be a left ideal of A and FL the Γ-filtration of L induced by FA. Using the inclusion map $L \hookrightarrow A$ (note that this is a strict Γ-filtered A-homomorphism), we may view $G^\Gamma(L)$ as a Γ-graded left ideal of $G^\Gamma(A)$ by Corollary 1.4. Thus, $G^\Gamma(L)$ is a projective $G^\Gamma(A)$-module, and it follows from Proposition 2.4 that L is a projective A-module. Therefore, A is left hereditary.

(iv). Note that any exact sequence

$$0 \to N \to L_n \to \cdots \to L_1 \to L_0 \to M \to 0$$

consisting of Γ-filtered free A-modules L_i and strict Γ-filtered A-homomorphisms yields an exact sequence

$$0 \to G^\Gamma(N) \to G^\Gamma(L_n) \to \cdots \to G^\Gamma(L_1) \to G^\Gamma(L_0) \to G^\Gamma(M) \to 0$$

consisting of Γ-graded free $G^\Gamma(A)$-modules $G^\Gamma(L_i)$ and Γ-graded $G^\Gamma(A)$-homomorphisms, where N has the Γ-filtration FN induced by FL_n. This assertion is an immediate consequences of Proposition 2.7.

Finally, both (v) and (vi) follow from (iv). □

2.3 The Γ-Graded Isomorphism $A^\Gamma_{\mathbf{LH}} \xrightarrow{\cong} G^\Gamma(A)$

In this section, we let Γ be a *totally ordered semigroup* with the total ordering $\prec$, and $R = \oplus_{\gamma\in\Gamma} R_\gamma$ a Γ-graded K-algebra. Consider the Γ-*grading filtration* $FR = \{F_\gamma R\}_{\gamma\in\Gamma}$ of R in the sense of (Ch.1, Section 3), that is, $F_\gamma R = \oplus_{\gamma'\preceq\gamma} R_{\gamma'}$, $\gamma \in \Gamma$. If I is an arbitrary ideal of R and $A = R/I$ is the corresponding quotient algebra, then A can be equipped with the Γ-filtration

$$FA = \{F_\gamma A = (F_\gamma R + I)/I\}_{\gamma\in\Gamma}$$

induced by FR which defines the associated Γ-graded algebra $G^\Gamma(A) = \oplus_{\gamma\in\Gamma} G^\Gamma(A)_\gamma$ of A with

$$G^\Gamma(A)_\gamma = F_\gamma A/F^*_\gamma A, \text{ where } F^*_\gamma A = \cup_{\gamma'\prec\gamma} F_{\gamma'} A.$$

Our aim is to show that there is a Γ-graded K-algebra isomorphism $G^\Gamma(A) \cong R/\langle \mathbf{LH}(I)\rangle$ (see the definition of $\mathbf{LH}(I)$ below).

To begin with, note that each element $f \in R$ can be uniquely written as a sum of finitely many Γ-homogeneous elements, say $f = \sum_{i=1}^{s} r_{\gamma_i}$, $r_{\gamma_i} \in R_{\gamma_i}$.

Assuming $\gamma_1 \succ \gamma_2 \succ \cdots \succ \gamma_s$, we define the Γ-*leading homogeneous element* of f, denoted $\mathbf{LH}(f)$, to be r_{γ_1}, that is, $\mathbf{LH}(f) = r_{\gamma_1}$, and say that f is of *degree* γ_1, denoted $d(f) = \gamma_1$. Thus, for a subset $S \subset R$, we write

$$\mathbf{LH}(S) = \{\mathbf{LH}(f) \mid f \in S\}$$

for the set of Γ-leading homogeneous elements of S. Considering an ideal I of R, since $\mathbf{LH}(I)$ consists of Γ-homogeneous elements, the ideal $\langle\mathbf{LH}(I)\rangle$ generated by $\mathbf{LH}(I)$ in R is Γ-graded, and consequently, the quotient algebra $R/\langle\mathbf{LH}(I)\rangle$ is a Γ-graded K-algebra, namely

$$R/\langle\mathbf{LH}(I)\rangle = \bigoplus_{\gamma\in\Gamma}(R/\langle\mathbf{LH}(I)\rangle)_\gamma$$
$$\text{with } (R/\langle\mathbf{LH}(I)\rangle)_\gamma = (R_\gamma + \langle\mathbf{LH}(I)\rangle)/\langle\mathbf{LH}(I)\rangle.$$

Definition 3.1. We call $R/\langle\mathbf{LH}(I)\rangle$ the Γ-*leading homogeneous algebra* of the quotient algebra $A = R/I$ and denote it by $A^{\Gamma}_{\rm LH}$, that is,

$$A^{\Gamma}_{\rm LH} = R/\langle\mathbf{LH}(I)\rangle.$$

Theorem 3.2. *With notation as fixed above, there is a* Γ*-graded* K*-algebra isomorphism*

$$A^{\Gamma}_{\rm LH} = R/\langle\mathbf{LH}(I)\rangle \xrightarrow{\cong} G^{\Gamma}(A).$$

Proof. We first note that Γ is ordered by the total ordering $\prec$, and that $G^{\Gamma}(A) = \oplus_{\gamma\in\Gamma}G^{\Gamma}(A)_\gamma$, where for each $\gamma \in \Gamma$, $G^{\Gamma}(A)_\gamma = F_\gamma A/F^*_\gamma A$ with $F_\gamma A = (F_\gamma R + I)/I$ and, as a K-subspace,

$$F^*_\gamma A = \bigcup_{\gamma'\prec\gamma} F_{\gamma'}A = \bigcup_{\gamma'\prec\gamma}\frac{F_{\gamma'}R + I}{I}$$
$$= \frac{\bigcup_{\gamma'\prec\gamma}F_{\gamma'}R + I}{I} = \frac{F^*_\gamma R + I}{I}.$$

It turns out that there are canonical isomorphisms of K-subspaces

$$\frac{R_\gamma \oplus F^*_\gamma R}{(I\cap F_\gamma R) + F^*_\gamma R} = \frac{F_\gamma R}{(I\cap F_\gamma R) + F^*_\gamma R} \xrightarrow{\cong} G^{\Gamma}(A)_\gamma, \quad \gamma\in\Gamma,$$

and consequently, we can extend the natural epimorphisms of K-subspaces

$$\phi_\gamma :\ R_\gamma \longrightarrow \frac{R_\gamma \oplus F^*_\gamma R}{(I\cap F_\gamma R) + F^*_\gamma R}, \quad \gamma\in\Gamma,$$

to define a Γ-graded K-algebra epimorphism $\phi: R \longrightarrow G^{\Gamma}(A)$. We claim that $\mathrm{Ker}\phi = \langle \mathbf{LH}(I)\rangle$. To see this, noticing $\langle \mathbf{LH}(I)\rangle$ is a Γ-graded ideal, it is sufficient to prove the equalities

$$\mathrm{Ker}\phi_\gamma = \langle \mathbf{LH}(I)\rangle \cap R_\gamma, \quad \gamma \in \Gamma.$$

Suppose $r_\gamma \in \mathrm{Ker}\phi_\gamma \subset R_\gamma$. Then $r_\gamma \in (I \cap F_\gamma R) + F_\gamma^* R$. If $r_\gamma \neq 0$, then, as a homogeneous element of degree γ, $r_\gamma = \mathbf{LH}(f)$ for some $f \in I \cap F_\gamma R$. This shows that $r_\gamma \in \langle \mathbf{LH}(I)\rangle \cap R_\gamma$. Hence $\mathrm{Ker}\phi_\gamma \subseteq \langle \mathbf{LH}(I)\rangle \cap R_\gamma$. Conversely, suppose $r_\gamma \in \langle \mathbf{LH}(I)\rangle \cap R_\gamma$. Then, as a homogeneous element of degree γ, $r_\gamma = \sum_{i=1}^{s} v_i \mathbf{LH}(f_i) w_i$, where v_i, w_i are homogeneous elements of R and $f_i \in I$. Write $f_i = \mathbf{LH}(f_i) + f_i'$ such that $d(f_i') \prec d(f_i)$, $i = 1, \ldots, s$. By the fact that Γ is ordered by the total ordering $\prec$, we may see that the expression

$$r_\gamma = \sum_{i=1}^{s} v_i f_i w_i - \sum_{i=1}^{s} v_i f_i' w_i$$

satisfies $\sum_{i=1}^{s} v_i f_i w_i \in I \cap F_\gamma R$ and $\sum_{i=1}^{s} v_i f_i' w_i \in F_\gamma^* R$. This shows that $r_\gamma \in (I \cap F_\gamma R) + F_\gamma^* R$, i.e., $r_\gamma \in \mathrm{Ker}\phi_\gamma$. Hence, $\langle \mathbf{LH}(I)\rangle \cap R_\gamma \subseteq \mathrm{Ker}\phi_\gamma$. Summing up, we conclude the desired equalities $\mathrm{Ker}\phi_\gamma = \langle \mathbf{LH}(I)\rangle \cap R_\gamma$, $\gamma \in \Gamma$. □

Remark Obviously, if I is a Γ-graded ideal of R, then $A = R/I = G^{\Gamma}(A)$ with respect to FA induced by the Γ-grading filtration FR of R.

As a primary illustration of Theorem 3.2, let us look at some classical examples reviewed in (Ch.1, Section 1).

Example 1. Let g be a K-Lie algebra with the K-basis $\{x_1, \ldots, x_n\}$ and the bracket product

$$[x_i,\, x_j] = \sum_{\ell=1}^{n} \lambda_{ij}^{\ell} x_\ell, \quad 1 \le i < j \le n,\ \lambda_{ij}^{\ell} \in K,$$

and let $U(\mathrm{g})$ be the universal enveloping algebra of g. Then by (Ch.1, Section 1), $U(\mathrm{g}) = K\langle X\rangle/I$, where $K\langle X\rangle = K\langle X_1, \ldots, X_n\rangle$ is the free K-algebra generated by $X = \{X_1, \ldots, X_n\}$, and the ideal I is generated by $\{X_jX_i - X_iX_j - \sum_{\ell=1}^{n} \lambda_{ij}^{\ell} X_\ell \mid 1 \le i < j \le n\}$. If we consider the natural $\mathbb{N}$-filtration $FU(\mathrm{g})$ of $U(\mathrm{g})$, then by Theorem 3.2, $U(\mathrm{g})$ has the associated $\mathbb{N}$-graded algebra $G^{\mathbb{N}}(U(\mathrm{g})) \cong K\langle X\rangle/\langle \mathbf{LH}(I)\rangle$, where $\mathbf{LH}(I)$ is the set of $\mathbb{N}$-leading homogeneous elements of I with respect to the natural $\mathbb{N}$-gradation of $K\langle X\rangle$. Note that $\Omega = \{X_jX_i - X_iX_j \mid 1 \le i < j \le n\} \subset \mathbf{LH}(I)$. By the PBW theorem concerning $U(\mathbf{g})$ (cf. [Hu]), $\langle\mathbf{LH}(I)\rangle$ is generated by Ω, i.e., $G^{\mathbb{N}}(U(\mathbf{g}))$ is isomorphic to the commutative polynomial K-algebra $K[x_1, \ldots, x_n]$ in variables $x_1, \ldots, x_n$. In later (Ch.4, Section 3) we will recapture this fact by showing that the defining relations of $U(\mathbf{g})$ form a Gröbner basis of I and moreover using (Ch.4, Proposition 2.2).

Example 2. Let $A_n(K)$ be the n-th Weyl algebra defined in (Ch.1, Section 1), that is, $A_n(K) = K\langle X, Y\rangle/I$, where $K\langle X,\ Y\rangle = K\langle X_1, \ldots, X_n, Y_1, \ldots, Y_n\rangle$ is the free K-algebra generated by $\{X, Y\} = \{X_1, \ldots, X_n, Y_1, \ldots, Y_n\}$ and the ideal I is generated by the elements

$$\begin{array}{ll} X_iX_j - X_jX_i,\ Y_iY_j - Y_jY_i, & 1 \le i < j \le n, \\ Y_jX_i - X_iY_j - \delta_{ij} \text{ (the Kronecker delta)}, & 1 \le i, j \le n. \end{array}$$

Considering the natural $\mathbb{N}$-filtration $FA_n(K)$ of $A_n(K)$, it follows from Theorem 3.2 that $A_n(K)$ has the associated $\mathbb{N}$-graded K-algebra $G^{\mathbb{N}}(A_n(K)) \cong K\langle X,\ Y\rangle/\langle\mathbf{LH}(I)\rangle$, where $\mathbf{LH}(I)$ is the set of $\mathbb{N}$-leading homogeneous elements of I with respect to the natural $\mathbb{N}$-gradation of $K\langle X,\ Y\rangle$. It is clear that $\mathbf{LH}(I)$ contains

$$\Omega = \left\{ \begin{array}{ll} X_iX_j - X_jX_i,\ Y_iY_j - Y_jY_i, & 1 \le i < j \le n \\ Y_jX_i - X_iY_j, & 1 \le i,\ j \le n \end{array} \right\}.$$

Indeed, a well-known result on $A_n(K)$ (cf. [Bj]) asserts that $\langle\mathbf{LH}(I)\rangle$ is generated by Ω, i.e., $G^{\mathbb{N}}(A_n(K))$ is isomorphic to the commutative polynomial K-algebra $K[x_1, \ldots, x_n, y_1, \ldots, y_n]$ in variables $x_1, \ldots, x_n, y_1, \ldots, y_n$. In (Ch.4, Section 3) we will recapture this fact by using (Ch.4, Proposition 2.2) after showing that the defining relations of $A_n(K)$ form a Gröbner basis of I.

Example 3. Let $\mathsf{C} = K[T]$ be the Clifford algebra defined in (Ch.1, Section 1), where $T = \{x_i\}_{i\in J}$, that is, $\mathsf{C} = K\langle X\rangle/I$, where $K\langle X\rangle$ is the free K-algebra generated by $X = \{X_i\}_{i\in J}$ and the ideal I is generated by the elements

$$\begin{array}{ll} X_i^2 - q_i & i \in J, \\ X_jX_i + X_iX_j - q_{ji} & i, j \in J,\ i \ne j, \end{array} \quad q_i, q_{ji} \in K.$$

If we consider the natural $\mathbb{N}$-filtration $F\mathsf{C}$ of C, then it follows from Theorem 3.2 that C has the associated $\mathbb{N}$-graded K-algebra $G^{\mathbb{N}}(\mathsf{C}) \cong K\langle X\rangle/\langle\mathbf{LH}(I)\rangle$, where $\mathbf{LH}(I)$ is the set of $\mathbb{N}$-leading homogeneous elements of I with respect to the natural $\mathbb{N}$-gradation of $K\langle X\rangle$. Note that $\mathbf{LH}(I)$ contains

$$\Omega = \{X_i^2,\ X_jX_i + X_iX_j \mid i, j \in J,\ i \ne j\},$$

which is clearly the set of defining relations of the exterior algebra E (see Ch.1, Section 1). In (Ch.3, Section 5), we will prove that the defining relations of C form a Gröbner basis, and consequently, we will show by using (Ch.4, Proposition 2.2) that $\langle \mathbf{LH}(I)\rangle$ is generated by Ω, i.e., $G^{\mathbb{N}}(\mathsf{C})$ is isomorphic to the exterior algebra E, recapturing another well-known fact.

Let $A = R/I$ be as before and L a left ideal of R such that $I \subset L$. If we equip $M = R/L$ with the Γ-filtration $FM = \{F_\gamma M = (F_\gamma R + L)/L\}_{\gamma\in\Gamma}$ induced by FR, M becomes a Γ-filtered A-module with the associated Γ-graded $G^\Gamma(A)$-module $G^\Gamma(M) = \oplus_{\gamma\in\Gamma} G^\Gamma(M)_\gamma$, where $G^\Gamma(M)_\gamma = F_\gamma M/F^*_\gamma M$ and $F^*_\gamma M = (F^*_\gamma R + L)/L$ for all $\gamma \in \Gamma$. Here we point out that the proof of Theorem 3.2 may be carried out to deal with the A-module $M = R/L$ directly so long as L is used in place of I and only left-hand side action is considered. We mention the result below but will not explore this module theory in detail in this book.

Theorem 3.3. *Let $M = R/L$ be as fixed above. Then there is an isomorphism of Γ-graded $G^\Gamma(A)$-modules:*

$$G^\Gamma(M) \cong R/\langle \mathbf{LH}(L)],$$

where $\langle \mathbf{LH}(L)]$ denotes the Γ-graded left ideal of R generated by $\mathbf{LH}(L)$. □

2.4 Recognizing A via $A^\Gamma_{\rm LH}$

Let Γ be a *well-ordered monoid* with the well-ordering $\prec$, and let $R = \oplus_{\gamma\in\Gamma} R_\gamma$ be a Γ-graded K-algebra. Consider an arbitrary ideal I of R, the corresponding quotient algebra $A = R/I$, and the Γ-filtration FA of A induced by the Γ-grading filtration FR of R. Then, in the course of lifting structural properties from $G^\Gamma(A)$ to A (Sections 1, 2), Theorem 3.2 allows us to replace the role of $G^\Gamma(A)$ by the Γ-leading homogeneous algebra $A^\Gamma_{\rm LH} = R/\langle \mathbf{LH}(I)\rangle$, namely, A can be studied more effectively in terms of the structure of $A^\Gamma_{\rm LH}$, as indicated by the following diagram:

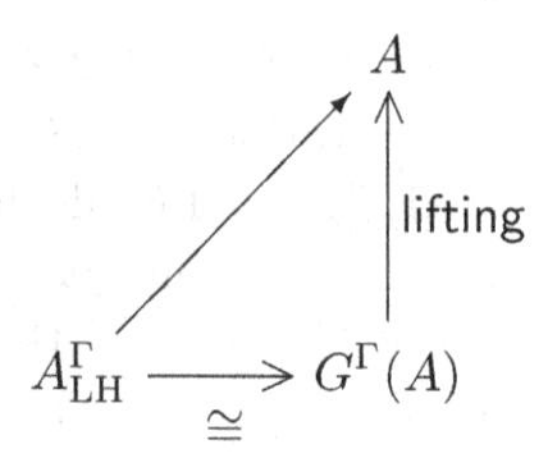

In this section, we translate explicitly the lifting properties from $G^{\Gamma}(A)$ to A obtained in Sections 1, 2 into the transferring properties from $A^{\Gamma}_{\rm LH}$ to A via the above diagram. Notations and conventions are maintained as before.

To better understand some of the transferring properties, let us first note a useful fact on the σ-elements associated to elements of A (see Section 1), that is, if $f \in R$, then

$$\sigma(\overline{f}) = \sigma(\ \overline{\mathbf{LH}(f)}\ ),$$

where $\overline{f}$ and $\overline{\mathbf{LH}(f)}$ denote the canonical images of f and $\mathbf{LH}(f)$ in A, respectively.

Theorem 4.1. *Let $A^{\Gamma}_{\rm LH} = R/\langle \mathbf{LH}(I)\rangle$ be the Γ-leading homogeneous algebra of the algebra $A = R/I$. The following statements hold.*
(i) *Suppose that $\{f_i\}_{i\in J}$ is a subset of R such that the image of $\{\mathbf{LH}(f_i)\}_{i\in J}$ in $A^{\Gamma}_{\rm LH}$ forms a K-basis for $A^{\Gamma}_{\rm LH}$, then the image of $\{f_i\}_{i\in J}$ in A forms a K-basis for A. Hence, if $A^{\Gamma}_{\rm LH}$ is finite dimensional over K, then A has the same property.*
(ii) *If $A^{\Gamma}_{\rm LH}$ is a domain, then so too is A.*
(iii) *If $A^{\Gamma}_{\rm LH}$ is a (semi-)prime ring, then so too is A.*
(iv) *If $A^{\Gamma}_{\rm LH}$ is a Noetherian domain and a maximal order in its quotient ring, then so too is A.*

□

Theorem 4.2. *Let $A = R/I$ and $A^{\Gamma}_{\rm LH} = R/\langle \mathbf{LH}(I)\rangle$ be as in Theorem 4.1. The following statements hold.*
(i) *Suppose that $A^{\Gamma}_{\rm LH}$ is Γ-graded left Noetherian, that is, every Γ-graded left ideal of $A^{\Gamma}_{\rm LH}$ is finitely generated. Then A is left Noetherian.*
(ii) *Suppose that $A^{\Gamma}_{\rm LH}$ is Γ-graded left Artinian, that is, $A^{\Gamma}_{\rm LH}$ satisfies the descending chain condition for Γ-graded left ideals. Then A is left Artinian.*
(iii) *If $A^{\Gamma}_{\rm LH}$ is a Γ-graded simple K-algebra, that is, $A^{\Gamma}_{\rm LH}$ does not have nontrivial Γ-graded ideal, then A is a simple K-algebra.*
(iv) *If the Krull dimension (in the sense of Gabriel and Rentschler) of $A^{\Gamma}_{\rm LH}$ is well defined, then the Krull dimension of A is defined and $K.dimA \le K.dimA^{\Gamma}_{\rm LH}$.*
(v) *Let L be a left ideal of R containing I. If $\langle \mathbf{LH}(L)]/\langle \mathbf{LH}(I)\rangle$ is a Γ-graded maximal left ideal of $A^{\Gamma}_{\rm LH}$, then L/I is a maximal left ideal of A, that is, R/L is a simple A-module.*
(vi) *If $A^{\Gamma}_{\rm LH}$ is semisimple (simple) Artinian, then so too is A.*

□

Remark Let A be a finitely generated K-algebra and let $\varphi\colon K[X] \to A$ be an epimorphism, where X is a free semigroup with free generators $x_1, \ldots, x_n$. If $f \in K[X]$ and $\overline{f}$ denotes the dominant (i.e. maximal) word in the support of f with respect to the graded lexicographic ordering on X (see (Ch.3, Section 2) for the definition), then $I = \{g \in X \mid g = \overline{f}$ for some $f \in K[X]$ with $\varphi(f) = 0\}$ is an ideal of the semigroup X. In [Ok], Jan Okninski proved that the Noetherian property, Artinian property, and the (semi-)primeness of the quotient semigroup X/I are equivalent to the same properties of the semigroup algebra $K[X/I]$ and thereby may be lifted to A. With notation used in this book, let us note that the algebra $K[X/I]$ is nothing but $K[X]/\langle \mathbf{LH}(\mathrm{Ker}\varphi)\rangle$. So, our results transferring Noetherian property, Artinian property, and the (semi-)primeness are just extensions of Okninski's results to general Γ-filtered algebras.

Theorem 4.3. *Let $A = R/I$ and $A^{\Gamma}_{\mathrm{LH}} = R/\langle \mathbf{LH}(I)\rangle$ be as in Theorem 4.2. The following statements hold.*
(i) *$gl.dimA \le gl.domA^{\Gamma}_{\mathrm{LH}}$.*
(ii) *If A^{Γ}_{LH} is left hereditary, then A is left hereditary.*
(iii) *$gl.wdimA \le gl.wdimA^{\Gamma}_{\mathrm{LH}}$.*
(iv) *If A^{Γ}_{LH} is a von Neuman regular ring, then so too is A.*

□

Remark (i) In dealing with Γ-graded projective modules, we have reminded the reader to notice a delicate point, namely Proposition 2.1 requires Γ satisfying the left cancelation law and this property is guaranteed by the fact that Γ is a totally ordered monoid. We take this place to point out that in ([Li3], Section 4), when Γ-graded projective module was involved, a version of Proposition 2.1 and such an indication were missing.
(ii) Note that the (left) *graded global dimension* of a Γ-graded algebra R, denoted gr.gl.dim$R$, may be well defined in terms of Γ-graded projective objects in the graded context (cf. [NVO], [LVO2]). In light of Proposition 2.1 and Proposition 2.4, we can modify Theorem 2.8(ii) as

$$\mathrm{gl.dim}A \le \ \mathrm{gr.gl.dim}G^{\Gamma}(A) \le \ \mathrm{gl.dim}G^{\Gamma}(A),$$

and modify Theorem 4.3(i) as

$$\mathrm{gl.dim}A \le \ \mathrm{gr.gl.dim}A^{\Gamma}_{\mathrm{LH}} \le \ \mathrm{gl.dim}A^{\Gamma}_{\mathrm{LH}},$$

that may be useful whenever gr.gl.dim can be controlled in an effective way.

Chapter 3

Gröbner Bases: Conception and Construction

Let $R = \oplus_{\gamma\in\Gamma}R_\gamma$ be a Γ-graded K-algebra, where Γ is a totally ordered semigroup, and let I be an arbitrary proper ideal of R. In view of (Ch.2, Sections 3, 4), we are naturally concerned with how to study the algebra $A = R/I$ by finding a "better generating set" F of I such that $\langle \mathbf{LH}(I)\rangle = \langle \mathbf{LH}(F)\rangle$, or more intrinsically, such that

$$R/\langle \mathbf{LH}(F)\rangle = R/\langle \mathbf{LH}(I)\rangle \xrightarrow{\cong} G^\Gamma(A).$$

To realize such a core idea in the subsequent chapters, this chapter is devoted to introducing the structure of Gröbner bases in a quite extensive context, but more attention is paid to Gröbner bases with respect to a skew multiplicative K-basis (see the definition in Section 4). Although two-sided Gröbner bases for two-sided ideals will play the leading role in this book, left Gröbner bases for left ideals are also introduced for completeness of the theory and a later use in Ch.8 (a right Gröbner basis theory for right ideals can be stated in a similar way). The general theory is developed through Sections 1 – 5 by employing the principle and methods adapted from [BW], [Gr1, 2], [Mor1, 2], and [Uf3]. In Sections 6, 7, we specialize the construction of (de)homogenized Gröbner bases for the reason that most well-developed computer algebra systems have the superiority of producing homogeneous Gröbner bases, and that this topic provides a useful computational channel between affine and projective algebraic geometry in both the commutative and noncommutative cases.

We maintain the conventions and notations used in Ch.1, Ch.2.

3.1 Monomial Ordering and Admissible System

In this section, we introduce the notions of (left) monomial ordering and (left) admissible system which are essential for developing a (left) Gröbner basis theory.

Let R be a K-algebra with the K-basis $\mathcal{B}$, and let $\prec$ be a total ordering on $\mathcal{B}$. Adopting the commonly used terminology in computational algebra, we call an element $u \in \mathcal{B}$ a *monomial*; and for each $f \in R$, say

$$f = \sum_{i=1}^{s} \lambda_i u_i, \quad \lambda_i \in K^*, \ u_i \in \mathcal{B}, \ u_1 \prec u_2 \prec \cdots \prec u_s,$$

we define

$$\begin{aligned} &\mathbf{LM}(f) = u_s, && \text{the } \textit{leading monomial} \text{ of } f; \\ &\mathbf{LC}(f) = \lambda_s, && \text{the } \textit{leading coefficient} \text{ of } f; \\ &\mathbf{LT}(f) = \lambda_s u_s, && \text{the } \textit{leading term} \text{ of } f. \end{aligned}$$

If $S \subset R$, then we write $\mathbf{LM}(S) = \{\mathbf{LM}(f) \mid f \in S\}$ for the set of leading monomials of S.

Definition 1.1. Let $\prec$ be a well-ordering on $\mathcal{B}$.
(i) $\prec$ is said to be a *monomial ordering* on $\mathcal{B}$ (or on R) if the following conditions are satisfied.

- **(MO1)** If $w, u, v, s \in \mathcal{B}$ are such that $w \prec u$ and $\mathbf{LM}(vws)$, $\mathbf{LM}(vus) \notin K^* \cup \{0\}$, then $\mathbf{LM}(vws) \prec \mathbf{LM}(vus)$.
- **(MO2)** For $w, u \in \mathcal{B}$, if $u = \mathbf{LM}(vws)$ for some $v, s \in \mathcal{B}$ with $v \neq 1$ or $s \neq 1$ (in the case $1 \in \mathcal{B}$), then $w \prec u$.

(ii) $\prec$ is said to be a *left monomial ordering* on $\mathcal{B}$ (or on R) if the following conditions are satisfied.

- **(LMO1)** If $w, u, v \in \mathcal{B}$ are such that $w \prec u$ and $\mathbf{LM}(vw)$, $\mathbf{LM}(vu) \notin K^* \cup \{0\}$, then $\mathbf{LM}(vw) \prec \mathbf{LM}(vu)$.
- **(LMO2)** For $w, u \in \mathcal{B}$, if $u = \mathbf{LM}(vw)$ for some $v \in \mathcal{B}$ with $v \neq 1$ (in the case $1 \in \mathcal{B}$), then $w \prec u$.

(iii) If $\prec$ is a (left) monomial ordering on $\mathcal{B}$, then we call the pair $(\mathcal{B}, \prec)$ a *(left) admissible system* of R.

Remark (i) Note that if KQ is the path algebra defined by a finite directed graph Q with finitely $n \geq 2$ edges $e_1, \ldots, e_n$, then the multiplicative K-basis

$\mathcal{B}$ of Λ does not contain the identity element $1 = e_1 + \cdots + e_n$. That is why we specified the case $1 \in \mathcal{B}$.

(ii) In the case that $1 \in \mathcal{B}$, (LMO2) and (MO2) turn out that $1 \prec u$ for all $u \in \mathcal{B}$ with $u \neq 1$. This shows that our definition of a (left, two-sided) monomial ordering is consistent with that used in the classical commutative and noncommutative cases.

(iii) If $1 \in \mathcal{B}$ and $u, v \in \mathcal{B} - \{1\}$ satisfy $uv = \lambda \in K^*$, then $\mathbf{LM}(uv) = 1$. Noticing (ii) above, we require that $\mathbf{LM}(uv) \notin K^* \cup \{0\}$ in order to make (LMO1) and (MO1) valid.

(iv) From the definition it is clear that if the K-basis $\mathcal{B}$ of R contains the identity element 1, then any monomial ordering on $\mathcal{B}$ is also a left monomial ordering, and hence any admissible system $(\mathcal{B}, \prec)$ of R is also a left admissible system of R. In (Ch.8, Section 3) we will see that the converse of this observation is not true in general.

Let $R = \oplus_{\gamma \in \Gamma} R_\gamma$ be a Γ-graded K-algebra, where Γ is a totally ordered semigroup with the total ordering $<$. If R has a K-basis $\mathcal{B}$ *consisting of Γ-homogeneous elements*, and if $\prec$ is a well-ordering on $\mathcal{B}$, then we may define the ordering $\prec_{gr}$ for u, $v \in \mathcal{B}$ subject to the rule:

$$u \prec_{gr} v \Leftrightarrow d(u) < d(v) \\ \text{or } d(u) = d(v) \text{ and } u \prec v,$$

where $d(\ )$ refers to the degree function on Γ-homogeneous elements of R. If $\prec_{gr}$ is a (left) monomial ordering on $\mathcal{B}$, then we call $\prec_{gr}$ a (left) *Γ-graded monomial ordering* on R.

It is straightforward to verify the monomial orderings given in the following examples.

Example 1. Let $R = K[x_1, \ldots, x_n]$ be the commutative polynomial K-algebra in n variables $x_1, \ldots, x_n$, and let

$$\mathcal{B} = \left\{ x^\alpha = x_1^{\alpha_1} \cdots x_n^{\alpha_n} \;\middle|\; \alpha = (\alpha_1, \ldots, \alpha_n) \in \mathbb{N}^n \right\}$$

be the standard K-basis of R. Then the following monomial orderings on $\mathcal{B}$ are commonly used in the literature.

(a) The lexicographic ordering $\prec_{lex}$. Ordering $x_1, \ldots, x_n$ arbitrarily, say

$$x_{i_1} \prec_{lex} x_{i_2} \prec_{lex} \cdots \prec_{lex} x_{i_n},$$

then $\prec_{lex}$ is defined subject to the rule: For $x^\alpha = x_{i_1}^{\alpha_1}\cdots x_{i_n}^{\alpha_n}$, $x^\beta = x_{i_1}^{\beta_1}\cdots x_{i_n}^{\beta_n} \in \mathcal{B}$,

$$x^\alpha \prec_{lex} x^\beta \Leftrightarrow \text{there is some } 1 \le j \le n \text{ such that } \alpha_\ell = \beta_\ell \text{ for } \ell < j \text{ and } \alpha_j < \beta_j.$$

Obviously, there are $n!$ different lexicographic orderings.

(b) The $\mathbb{N}$-graded lexicographic ordering $\prec_{grlex}$. Ordering $x_1,\dots,x_n$ arbitrarily, say

$$x_{i_1} \prec_{lex} x_{i_2} \prec_{lex} \cdots \prec_{lex} x_{i_n},$$

then $\prec_{grlex}$ is defined subject to the rule: For $x^\alpha = x_{i_1}^{\alpha_1}\cdots x_{i_n}^{\alpha_n}$, $x^\beta = x_{i_1}^{\beta_1}\cdots x_{i_n}^{\beta_n} \in \mathcal{B}$,

$$x^\alpha \prec_{grlex} x^\beta \Leftrightarrow d(x^\alpha) < d(x^\beta) \text{ or } d(x^\alpha) = d(x^\beta) \text{ and } x^\alpha \prec_{lex} x^\beta \text{ for the chosen } \prec_{lex},$$

where $d(\ )$ denotes the degree function on $\mathcal{B}$ with respect to the natural $\mathbb{N}$-gradation of R.

(c) Similarly, the $\mathbb{N}$-graded reverse lexicographic ordering $\prec_{grevlex}$ on R is obtained by using degrees of monomials in the natural $\mathbb{N}$-gradation of R and a chosen reverse lexicographic ordering $\prec_{revlex}$, which is defined subject to a fixed ordering

$$x_{i_1} \prec_{revlex} x_{i_2} \prec_{revlex} \cdots \prec_{revlex} x_{i_n}$$

and the rule: For $x^\alpha = x_{i_1}^{\alpha_1}\cdots x_{i_n}^{\alpha_n}$, $x^\beta = x_{i_1}^{\beta_1}\cdots x_{i_n}^{\beta_n} \in \mathcal{B}$,

$$x^\alpha \prec_{revlex} x^\beta \Leftrightarrow \text{there is some } 1 \le j \le n \text{ such that } \alpha_\ell = \beta_\ell \text{ for } \ell > j \text{ and } \alpha_j > \beta_j.$$

Note that $\prec_{revlex}$ itself is not a monomial ordering (Check it!). There are also $n!$ different $\mathbb{N}$-graded reverse lexicographic orderings.

So, $R = K[x_1,\dots,x_n]$ has the admissible systems

$$(\mathcal{B}, \prec_{lex}),\quad (\mathcal{B}, \prec_{grlex}), \text{ and } (\mathcal{B}, \prec_{grevlex}).$$

Example 2. Let $R = \mathcal{O}_n(\lambda_{ji})$ be the skew polynomial K-algebra generated by $x_1,\dots,x_n$ subject to the relations $x_jx_i = \lambda_{ji}x_ix_j$ with $\lambda_{ji} \in K^*$, $1 \le i < j \le n$ (see Ch.1, Section 1). Then R has the K-basis

$$\mathcal{B} = \left\{ x^\alpha = x_1^{\alpha_1}\cdots x_n^{\alpha_n} \;\middle|\; \alpha = (\alpha_1,\dots,\alpha_n) \in \mathbb{N}^n \right\}.$$

Similar argument as with a commutative polynomial algebra in (1) above shows that R has the admissible systems

$$(\mathcal{B}, \prec_{lex}),\quad (\mathcal{B}, \prec_{grlex}),\ \text{and}\ (\mathcal{B}, \prec_{grevlex}).$$

Example 3. Let $R = \mathsf{E}$ be the exterior algebra generated by $T = \{x_i\}_{i\in J}$ subject to the relations $x_i^2 = 0,\ x_jx_i + x_ix_j = 0,\ i \in J$ (see Ch.1, Section 1). If J is well ordered by some $<$, then R has the K-basis

$$\mathcal{B} = \{1,\ x_{i_1}\cdots x_{i_s} \mid i_1 < i_2 < \cdots < i_s,\ i_\ell \in J\}.$$

As with a commutative polynomial algebra, R has the admissible systems

$$(\mathcal{B}, \prec_{lex}),\quad (\mathcal{B}, \prec_{grlex}),\ \text{and}\ (\mathcal{B}, \prec_{grevlex}).$$

Example 4. Let $R = K\langle X_1,\ldots,X_n\rangle$ be the free K-algebra generated by $X = \{X_1,\ldots,X_n\}$. Then R has the standard K-basis

$$\mathcal{B} = \left\{1,\ w = X_{i_1}\cdots X_{i_s}\ \middle|\ X_{i_j} \in X,\ s \geq 1\right\}.$$

Ordering $X_1,\ldots,X_n$ arbitrarily, say

$$X_{i_1} \prec_{lex} X_{i_2} \prec_{lex} \cdots \prec_{lex} X_{i_n},$$

then for $u = X_{\ell_1}\cdots X_{\ell_s},\ v = X_{t_1}\cdots X_{t_m} \in \mathcal{B}$, the lexicographic ordering $\prec_{lex}$ on B may be defined as

$$u \prec_{lex} v \Leftrightarrow \text{there is some } 1 \leq p \leq \min\{s,\ m\} \text{ such that}$$
$$X_{\ell_k} = X_{t_k} \text{ for } k < p \text{ and } X_{\ell_p} \prec_{lex} X_{t_p}.$$

Similarly, we may define the reverse lexicographic ordering $\prec_{revlex}$ on $\mathcal{B}$. It is easy to see that both $\prec_{lex}$ and $\prec_{revlex}$ are total ordering on $\mathcal{B}$. However, *they are not monomial orderings* on R. For instance, look at the free algebra $K\langle X_1, X_2\rangle$ with $X_1 \prec_{lex} X_2$. Then we have

$$X_2 \succ_{lex} X_1X_2 \succ_{lex} X_1X_1X_2 \succ_{lex} X_1X_1X_1X_2 \succ_{lex} \cdots.$$

This shows that $\prec_{lex}$ *is not a well-ordering on* R. Similarly, $\prec_{revlex}$ *is not a monomial ordering on* R. Nevertheless, we do have the following commonly used monomial orderings on R.

(a) The $\mathbb{N}$-graded lexicographic ordering $\prec_{grlex}$. Ordering $X_1,\ldots,X_n$ arbitrarily, say

$$X_{i_1} \prec_{grlex} X_{i_2} \prec_{grlex} \cdots \prec_{grlex} X_{i_n},$$

then for $u = X_{\ell_1} \cdots X_{\ell_s}$, $v = X_{t_1} \cdots X_{t_m} \in \mathcal{B}$, $\prec_{grlex}$ is defined as

$$u \prec_{grlex} v \Leftrightarrow d(u) < d(v)$$
$$\text{or } d(u) = d(v) \text{ and } u \prec_{lex} v \text{ for the chosen } \prec_{lex},$$

where $d(\ )$ denotes the degree function on $\mathcal{B}$ with respect to the natural $\mathbb{N}$-gradation of R. Note that in the definition of $\prec_{grlex}$, comparing degree takes priority and there are only finitely many $u \in \mathcal{B}$ with $d(u) \le d(v)$. Thus, $\prec_{grlex}$ is a well-ordering and hence a monomial ordering on R.

(b) Similarly, the $\mathbb{N}$-graded reverse lexicographic ordering $\prec_{grevlex}$ is defined on R by using degrees of monomials and a chosen reverse lexicographic ordering.

Hence, $R = K\langle X_1, \ldots, X_n\rangle$ has the admissible systems

$$(\mathcal{B}, \prec_{grlex}) \text{ and } (\mathcal{B}, \prec_{grevlex}).$$

Example 5. Let $R = KQ$ be the path algebra defined by a finite directed graph Q. Then the standard K-basis

$$\mathcal{B} = \{\text{finite directed paths of } Q, \text{ including the vertices}\}$$

is a skew multiplicative K-basis of R. A (reverse) lexicographic ordering on $\mathcal{B}$ may be defined as follows. Suppose the set of vertices $Q_0 = \{v_1, \ldots, v_s\}$, and the set of arrows $Q_1 = \{a_1, \ldots, a_m\}$. Ordering the vertices and arrows arbitrarily in which the vertices smaller than the arrows, say

$$v_{i_1} \prec_{lex} v_{i_2} \prec_{lex} \cdots \prec_{lex} v_{i_s} \prec_{lex} a_{j_1} \prec_{lex} a_{j_2} \prec_{lex} \cdots \prec_{lex} a_{j_m},$$

then the (reverse) lexicographic ordering $\prec_{lex}$ ($\prec_{revlex}$) on $\mathcal{B}$ is defined exactly as in Example 4 for a free algebra. Since the path algebra defined by a directed graph with one vertex and two loops is isomorphic to the free algebra $K\langle X_1, X_2\rangle$ (see Ch.1, Section 1), it follows that in general $\prec_{lex}$, respectively $\prec_{revlex}$, is not a monomial ordering on R. But actually, as with a free algebra, we have the $\mathbb{N}$-graded (reverse) lexicographic monomial ordering $\prec_{grlex}$ ($\prec_{grevlex}$) on R.

Therefore, $R = KQ$ has the admissible systems

$$(\mathcal{B}, \prec_{grlex}) \text{ and } (\mathcal{B}, \prec_{grevlex}).$$

Remark If each algebra considered in Examples 1 – 5 is endowed respectively with a weight $\mathbb{N}$-gradation in the sense of (Ch.1, Section 2), then the obtained $\mathbb{N}$-graded monomial ordering is called a *weight* $\mathbb{N}$*-graded monomial ordering*.

Some other monomial orderings defined for free algebras and path algebras may be found in [Gr2]. For more monomial orderings and admissible systems, such as those for Clifford algebras (exterior algebras), Weyl algebras and enveloping algebras of Lie algebras, we refer to [HT], [KR-W], [Li1], and [Lev] for details.

New convention Henceforth in this book, unless otherwise stated, a graded monomial ordering of any type will be denoted by $\prec_{gr}$.

3.2 Division Algorithm and Gröbner Basis

Let R be a K-algebra, and let $(\mathcal{B}, \prec)$ be a (left) admissible system of R as defined in the last section, that is, $\mathcal{B}$ is a K-basis of R, and $\prec$ is a (left) monomial ordering on $\mathcal{B}$. In this section, after demonstrating a (left) division algorithm we introduce the notion of a (left) Gröbner basis and, in the case that the (left) division of monomials is transitive, we prove that every (left) ideal of R has a (minimal) Gröbner basis.

The (left) division algorithm

Let $(\mathcal{B}, \prec)$ be an admissible system of R. For $u, v \in \mathcal{B}$, we say that u *divides* v, denoted $u|v$, provided there are $w, s \in \mathcal{B}$ such that $v = \mathbf{LM}(wus)$; similarly if $(\mathcal{B}, \prec)$ is a left admissible system of R, then, for $u, v \in \mathcal{B}$, we say that u *divides* v *from left-side*, denoted also by $u|v$, provided there is some $w \in \mathcal{B}$ such that $v = \mathbf{LM}(wu)$. Clearly, if $1 \in \mathcal{B}$ then $1|u$ for all $u \in \mathcal{B}$.

Let $f,\ g \in R - \{0\}$. The (left) division of monomials defined above extends naturally to a (left) division of f by g as follows.

If $\mathbf{LM}(g)|\mathbf{LM}(f)$, then there are $w, s \in \mathcal{B}$ such that $\mathbf{LM}(f) = \mathbf{LM}(w\mathbf{LM}(g)s)$; furthermore, writing

$$\begin{aligned} &g = \mathbf{LC}(g)\mathbf{LM}(g) + g' \text{ with } \mathbf{LM}(g') \prec \mathbf{LM}(g), \\ &\lambda = \mathbf{LC}(wgs - wg's), \end{aligned}$$

we have

$$f = \frac{\mathbf{LC}(f)}{\lambda \mathbf{LC}(g)} wgv + f'$$

satisfying $\mathbf{LM}(f) = \mathbf{LM}(w\mathbf{LM}(g)s)$, $\mathbf{LM}(f') \prec \mathbf{LM}(f)$. If $\mathbf{LM}(g)|\mathbf{LM}(f')$, repeat the same procedure for f'. While in the case that $\mathbf{LM}(g) \nmid \mathbf{LM}(f)$, consider the divisibility of $\mathbf{LM}(f_1)$ by $\mathbf{LM}(g)$, where $f_1 = f - \mathbf{LT}(f)$. Since

$\prec$ is a well-ordering, the division process stops after a finite number of steps and it turns out an expression

$$f = \sum_i \lambda_i w_i g s_i + r_{_f}$$

with $\lambda_i \in K^*$, $w_i, s_i \in \mathcal{B}$, and either $r_{_f} = 0$ or $r_{_f} = \sum_m \lambda_m v_m$, $\lambda_m \in K^*$, $v_m \in \mathcal{B}$, satisfying $w_i\mathbf{LM}(g)s_i \neq 0$, $\mathbf{LM}(w_i g s_i) \preceq \mathbf{LM}(f)$, $\mathbf{LM}(r_{_f}) \preceq \mathbf{LM}(f)$, and no v_m can be divided by $\mathbf{LM}(g)$ provided $r_{_f} \neq 0$.

The division procedure of f by g from left-side can be performed in a similar way, and the output result expresses f as

$$f = \sum_i \lambda_i w_i g + r_{_f}$$

with $\lambda_i \in K^*$, $w_i \in \mathcal{B}$, and either $r_{_f} = 0$ or $r_{_f} = \sum_m \lambda_m v_m$, $\lambda_m \in K^*$, $v_m \in \mathcal{B}$, satisfying $w_i\mathbf{LM}(g) \neq 0$, $\mathbf{LM}(w_i g) \preceq \mathbf{LM}(f)$, $\mathbf{LM}(r_{_f}) \preceq \mathbf{LM}(f)$, and no v_m can be divided by $\mathbf{LM}(g)$ provided $r_{_f} \neq 0$.

Now, let S be a nonempty subset of $R - K$. Actually the (left) division of f by g demonstrated above gives rise to a (left) division procedure manipulating the reduction of f by S. That is, if we start the (left) division by detecting whether or not $\mathbf{LM}(f)$ can be divided by some $\mathbf{LM}(g_i)$ with $g_i \in S$, then, after a finite number of reductions as described above, f has a representation

$$f = \sum_{i,j} \lambda_{ij} w_{ij} g_i s_{ij} + r_{_f}$$

with $\lambda_{ij} \in K^*$, $g_i \in S$, $w_{ij}, s_{ij} \in \mathcal{B}$, and either $r_{_f} = 0$ or $r_{_f} = \sum_m \lambda_m v_m$, $\lambda_m \in K^*$, $v_m \in \mathcal{B}$, satisfying $w_{ij}\mathbf{LM}(g_i)s_{ij} \neq 0$, $\mathbf{LM}(w_{ij} g_i s_{ij}) \preceq \mathbf{LM}(f)$, $\mathbf{LM}(r_{_f}) \preceq \mathbf{LM}(f)$, and no v_m can be divided by any $\mathbf{LM}(g_j)$ with $g_j \in S$ (provided $r_{_f} \neq 0$), respectively

$$f = \sum_{ij} \lambda_{ij} w_{ij} g_i + r_{_f}$$

with $\lambda_{ij} \in K^*$, $g_i \in S$, $w_{ij} \in \mathcal{B}$, and either $r_{_f} = 0$ or $r_{_f} = \sum_m \lambda_m v_m$, $\lambda_m \in K^*$, $v_m \in \mathcal{B}$, satisfying $w_{ij}\mathbf{LM}(g_i) \neq 0$, $\mathbf{LM}(w_{ij} g_i) \preceq \mathbf{LM}(f)$, $\mathbf{LM}(r_{_f}) \preceq \mathbf{LM}(f)$, and no v_m can be divided by any $\mathbf{LM}(g_j)$ with $g_j \in S$ (provided $r_{_f} \neq 0$).

If $S = \{g_1, \ldots, g_s\} \subset R - K$ is an *ordered finite set*, then the division procedure performed above may be written (in pseudo-code) into an algorithm (similar algorithm may be written for the left division procedure).

Algorithm-DIV

Input: $g_1, \ldots, g_s$ (ordered), f
Output: $q_{ij} = \lambda_{ij} w_{ij} g_i s_{ij}$, $r = r_f$ such that $f = \sum_{ij} q_{ij} + r_f$ as described above.
Initialization: $q_{ij} = 0$, $r = 0$, $h = f$, Divoccur: = false
While ($h \neq 0$ and Divoccur == false) do
  For (i = 1) to s do
    If $\mathbf{LM}(h) = \mathbf{LM}(s_{ij}\mathbf{LM}(g_i)w_{ij})$ for $s_{ij}, w_{ij} \in \mathcal{B}$ then
    $q_{ij} \colon = q_{ij} + \dfrac{\mathbf{LC}(h)}{\lambda_i \mathbf{LC}(g_i)} s_{ij} g_i w_{ij}$
    $h \colon = h - \dfrac{\mathbf{LC}(h)}{\lambda_i \mathbf{LC}(g_i)} s_{ij} g_i w_{ij}$
    (where $g_i = \mathbf{LC}(g_i)\mathbf{LM}(g_i) + g_i'$ with $\mathbf{LM}(g_i') \prec \mathbf{LM}(g_i)$
     and $\lambda_i = \mathbf{LC}(s_{ij} g_i w_{ij} - s_{ij} g_i' w_{ij})$.
    Divoccur: = true
    Else i: = i+1
  If Divoccur == false then
    $r \colon = r + \mathbf{LC}(h)\mathbf{LM}(h)$
    $h \colon = h - \mathbf{LC}(h)\mathbf{LM}(h)$
    End
  End
End

The element r obtained in **Algorithm-DIV** is called *a remainder of* f *on division by* S. As illustrated by the following examples, r is in general not unique (if $s \geq 2$), that is, it depends on the order of elements used in the division process. Nevertheless, bearing in mind this remark, from now on we adopt the commonly used notation $\overline{f}^S$ to denote *any* remainder of f on division by S and understand in the context when it is unique by referring to the forthcoming Theorem 3.2.

Example 1. Let $R = K[x, y]$ be the commutative polynomial algebra in x, y, and let $S = \{g_1 = x^2 + xy + 1,\ g_2 = xy^2 + y\}$. Consider the $\mathbb{N}$-graded lexicographic ordering $y \prec_{gr} x$ with respect to the natural $\mathbb{N}$-gradation of R. Then $\mathbf{LM}(g_1) = x^2$, $\mathbf{LM}(g_2) = xy^2$. For $f = x^2y^3 + x^2y + x$, let us divide f by S in two different orders.

(a) At each step, g_1 always takes the priority for division. In this order we have

$$f = (y^3 + y)g_1 - (y^2 + 1)g_2 + x,$$
$$\overline{f}^S = x.$$

(b) At each step, g_2 always takes the priority for division. In this order we get

$$f = (xy - 2)g_2 + yg_1 + x + y,$$
$$\overline{f}^S = x + y.$$

Example 2. Let $R = K\langle X_1, X_2, X_3\rangle$ be the free K-algebra generated by X_1, X_2, X_3 over K, and let $S = \{g_1 = X_1X_2 - X_1,\ g_2 = X_2^2 - X_1X_3\}$. Consider the $\mathbb{N}$-graded lexicographic ordering $X_3 \prec_{gr} X_1 \prec_{gr} X_2$ with respect to the natural $\mathbb{N}$-gradation of R. Then $\mathbf{LM}(g_1) = X_1X_2$, $\mathbf{LM}(g_2) = X_2^2$. For $f = X_1^2X_2^3X_3$, let us divide f by S in two different orders.

(a) At each step, g_1 always takes the priority for division. In this order we have

$$f = X_1g_1X_2^2X_3 + X_1g_1X_2X_3 + X_1g_1X_3 + X_1^2X_3,$$
$$\overline{f}^S = X_1^2X_3.$$

(b) At each step, g_2 always takes the priority for division. In this order we get

$$f = X_1^2g_2X_3 + X_1^3X_3^2,$$
$$\overline{f}^S = X_1^3X_3^2.$$

In the forthcoming Section 3 we will see that (left) Gröbner basis provides a perfect solution to the uniqueness of $\overline{f}^S$ and the related "membership problem" for a (left) ideal. We introduce below the notion of a (left) Gröbner basis and prove the existence of a minimal (left) Gröbner basis under appropriate condition.

(*Left*) *Gröbner bases*

In what follows, R is a K-algebra with the K-basis $\mathcal{B}$.

Definition 2.1. (i) Let $(\mathcal{B}, \prec)$ be an admissible system of the algebra R, and I an ideal of R. A subset $\mathcal{G} \subset I$ is said to be a *Gröbner basis* of I if for each nonzero $f \in I$, there is some $g \in \mathcal{G}$ such that $\mathbf{LM}(g)|\mathbf{LM}(f)$.

(ii) Let $(\mathcal{B}, \prec)$ be a left admissible system of the algebra R, and L a left ideal of R. A subset $\mathcal{G} \subset L$ is said to be a *left Gröbner basis* of L if for each nonzero $f \in L$, there is some $g \in \mathcal{G}$ such that $\mathbf{LM}(g)|\mathbf{LM}(f)$ from left-side.
(iii) If in (i) and (ii) above $R = \oplus_{g\in \mathsf{S}} R_g$ is an S-graded K-algebra, where S is a semigroup, then a (left) Gröbner basis consisting of S-homogeneous elements is called a *homogeneous Gröbner basis.*
(iv) A (left) Gröbner basis $\mathcal{G}$ is said to be a *minimal Gröbner basis* if any proper subset of $\mathcal{G}$ cannot be a (left) Gröbner basis for the (left) idea generated by $\mathcal{G}$.

Let $\mathcal{G}$ be a (left) Gröbner basis as defined above. By executing the division algorithm, it is clear that $\mathcal{G}$ is first of all a generating set for the (left) ideal considered; moreover, it is easy to see that if the division of monomials is transitive, then $\mathcal{G}$ is minimal if and only if $\mathbf{LM}(g_1) \nmid \mathbf{LM}(g_2)$ for all $g_1, g_2 \in \mathcal{G}$ with $g_1 \neq g_2$.

Let $(\mathcal{B}, \prec)$ be a (left) admissible system of the algebra R. If J is a nonzero (left) ideal of R, then $J^* = J - \{0\}$ is trivially a (left) Gröbner basis of J. Although we do not know in general if there is an effective way to construct a "smaller" (left) Gröbner basis for J, the next proposition tells us that the *transitivity* of the division of monomials may guarantee the existence of a *minimal* (left) Gröbner basis $\mathcal{G}$ for J.

Before stating the proposition concerning the existence of a minimal (left) Gröbner basis, let us introduce some more notions needed not only by the present section but also by the subsequent contents. Let $(\mathcal{B}, \prec)$ be a (left) admissible system of the algebra R. A subset $\Omega \subset \mathcal{B}$ is said to be *reduced* if $u, v \in \Omega$ and $u \neq v$ implies $u \nmid v$. If S is a subset in R such that $\mathbf{LM}(S)$ is reduced, then we say that S is *LM-reduced* (In [Gr2], such a subset S in a path algebra is called *tip reduced*). It is easy to see that a (left) Gröbner basis $\mathcal{G}$ of a (left) ideal J is LM-reduced if and only if $\mathcal{G}$ is a minimal (left) Gröbner basis.

Proposition 2.2. (i) *Suppose that the algebra R has an admissible system $(\mathcal{B}, \prec)$, and that the division of monomials is transitive. Then every ideal I of R has a Gröbner basis*

$$\mathcal{G} = \{g \in I \mid \textit{if } g' \in I \textit{ and } \mathbf{LM}(g') \neq \mathbf{LM}(g), \textit{ then } \mathbf{LM}(g') \nmid \mathbf{LM}(g)\},$$

and $\mathcal{G}$ is a minimal Gröbner basis if it satisfies: g_1, $g_2 \in \mathcal{G}$ and $g_1 \neq g_2$ imply $\mathbf{LM}(g_1) \neq \mathbf{LM}(g_2)$.

(ii) *Suppose that the algebra R has a left admissible system $(\mathcal{B}, \prec)$, and that the division of monomials from left-side is transitive. Then every left ideal L of R has a left Gröbner basis*

$$\mathcal{G} = \{g \in L \mid \text{if } g' \in L \text{ and } \mathbf{LM}(g') \neq \mathbf{LM}(g), \text{ then } \mathbf{LM}(g') \nmid \mathbf{LM}(g)\},$$

and $\mathcal{G}$ is a minimal left Gröbner basis if it satisfies: $g_1, g_2 \in \mathcal{G}$ and $g_1 \neq g_2$ imply $\mathbf{LM}(g_1) \neq \mathbf{LM}(g_2)$.
(iii) *If in* (i) *and* (ii) *above $R = \oplus_{g\in S} R_g$ is an S-graded K-algebra, where S is a semigroup, and if $\mathcal{B}$ consists of S-homogeneous elements, then every S-graded (left) ideal has a minimal homogeneous (left) Gröbner basis.*

Proof. By the transitivity of the division of monomials, i.e., $u, v, w \in \mathcal{B}$ and $u|v$, $v|w$ implies $u|w$, the verification of (i) – (ii) is straightforward by referring to the remark on the LM-reduced (left) Gröbner bases given preceding the proposition. To prove (iii), noticing that $\mathcal{B}$ consists of homogeneous elements, by (Ch.1, Proposition 2.1), it is sufficient to consider only homogeneous elements in (i) and (ii). □

Definition 2.3. If the algebra R has a (left) admissible system $(\mathcal{B}, \prec)$, such that the (left) division of monomials with respect to $\prec$ is transitive, then we say that R has a *(left) Gröbner basis theory.*

If R has a (left) Gröbner basis theory such that every (left) ideal has a finite (left) Gröbner basis, then we say that R has a *finite (left) Gröbner basis theory.*

3.3 Gröbner Bases and Normal Elements

In this section we assume that the K-algebra R has a (left) Gröbner basis theory with respect to a (left) admissible system $(\mathcal{B}, \prec)$ in the sense of Definition 2.3. Given a (left) ideal I of R, after introducing normal elements in R (mod I), we show, for any $f \in R$, the uniqueness of the remainder $\overline{f}^{\mathcal{G}}$ of f on division by a (left) Gröbner basis $\mathcal{G}$ of I, and then we derive the fundamental decomposition theorem of the vector space R by I. It turns out that the "membership problem" for I is solved, and a K-basis for the quotient algebra (quotient module) R/I is obtained.

Notations are maintained as before.

Definition 3.1. Let I be a (left) ideal of R. Put

$$N(I) = \{u \in \mathcal{B} \mid \mathbf{LM}(f) \nmid u \text{ for any } f \in I\}.$$

(i) A monomial $u \in N(I)$ is called a *normal monomial* (mod I).
(ii) An element $f \in R$ is called a *normal element* (mod I) if f is a K-linear combination of normal monomials (mod I), or equivalently, if $f \in K$-span$N(I)$.

Theorem 3.2. (i) *Let I be an ideal of R and $\mathcal{G}$ a Gröbner basis of I. Then*

$$N(I) = \{u \in \mathcal{B} \mid \mathbf{LM}(g) \nmid u \textit{ for any } g \in \mathcal{G}\},$$

and each nonzero $f \in R$ has a finite representation

$$f = \sum_{i,j} \lambda_{ij} s_{ij} g_i w_{ij} + \overline{f}^{\mathcal{G}}, \quad \lambda_{ij} \in K,\ s_{ij}, w_{ij} \in \mathcal{B},\ g_i \in \mathcal{G},$$

in which $\mathbf{LM}(s_{ij} g_i w_{ij}) \preceq \mathbf{LM}(f)$ whenever $\lambda_{ji} s_{ij} \mathbf{LM}(g_i) w_{ij} \neq 0$, and either $\overline{f}^{\mathcal{G}} = 0$ or $\overline{f}^{\mathcal{G}}$ is a unique normal element (mod I). Hence, $f \in I$ if and only if $\overline{f}^{\mathcal{G}} = 0$, solving the "membership problem" for I.
(ii) *Let L be a left ideal of R and $\mathcal{G}$ a left Gröbner basis of I. Then*

$$N(L) = \{u \in \mathcal{B} \mid \mathbf{LM}(g) \nmid u \textit{ for any } g \in \mathcal{G}\},$$

and each nonzero $f \in R$ has a finite representation

$$f = \sum_{i,j} \lambda_{ij} s_{ij} g_i + \overline{f}^{\mathcal{G}}, \quad \lambda_{ij} \in K,\ s_{ij} \in \mathcal{B},\ g_i \in \mathcal{G},$$

in which $\mathbf{LM}(s_{ij} g_i) \preceq \mathbf{LM}(f)$ whenever $\lambda_{ji} s_{ij} \mathbf{LM}(g_i) \neq 0$, and either $\overline{f}^{\mathcal{G}} = 0$ or $\overline{f}^{\mathcal{G}}$ is a unique normal element (mod L). Hence, $f \in L$ if and only if $\overline{f}^{\mathcal{G}} = 0$, solving the "membership problem" for L.

Proof. (i). The determination of $N(I)$ by $\mathcal{G}$ follows from the definition of a Gröbner basis. The existence of a representation of f as described in the theorem follows from the division of f by $\mathcal{G}$. Suppose somehow we also have $f = \sum_{t,j} \lambda_{tj} s_{tj} g_t w_{tj} + r$, where r is normal (mod I). Then $r - \overline{f}^{\mathcal{G}} \in I$ and hence there is some $g \in \mathcal{G}$ such that $\mathbf{LM}(g) | \mathbf{LM}(r - \overline{f}^{\mathcal{G}})$. But by the definition of a remainder this is possible only if $r = \overline{f}^{\mathcal{G}}$.
(ii). This is similar to the proof of (i). □

Now, the above discussion on normal elements allows us to mention the decomposition theorem of the vector space R by a (left) ideal.

Theorem 3.3. *Let I be a (left) ideal of R and $N(I)$ the set of normal monomials in $\mathcal{B}$ (mod I). The following statements hold.*
(i) *As a K-vector space,*

$$R = I \oplus K\text{-}spanN(I).$$

(ii) *The canonical image of $N(I)$ in R/I, that is the set of cosets $\overline{N(I)} = \{\overline{u} = u + I \mid u \in N(I)\}$, forms a K-basis for R/I and hence $dim_K(R/I) = |N(I)|$, where the latter is the cardinal number of $N(I)$.*

□

Example 1. Let $R = K[x_1, \ldots, x_n]$ be the commutative polynomial K-algebra in n variables, and let $(\mathcal{B}, \prec)$ be an admissible system of R as described in Section 1. If $\mathcal{G}$ is a Gröbner basis of the ideal $I = \langle \mathcal{G} \rangle$ in R with respect to the data $(\mathcal{B}, \prec)$, then it is well known (e.g., see [BW]) that the algebra $A = R/I$ is finite dimensional over K, i.e., $|N(I)|$ is finite, if and only if for each $i = 1, 2, \ldots, n$, there exists a $g_i \in \mathcal{G}$ such that $\mathbf{LM}(g_i) = x_i^{m_i}$ with $m_i \in \mathbb{N}$. Moreover, an analogue of this result holds for solvable polynomial algebras studied in [K-RW].

We finish this section by deriving the notion of a reduced (left) Gröbner basis in terms of normal elements.

By the definition of a (left) Gröbner basis and Theorem 3.2, the following proposition is straightforward.

Theorem 3.4. (i) *Let I be an ideal of R and $\mathcal{G} \subset I$. Then $\mathcal{G}$ is a Gröbner basis of I if and only if each nonzero $f \in I$ has a Gröbner representation by elements of $\mathcal{G}$, i.e.,*

$$f = \sum_{i,j} \lambda_{ij} w_{ij} g_j v_{ij}$$

with $\lambda_{ij} \in K^*$, $w_{ij}, v_{ij} \in \mathcal{B}$, $g_j \in \mathcal{G}$, satisfying $\mathbf{LM}(w_{ij} g_j v_{ij}) \preceq \mathbf{LM}(f)$, and $\mathbf{LM}(w_{ij^*} \mathbf{LM}(g_{j^*}) v_{ij^*}) = \mathbf{LM}(f)$ for some j^*.
(ii) *Let L be a left ideal of R and $\mathcal{G} \subset L$. Then $\mathcal{G}$ is a left Gröbner basis of L if and only if each nonzero $f \in L$ has a left Gröbner representation by elements of $\mathcal{G}$, i.e.,*

$$f = \sum_{i,j} \lambda_{ij} w_{ij} g_j$$

with $\lambda_{ij} \in K^*$, $w_{ij} \in \mathcal{B}$, $g_j \in \mathcal{G}$, satisfying $\mathbf{LM}(w_{ij} g_j) \preceq \mathbf{LM}(f)$, and $\mathbf{LM}(w_{ij^*} \mathbf{LM}(g_{j^*})) = \mathbf{LM}(f)$ for some j^*.

□

We will see in the next chapter that Theorem 3.4 makes an essential link between a constructive PBW theory and the Gröbner basis theory.

By Theorem 3.4(i), if $g, g_1 \in \mathcal{G}$, $g \neq g_1$ and $\mathbf{LM}(g_1)|\mathbf{LM}(g)$, then $\mathcal{G}-\{g\}$ is also a Gröbner basis for I. Furthermore, if $g \in \mathcal{G}$ satisfies $u \nmid \mathbf{LM}(g)$ for any $u \in \mathbf{LM}(\mathcal{G}-\{g\})$, then by the division of g by $S = \mathcal{G}-\{g\}$, there is a unique nonzero element $r_g = \sum_i \mu_i v_i$ with $\mu_i \in K^*$ and $v_i \in \mathcal{B}$, such that

$$g = \sum_{i,j} \lambda_{ij} s_{ij} g_i w_{ij} + r_g, \quad \lambda_{ij} \in K^*, \ s_{ij}, w_{ij} \in \mathcal{B}, \ g_i \in \mathcal{G}-\{g\},$$

and no v_i is divisible by any $\mathbf{LM}(g_j)$, $g_j \in \mathcal{G}-\{g\}$. Thus, we may use r_g in place of g in $\mathcal{G}$. By Theorem 3.4(ii), similar argument works for a left Gröbner basis.

The argument made above leads to the following fact.

Proposition 3.5. *Every (left) ideal I of R has a (left) Gröbner basis $\mathcal{G}$ with the following properties:*
(1) *If $g \in \mathcal{G}$ then $\mathbf{LC}(g) = 1$.*
(2) *If $g_1, g_2 \in \mathcal{G}$ and $g_1 \neq g_2$, then $\mathbf{LM}(g_1) \nmid \mathbf{LM}(g_2)$.*
(3) *If $g \in \mathcal{G}$, then $g - \mathbf{LM}(g)$ is a normal element (mod I).*

□

The (left) Gröbner basis $\mathcal{G}$ described in Proposition 3.5 is called a *reduced (left) Gröbner basis.*

It is clear that if $\mathcal{G}$ is a reduced (left) Gröbner basis for I, then $\mathcal{G}$ is a minimal (left) Gröbner basis for I. One may easily find (for instance in a commutative polynomial algebra $K[x_1, \ldots, x_n]$) a minimal Gröbner basis $\mathcal{G}$ which is not reduced.

3.4 Gröbner Bases w.r.t. Skew Multiplicative K-Bases

To demonstrate effective applications of (noncommutative) Gröbner bases in subsequent chapters, in this section we give a more detailed discussion on the Gröbner basis theory (in the sense of Definition 2.3) for K-algebras with a skew multiplicative K-basis defined below.

Definition 4.1. Let R be a K-algebra. If R has a K-basis $\mathcal{B}$ satisfying

$$u,\ v \in \mathcal{B} \text{ implies } \begin{cases} u \cdot v = \lambda w \text{ for some } \lambda \in K^*,\ w \in \mathcal{B}, \\ \text{or } u \cdot v = 0, \end{cases}$$

then we call $\mathcal{B}$ a *skew multiplicative K-basis* of R.

Obviously, a skew multiplicative K-basis generalizes the notion of a *multiplicative K-basis* used in representation theory and classical Gröbner basis theory of associative algebras (i.e., a K-basis $\mathcal{B}$ with the property that $u,\ v \in \mathcal{B}$ implies $uv = 0$ or $uv \in \mathcal{B}$). The class of algebras with a skew multiplicative K-basis includes not only ordered semigroup algebras, free algebras, commutative polynomial algebras, path algebras defined by finite directed graphs, but also many other popularly studied algebras such as exterior algebras, and the skew polynomial algebra $\mathcal{O}_n(\lambda_{ji})$ subject to the relations $x_jx_i - \lambda_{ji}x_ix_j = 0$ with $\lambda_{ji} \in K^*$ and $1 \le i < j \le n$ (see Ch.1, Section 1). More such algebras will be given in Ch.8.

Let $(\mathcal{B}, \prec)$ be an admissible system of R with $\mathcal{B}$ a skew multiplicative K-basis. Noticing the feature of $u \cdot v$ for $u, v \in \mathcal{B}$, the division of monomials in $\mathcal{B}$ (as defined in Section 2) may be simplified as follows:

- For $u, v \in \mathcal{B}$, we say that u divides v, denoted $u|v$, if there are $w, s \in \mathcal{B}$ and $\lambda \in K^*$ such that $v = \lambda wus$; similarly we say that u divides v from left-side, again denoted $u|v$, if there are $w \in \mathcal{B}$ and $\lambda \in K^*$ such that $v = \lambda wu$.

Thus, the following assertion is clear.

Proposition 4.2. *Let $(\mathcal{B}, \prec)$ be a (left) admissible system of R with $\mathcal{B}$ a skew multiplicative K-basis. Then the division of monomials is transitive and hence R has a (left) Gröbner basis theory.*

□

The advantage of working with a skew multiplicative K-basis is that we may construct a (left) Gröbner basis of a (left) ideal I via a monomial generating set of the (left) ideal generated by the set of leading monomials of I. We demonstrate below the details.

In what follows, R denotes a K-algebra with a (left) admissible system $(\mathcal{B}, \prec)$, where $\mathcal{B}$ is a skew multiplicative K-basis.

The Generating set of a monomial (left) ideal

If a (left) ideal I of R is generated by a subset $U \subset \mathcal{B}$, then I is called a *monomial (left) ideal.*

Proposition 4.3. *Let I be a monomial (left) ideal of R generated by $U \subset \mathcal{B}$. The following statements hold.*
(i) *For a monomial $v \in \mathcal{B}$, $v \in I$ if and only if there exists some $u \in U$ such that $u|v$.*
(ii) *For an element $f = \sum_\ell \lambda_\ell v_\ell \in R$, where $\lambda_\ell \in K^*$ and $v_\ell \in \mathcal{B}$, $f \in I$ if and only if all $v_\ell \in I$.*

Proof. We prove (i) and (ii) for a two-sided ideal I, for, a similar argument works for left ideals.

(i). For $v \in \mathcal{B}$, $u \in U$, if $u|v$, then $\lambda v = suw \in I$ for some $s, w \in \mathcal{B}$, $\lambda \in K^*$. Conversely, suppose $v \in I \cap \mathcal{B}$. Note that U generates I and each element of R is a linear combination of monomials in $\mathcal{B}$. Thus, v may be written as a finite sum $v = \sum_i \lambda_i s_i u_i w_i$, where $\lambda_i \in K$, $s_i, w_i \in \mathcal{B}$, and $u_i \in U$. Since $\mathcal{B}$ is a skew multiplicative K-basis, it follows that $\lambda v = s_i u_i w_i$ for some i and $\lambda \in K^*$, i.e., $u_i|v$, as desired.

(ii). The sufficiency is clear. Suppose now $f = \sum_\ell \lambda_\ell v_\ell \in I$, where $\lambda_\ell \in K$ and $v_\ell \in \mathcal{B}$. Then since U is a monomial generating set of I, f may be written as

$$\sum_\ell \lambda_\ell v_\ell = f = \sum_{i,j} \mu_{ij} s_{ij} u_j w_{ij}, \quad \mu_{ij} \in K,\ s_{ij}, w_{ij} \in \mathcal{B},\ u_j \in U.$$

As $s_{ij} u_j w_{ij} \neq 0$ implies $\lambda s_{ij} u_j w_{ij} \in \mathcal{B}$ for some $\lambda \in K^*$, it follows that all $v_\ell \in I$. □

By Proposition 4.3, any monomial generating set U of a monomial (left) ideal I in R is a (left) Gröbner basis for I. Furthermore, by Proposition 2.2, a unique minimal (monomial) Gröbner basis of I is obtained.

Theorem 4.4. *Let I be a (left) monomial ideal of R and put $\Lambda = \mathcal{B} \cap I$. The following statements hold.*

(i) *I has a unique minimal, or equivalently, a unique reduced monomial generating set* Ω *which is given by*

$$\Omega=\left\{w\in\Lambda\;\middle|\;\text{if } u\in\Lambda \text{ and } u\neq w,\ \text{then } u\nmid w\right\}.$$

(ii) *Let* Ω *be as in (i) above. If* T *is any monomial generating set of* I*, then* $\Omega\subseteq T$.

□

Remark In the case that R is a Noetherian ring, the minimal monomial generating set Ω in Theorem 4.4 is always finite. If $R=K[x_1,\ldots,x_n]$ is a commutative polynomial algebra, then there is a constructive proof for the finiteness of Ω that is known the consequence of using Dickson's Lemma instead of using the Noetherianity of R (e.g. see [BW]). As illustrated by the following example, however, in the non-Noetherian case, it is possible to have an infinite set Ω.

Example 1. Consider the free algebra $R=K\langle X_1,X_2\rangle$. Then the monomial ideal I generated by $\{X_1X_2^nX_1 \mid n\in\mathbb{N}\}$ has the unique minimal monomial generating set $\Omega=\{X_1X_2^nX_1 \mid n\in\mathbb{N}\}$.

LM(I) *and Gröbner basis*

Let I be a (left) ideal of R. Then we have the set of leading monomials of I, that is

$$\mathbf{LM}(I)=\{\mathbf{LM}(f)\mid f\in I\}.$$

We now proceed to show that any monomial generating set of the monomial (left) ideal generated by $\mathbf{LM}(I)$ gives rise to a (left) Gröbner basis of I.

Lemma 4.5. *If* I *is an ideal of* R*, then* $\langle\mathbf{LM}(I)\rangle\cap\mathcal{B}=\mathbf{LM}(I)$*; if* L *is a left ideal of* R*, then* $\langle\mathbf{LM}(L)]\cap\mathcal{B}=\mathbf{LM}(L)$*, where* $\langle\mathbf{LM}(L)]$ *denotes the left ideal generated by* $\mathbf{LM}(L)$.

Proof. Let I be an ideal of R. We need only to prove $\langle\mathbf{LM}(I)\rangle\cap\mathcal{B}\subseteq\mathbf{LM}(I)$. Let $v\in\langle\mathbf{LM}(I)\rangle\cap\mathcal{B}$. Then by Proposition 4.3, $v=\lambda u\mathbf{LM}(f)w$ for some $\lambda\in K^*$, $u,w\in\mathcal{B}$ and $f\in I$. Write $f=\mathbf{LC}(f)\mathbf{LM}(f)+f'$ such that $\mathbf{LM}(f')\prec\mathbf{LM}(f)$. Then $\lambda\mathbf{LC}(f)u\mathbf{LM}(f)w+\lambda uf'w=\lambda ufw=f_1\in I$, and $v=\lambda u\mathbf{LM}(f)w=\mathbf{LM}(f_1)\in\mathbf{LM}(I)$. This shows that $\langle\mathbf{LM}(I)\rangle\cap\mathcal{B}\subseteq\mathbf{LM}(I)$. Hence the desired equality holds.

The proof for a left ideal is similar. □

The proof of the next proposition tells us how to lift a monomial generating set of $\langle \mathbf{LM}(I)\rangle$, respectively, a monomial generating set of $\langle \mathbf{LM}(L)]$, to a Gröbner basis of I, respectively to a left Gröbner basis of L, and further obtain a minimal (left) Gröbner basis of I.

Proposition 4.6. (i) *Let I be an ideal of R and T a monomial generating set of $\langle \mathbf{LM}(I)\rangle$. Then $T \subset \mathbf{LM}(I)$. If $\mathcal{G} \subset I$ is such that $\mathbf{LM}(\mathcal{G}) = T$, i.e., $\langle \mathbf{LM}(\mathcal{G})\rangle = \langle T\rangle = \langle \mathbf{LM}(I)\rangle$, then $I = \langle \mathcal{G}\rangle$ and $\mathcal{G}$ is a Gröbner basis for I; moreover, $\mathcal{G}$ is a minimal Gröbner basis if and only if $T = \Omega$ is the unique reduced monomial generating set of $\langle \mathbf{LM}(I)\rangle$ in the sense of Theorem 4.4.*
(ii) *Let L be a left ideal of R and T a monomial generating set of $\langle \mathbf{LM}(L)]$. Then $T \subset \mathbf{LM}(L)$. If $\mathcal{G} \subset L$ is such that $\mathbf{LM}(\mathcal{G}) = T$, i.e., $\langle \mathbf{LM}(\mathcal{G})] = \langle T] = \langle \mathbf{LM}(L)]$, then $L = \langle \mathcal{G}]$ and $\mathcal{G}$ is a left Gröbner basis for L; moreover, $\mathcal{G}$ is a minimal left Gröbner basis if and only if $T = \Omega$ is the unique reduced monomial generating set of $\langle \mathbf{LM}(L)]$ in the sense of Theorem 4.4.*

Proof. (i). Firstly, the inclusion $T \subset \mathbf{LM}(I)$ follows from Lemma 4.5. So there is some subset $\mathcal{G} \subset I$ such that $\mathbf{LM}(\mathcal{G}) = T$. Next, if $f \in I$ and $f \neq 0$, then $\mathbf{LM}(f) \in \langle \mathbf{LM}(I)\rangle = \langle \mathbf{LM}(\mathcal{G})\rangle$. By Proposition 4.3, $\mathbf{LM}(g)|\mathbf{LM}(f)$ for some $g \in \mathcal{G}$. Hence $\mathcal{G}$ is a Gröbner basis of I by definition. The assertion on the minimality of $\mathcal{G}$ follows from Proposition 2.2(i).

(ii). The argument for a left ideal is similar to the proof of (i). □

Remark At this stage the reader is invited to prove, by using Proposition 3.5 and Proposition 4.6, that every proper (left) ideal of R has a *unique* reduced (left) Gröbner basis.

The above proposition leads to the following characterization of the Gröbner bases in a K-algebra R with a skew multiplicative K-basis $\mathcal{B}$.

Theorem 4.7. *Let $(\mathcal{B}, \prec)$ be an admissible system of R with $\mathcal{B}$ a skew multiplicative K-basis. The following statements hold.*
(i) *Let L be a left ideal of R. A subset $\mathcal{G} \subset L$ is a left Gröbner basis of L if and only if $\langle \mathbf{LM}(L)] = \langle \mathbf{LM}(\mathcal{G})]$; L has a finite left Gröbner basis if and only if $\langle \mathbf{LH}(L)]$ has a finite monomial generating set.*

(ii) *Let I be an ideal of R. A subset $\mathcal{G} \subset I$ is a Gröbner basis of I if and only if $\langle \mathbf{LM}(I) \rangle = \langle \mathbf{LM}(\mathcal{G}) \rangle$; I has a finite Gröbner basis if and only if $\langle \mathbf{LH}(I) \rangle$ has a finite monomial generating set.*

□

Example 2. Let $R = K[x_1, \ldots, x_n]$ be the commutative polynomial algebra in n variables, and let $g \in R$ be a nonconstant polynomial. Consider the principal ideal $I = \langle g \rangle = \{gf \mid f \in R\}$. Then, with respect to any monomial ordering $\prec$ on the standard K-basis $\mathcal{B} = \{x^{\alpha_1} \cdots x_n^{\alpha_n} \mid \alpha_i \in \mathbb{N}\}$ of R, $\mathbf{LM}(gf) = \mathbf{LM}(g)\mathbf{LM}(f)$ for all $0 \neq f \in R$, i.e., $\langle \mathbf{LM}(I) \rangle = \langle \mathbf{LM}(g) \rangle$. Hence, by Theorem 4.7, the ideal I has $\{g\}$ as a Gröbner basis with respect to any monomial ordering $\prec$ on $\mathcal{B}$.

The next three examples indicate that the notion of a Gröbner basis is by no means trivial, namely, not every generating set of an arbitrary ideal is a Gröbner basis, though each monomial generating set of a monomial (left) ideal is a (left) Gröbner basis.

Example 3. Let $R = K[x, y]$ be the commutative polynomial K-algebra in variables x and y, and $\mathcal{B} = \{x^i y^j \mid i, j \in \mathbb{N}\}$ the standard K-basis of R. Consider the ideal $I = \langle g_1, g_2 \rangle$ generated by $g_1 = xy - 1$, $g_2 = y^3 - x$. Then

$$f = y^2 g_1 - x g_2 = -y^2 + x^2 \in I.$$

If we use the $\mathbb{N}$-graded lexicographic ordering $x \prec_{gr} y$ on $\mathcal{B}$ with respect to the natural $\mathbb{N}$-gradation of R, then $\mathbf{LM}(g_1) = xy$, $\mathbf{LM}(g_2) = y^3$, and $\mathbf{LM}(f) = y^2$. But clearly $\mathbf{LM}(g_1) \nmid \mathbf{LM}(f)$ and $\mathbf{LM}(g_2) \nmid \mathbf{LM}(f)$. So, by Theorem 4.7, $\{g_1, g_2\}$ is not a Gröbner basis for I.

Example 4. Consider in the free K-algebra $R = K\langle X_1, X_2 \rangle$ the principal ideal $I = \langle g \rangle$ generated by $g = X_1^2 - X_1 X_2$, and let $\mathcal{B} = \{1, X_{i_1} \cdots X_{i_s} \mid X_{i_j} \in \{X_1, X_2\},\ s \geq 1\}$ be the standard K-basis of R. Note that $gX_2 = X_1^2 X_2 - X_1 X_2^2$, $g(-X_1) = -X_1^3 + X_1 X_2 X_1$, $X_1 g = X_1^3 - X_1^2 X_2$ $\in I$. So

$$f = gX_2 + g(-X_1) + X_1 g = X_1 X_2 X_1 - X_1 X_2^2 \in I.$$

If we use the $\mathbb{N}$-graded lexicographic ordering $X_2 \prec_{gr} X_1$ on $\mathcal{B}$ with respect to the natural $\mathbb{N}$-gradation of R, then $\mathbf{LM}(g) = X_1^2$ and $\mathbf{LM}(f) = X_1 X_2 X_1$. But clearly $\mathbf{LM}(g) \nmid \mathbf{LM}(f)$. Hence, by Theorem 4.7, $\{g\}$ cannot be a Gröbner basis of I.

Example 5. Consider in the free $\mathbb{Q}$-algebra $R = \mathbb{Q}\langle X_1, X_2, X_3\rangle$ the ideal $I = \langle G\rangle$ with G consisting of

$$\begin{aligned} g_1 &= -3X_1X_2X_3 + X_3^2 + 1, \\ g_2 &= X_2X_3^2 + X_1X_3 + 3, \\ g_3 &= X_1^2X_2 + X_2X_3, \end{aligned}$$

and let $\mathcal{B} = \{1,\ X_{i_1}\cdots X_{i_s} \mid X_{i_j} \in \{X_1, X_2, X_3\},\ s \geq 1\}$ be the standard K-basis of R. Then, one checks, by using the $\mathbb{N}$-graded lexicographic ordering $X_3 \prec_{gr} X_2 \prec_{gr} X_1$ on $\mathcal{B}$ with respect to the natural $\mathbb{N}$-gradation of R, that

$$f = X_1g_1X_3 + X_1^2g_2 + 2g_3X_3^2 \in I$$

but $\mathbf{LM}(f) = X_1^3X_3 \notin \langle\mathbf{LM}(G)\rangle$. It follows from Theorem 4.7 that G is not a Gröbner basis of I.

Remark Comparing Example 2 with Example 4 above, we see how big the difference between the commutative case and the noncommutative case is.

Let I be a (left) ideal of R. Finally we characterize normal elements in terms of $\mathbf{LM}(I)$. As in Section 3 we write $N(I)$ for the set of normal monomials in $\mathcal{B}$ (mod I).

Proposition 4.8. *Let $(\mathcal{B}, \prec)$ be a (left) admissible system of R with $\mathcal{B}$ a skew multiplicative K-basis, and let I be a (left) ideal of R. For $f = \sum_j \mu_j v_j \in R$ with $\mu_j \in K^*$, $v_j \in \mathcal{B}$, the following statements hold.*
(i) *f is normal (mod I) if and only if $v_j \notin \mathbf{LM}(I)$, or equivalently if and only if $v_j \in B - \mathbf{LM}(I)$ for all j. Consequently, we have*

$$N(I) = \mathcal{B} - \mathbf{LM}(I) = N(\langle\mathbf{LM}(I)\rangle)$$

(in the case that I is a left ideal we have $N(I) = \mathcal{B} - \mathbf{LM}(I) = N(\langle\mathbf{LM}(I)])$.
(ii) *If $\mathcal{G}$ is a (left) Gröbner basis of I, then*

$$N(I) = \{u \in \mathcal{B} \mid \mathbf{LM}(g) \nmid u,\ g \in \mathcal{G}\} = \mathcal{B} - \langle\mathbf{LM}(\mathcal{G})\rangle = N(\langle\mathbf{LM}(I)\rangle)$$

(in the case that $\mathcal{G}$ is a left Gröbner basis, $N(I) = \mathcal{B} - \langle\mathbf{LM}(\mathcal{G})] = N(\langle\mathbf{LM}(I)])$.

Proof. This follows from Theorem 3.2, Lemma 4.5 and Theorem 4.7. □

Remark A right Gröbner basis theory for right ideals may be stated exactly as the left Gröbner basis theory for left ideals we developed through Sections 1 – 4. As distinct from a two-sided Gröbner basis theory for two-sided ideals, a (left, right) Gröbner basis theory will be employed in (Ch.8, Sections 3 – 5).

3.5 Gröbner Bases in $K\langle X_1,\ldots,X_n\rangle$ and KQ

Let R be a K-algebra with a (left) admissible system $(\mathcal{B},\prec)$, where $\mathcal{B}$ is a skew multiplicative K-basis of R and $\prec$ is a monomial ordering on R. Theoretically we have known from Sections 3, 4 that every (left) ideal of R has a (left) Gröbner basis (which is even minimal). Yet, from a practical viewpoint, we are naturally concerned about obtaining a Gröbner basis in an algorithmic way. If $R=K[x_1,\ldots,x_n]$ is the commutative polynomial K-algebra in n variables, then the celebrated Buchberger Algorithm [Bu1, 2] produces a Gröbner basis for each ideal of R with a given generating set. Based on Bergman's diamond lemma, noncommutative Gröbner bases are constructed algorithmically for ideals in a free K-algebra $K\langle X_1,\ldots,X_n\rangle$ and a path algebra $KQ=K[v_1,\ldots,v_p,a_1,\ldots,a_q]$ with $Q_0=\{v_1,\ldots,v_p\}$ the set of vertices and $Q_1=\{a_1,\ldots,a_q\}$ the set of arrows ([Mor1, 2], [FFG]). It is also well known that there are analogues of the Buchberger Algorithm for producing Gröbner bases in other interesting algebras, such as solvable polynomial algebras (in the sense of [K-RW]) including enveloping algebras of Lie algebras and Weyl algebras as well as their analogues, exterior algebras and Clifford algebras (e.g., [GLS], [HT], [LSX]). To suit the extent of this book, in the present section we adapt Green's definition of an overlap element and the proof of the "termination theorem" concerning path algebras [Gr2] to demonstrate a unified approach to the construction of a *two-sided* Gröbner basis in both a free K-algebra and a path algebra.

Throughout this section R denotes a free K-algebra $K\langle X_1,\ldots,X_n\rangle$ or a path algebra KQ defined by a finite directed graph Q, and in both cases, $\mathcal{B}$ denotes the standard K-basis of R consisting of words, respectively consisting of finite directed paths.

Maintaining all notions and notations used before, we start this section by introducing some new concepts and notations.

Overlap elements

Why overlap elements? Exploring the classical commutative case, if $I = \langle \mathcal{G} \rangle$ is an ideal of the polynomial K-algebra $K[x_1, \ldots, x_n]$ generated by $\mathcal{G} = \{g_1, \ldots, g_s\}$, then, Buchberger's (termination) theorem [Bu1, 2] tells us that $\mathcal{G}$ is a Gröbner basis of I if and only if every S-polynomial $S(g_i, g_j)$, where $1 \leq i < j \leq s$, is reduced to zero on division by $\mathcal{G}$ (i.e., a remainder of $S(g_i, g_j)$ on division by $\mathcal{G}$, with respect to a fixed order on $\mathcal{G}$, is zero); and the core idea we learnt from the proof of Buchberger's theorem may be highlighted as follows:

- Reducing every $S(g_i, g_j)$ to zero guarantees that every nonzero $f \in I$ has a Gröbner representation by $\mathcal{G}$, or more precisely, in generating f by $\mathcal{G}$, if $f = \sum_{i,j} \lambda_{ij} w_{ij} g_j$, where the w_{ij}'s are monomials and $\lambda_j \in K^*$, which has a sub-sum $\sum_{k=1}^{m} \lambda_{ik} w_{ik} g_{j_k}$ with the property that

  $$\mathbf{LM}(f) \prec w_{i1}\mathbf{LM}(g_{j_1}) = w_{i2}\mathbf{LM}(g_{j_2}) = \cdots = w_{im}\mathbf{LM}(g_{j_m})$$

  and

  $$\sum_{j=1}^{m} \lambda_{ik} w_{ik} \mathbf{LM}(g_{j_k}) = 0,$$

  where $2 \leq m \leq s$ (e.g., check Example 3 of the last section), then $\sum_{k=1}^{m} \lambda_{ik} w_{ik} g_{j_k}$ has a Gröbner representation via the division of $S(g_{j_k}, g_{j_{k'}})$ by $\mathcal{G}$, $1 \leq k < k' \leq m$.

To have an analogue of Buchberger's (termination) theorem in the very noncommutative case here for ideals in R (a free algebra or a path algebra), we have to deal with the same problem of canceling leading monomials of multiples of generators, for instance, in Example 4 of the last section, we have

$$\mathbf{LM}(f) = X_1X_2X_1 \prec_{gr} X_1^3 = \mathbf{LM}(g)X_1 = X_1\mathbf{LM}(g)$$

which leads to the cancelation in the representation of f:

$$-\mathbf{LM}(g)X_1 + X_1\mathbf{LM}(g) = (-1+1)X_1^3 = 0;$$

while in Example 5 of the last section, we have

$$\begin{aligned}\mathbf{LM}(f) = X_1^3X_3 &\prec_{gr} X_1^2X_2X_3^2 \\ &= X_1\mathbf{LM}(g_1)X_3 = X_1^2\mathbf{LM}(g_2) = \mathbf{LM}(g_3)X_3^3\end{aligned}$$

which leads to the cancelation in the representation of f:

$$-3X_1\mathbf{LM}(g_1)X_3 + X_1^2\mathbf{LM}(g_2) + 2\mathbf{LM}(g_3)X_3^3 = (-3+1+2)X_1^2X_2X_3^2 = 0.$$

Thus, to tackle the cancelation problem as indicated above, overlap element, the noncommutative analogue of an S-element, comes into play. Before giving the definition of an overlap element, let us also point out that the proof of Buchberger's (termination) theorem uses, for each pair (g_i, g_j), only one S-polynomial

$$S(g_i, g_j) = \frac{u}{\mathbf{LC}(g_i)\mathbf{LM}(g_i)} \cdot g_i - \frac{u}{\mathbf{LC}(g_j)\mathbf{LM}(g_j)} \cdot g_j$$

which is determined by u=lcm$(\mathbf{LM}(g_i), \mathbf{LM}(g_j))$. This is because, if $u_1\mathbf{LM}(g_i) = w = u_2\mathbf{LM}(g_j)$ where u_1, u_2, w are monomials, then $w = v \cdot u$ for some monomial v. But for a subset G in a free algebra or a path algebra, such an analogue does not exists, and consequently for $g_i, g_j \in G$, all overlap elements of the ordered pair (g_i, g_j) and the ordered pair (g_j, g_i), including the case of $g_i = g_j$, must be considered (see Example 4 of the last section, the proof of Theorem 5.1 (Case A.2.2) and the similar (Case C.2.2) given later).

We start to define an overlap element with monomials. Let $(v, u) \in \mathcal{B} \times \mathcal{B}$. If there are $v_1, w, u_1 \in \mathcal{B}$ with $w \neq 1$ (if $\mathcal{B}$ contains 1), such that

$$(*) \qquad v = v_1 w \text{ and } u = w u_1,$$

then we say that the ordered pair (v, u) has an *overlap relation.*

Let $(f, g) \in R \times R$ and suppose that the ordered pair $(\mathbf{LM}(f), \mathbf{LM}(g))$ has an overlap relation, i.e., there are $v, w, u \in \mathcal{B}$ with $w \neq 1$, such that

$$(**) \qquad \mathbf{LM}(f) = vw \text{ and } \mathbf{LM}(g) = wu,$$

then we call the element

$$o(f, u;\ v, g) = \frac{1}{\mathbf{LC}(f)}(f \cdot u) - \frac{1}{\mathbf{LC}(g)}(v \cdot g)$$

an *overlap element* of the ordered pair (f, g).

Observation For the purpose of doing an algorithmic job later, we observe that

(i) any overlap element has the property that

$$\mathbf{LM}(o(f, u;\ v, g)) \prec \mathbf{LM}(fu) = \mathbf{LM}(vg);$$

(ii) since R is a free algebra or a path algebra, and the standard K-basis $\mathcal{B}$ of R is used, so for each ordered pair $(f, g) \in R \times R$, the set of overlap elements of (f, g) is always finite.

Example 1. Consider the principle ideal $I = \langle g \rangle$ with $g = X_1^2 - X_1X_2$ in the free K-algebra $K\langle X \rangle = K\langle X_1, X_2 \rangle$, as in Example 4 of Section 4. Note that for $n = 0$, $X_1X_2^0X_1 - X_1X_2^{0+1} = g \in I$, and for $n \geq 1$ we have

$$\begin{aligned} gX_2^n + g(-X_2^{n-1}X_1) &= (X_1^2X_2^n - X_1X_2^{n+1}) \\ &\quad + (-X_1^2X_2^{n-1}X_1 + X_1X_2^nX_1) \\ &= -X_1(X_1X_2^{n-1}X_1 - X_1X_2^n) \\ &\quad + (X_1X_2^nX_1 - X_1X_2^{n+1}) \\ &\in I. \end{aligned}$$

By making the hypothesis $X_1X_2^{n-1}X_1 - X_1X_2^n \in I$, an induction on n shows that $X_1X_2^nX_1 - X_1X_2^{n+1} \in I$. Put $\mathcal{G} = \{g_n = X_1X_2^nX_1 - X_1X_2^{n+1} \mid n \in \mathbb{N}\}$. Then $g = g_0$ and $I = \langle \mathcal{G} \rangle$. If we use the natural $\mathbb{N}$-gradation of $K\langle X \rangle$ and the $\mathbb{N}$-graded lexicographic ordering $X_2 \prec_{gr} X_1$ on the standard K-basis $\mathcal{B}$ of $K\langle X \rangle$, then $\mathbf{LM}(g_n) = X_1X_2^nX_1$, $n \in \mathbb{N}$. For ordered pairs (g_n, g_k), $(g_k, g_n) \in \mathcal{G} \times \mathcal{G}$, including $n = k$, we list all possible overlap elements as follows:

$$\begin{aligned} o(g_n, X_2^kX_1;\ X_1X_2^n, g_k) &= g_n \cdot X_2^kX_1 - X_1X_2^n \cdot g_k \\ &= X_1X_2^nX_1X_2^{k+1} - X_1X_2^{n+k+1}X_1. \end{aligned}$$

$$\begin{aligned} o(g_k, X_2^nX_1;\ X_1X_2^k, g_n) &= g_k \cdot X_2^nX_1 - X_1X_2^k \cdot g_n \\ &= X_1X_2^kX_1X_2^{n+1} - X_1X_2^{n+k+1}X_1. \end{aligned}$$

Example 2. Let $R = KQ$ be the path algebra defined by the finite directed graph

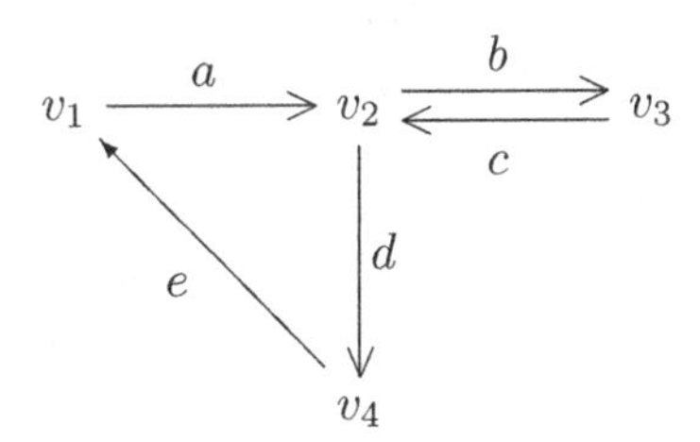

and let $\mathcal{B}$ be the standard K-basis of KQ consisting of finite directed paths (including vertices). Consider the ideal $I = \langle f, g \rangle$, where $f = abcdead - abc$, $g = deadea - dead$. If we use the $\mathbb{N}$-graded lexicographic ordering

$$v_1 \prec_{gr} v_2 \prec_{gr} v_3 \prec_{gr} v_4 \prec_{gr} a \prec_{gr} b \prec_{gr} c \prec_{gr} d \prec_{gr} e$$

on $\mathcal{B}$ with respect to the natural $\mathbb{N}$-gradation of KQ, then $\mathbf{LM}(f) = abcdead$, $\mathbf{LM}(g) = deadea$ and the ordered pairs (f, g), (g, f) and (g, g)

have overlap elements given below

$$\begin{aligned}
&o(f, ea;\ abc, g) = f \cdot ea - abc \cdot g = abcdead,\\
&o(f, eadea;\ abcdea, g) = f \cdot eadea - abcdea \cdot g = abcdeadead,\\
&o(g, bcdead;\ deade, f) = g \cdot bcdead - deade \cdot f = deadeabc,\\
&o(g, dea;\ dea, g) = g \cdot dea - dea \cdot g = deadead.
\end{aligned}$$

Let S be a nonempty subset of R. Recall from Section 2 that S is called LM-reduced if $f, g \in S$ and $f \neq g$ implies $\mathbf{LM}(f) \nmid \mathbf{LM}(g)$.

Consider an arbitrary ideal I of R. Starting with any generating set S of I, an LM-reduced generating set of I may always be obtained. To see this, suppose $\mathbf{LM}(f_j)|\mathbf{LM}(f_i)$ for some $f_j \neq f_i$ in S. Then $\mathbf{LM}(f_i) = u\mathbf{LM}(f_j)w$ for some $u, w \in \mathcal{B}$, the K-basis of R. Thus, $f_i - \frac{\mathbf{LC}(f_i)}{\mathbf{LC}(f_j)}uf_jw = f_i' \in I$ and we may consider f_i' in place of f_i. Note that $\mathbf{LM}(f_i') \prec \mathbf{LM}(f_i)$. We can repeat the same procedure for f_i' if $\mathbf{LM}(f_i')$ is divided by some $\mathbf{LM}(f_k)$, $f_k \neq f_i'$. By the well-ordering property of $\prec$, after a finite number of repetitions we find some $f \in I$ with the property that $\mathbf{LM}(f_\ell) \nmid \mathbf{LM}(f)$ for any $f_\ell \in S$, and thereby we can replace f_i by this f.

The above argument shows that we may always assume, if necessary, that *a generating set of an ideal I is LM-reduced.*

Uniform elements

The reason for introducing such elements is that a path algebra may have divisors of zero, while using uniform element may get rid of the annoying situation that $uf \neq 0$ ($fu \neq 0$) but $\mathbf{LM}(uf) \neq u\mathbf{LM}(f)$ ($\mathbf{LM}(fu) \neq \mathbf{LM}(f)u$) for $u \in \mathcal{B}$, $f \in R$.

Let $f \in R$, $f = \sum_{i=1}^{s} \lambda_i v_i$ with $\lambda_i \in K^*$ and $v_i \in \mathcal{B}$. If for each $u \in \mathcal{B}$, either $uv_i = 0$ for all $i = 1, 2, \ldots, s$, or $uv_i \neq 0$ for all $i = 1, 2, \ldots, s$, then we say that f is *uniform* in R.

Obviously, if R is a free algebra, then every element of R is uniform.

If $R = KQ$ is the path algebra defined by a finite directed graph Q over K, then R has elements which are not uniform, for example $f = v_1 + v_2$, where v_1 and v_2 are distinct vertices. However, as argued in [Gr2], every element of $R = KQ$ is a sum of uniform elements. For, if $Q_0 = \{v_1, \ldots, v_n\}$

is the set of vertices, then, for each pair $v_i, v_j \in Q_0$ with $i \neq j$, and each $f \in R$, it is easy to see that $v_i f v_j$ is a uniform element. Note that $1 = v_1 + \cdots + v_n$. It turns out that

$$f = \left(\sum_{i=1}^{n} v_i\right) f \left(\sum_{j=1}^{n} v_j\right) = \sum_{i,j=1}^{n} v_i f v_j$$

is a sum of uniform elements. Thus, we may always assume, if necessary, that *an ideal of $R = KQ$ is generated by uniform elements.*

Termination theorem

Now, we mention and prove the termination theorem for both a free algebra and a path algebra. We refer the reader to [Ber3], [Mor2], and [Gr2] for the respective version and the proof concerning this theorem.

Theorem 5.1. *Let R be a free algebra or a path algebra with the standrd K-basis $\mathcal{B}$. If $\mathcal{G}$ is an LM-reduced subset of R consisting of uniform elements, then $\mathcal{G}$ is a Gröbner basis for the ideal $I = \langle \mathcal{G} \rangle$ if and only if for each ordered pair $(g_t, g_\ell) \in \mathcal{G} \times \mathcal{G}$, including $g_t = g_\ell$, every overlap element $o(g_t, u;\ v, g_\ell)$ of (g_t, g_ℓ) (if it exists) has the property*

$$\overline{o(g_t, u;\ v, g_\ell)}^{\mathcal{G}} = 0,$$

that is, a remainder of $o(g_t, u;\ v, g_\ell)$ on the division by $\mathcal{G}$ is equal to 0. (To better under this theorem, the reader is referred to the remark on a remainder given after **Algorithm-DIV** in Section 2.)

Proof. Since $o(g_t, u;\ v, g_\ell) \in I$, the necessity follows from Theorem 3.4(i).

Under the assumption on overlap elements we prove the sufficiency by showing that if $f \in I$ then $\mathbf{LM}(g) | \mathbf{LM}(f)$ for some $g \in \mathcal{G}$. Suppose the contrary that $\mathbf{LM}(g) \nmid \mathbf{LM}(f)$ for any $g \in \mathcal{G}$. Then we proceed to derive a contradiction.

Since $I = \langle \mathcal{G} \rangle$, f may be represented as a finite sum

$$(1) \quad f = \sum_{i,j} \lambda_{ij} v_{ij} g_i w_{ij}, \text{ where } \lambda_{ij} \in K^*,\ v_{ij}, w_{ij} \in \mathcal{B}, \text{ and } g_i \in \mathcal{G}.$$

Let u be the largest monomial occurring on the right hand side of (1). Then u occurs as some $v_{ij}\mathbf{LM}(g_i)w_{ij}$ for all g_i's are uniform elements. Since $\mathbf{LM}(g_i) \nmid \mathbf{LM}(f)$ for the g_i occurring in (1), it follows that $\mathbf{LM}(f) \prec u$ and

u must occur at least twice on the right hand side of (1) for a cancelation, that is, we may have

(2) $$u = v_{ij}\mathbf{LM}(g_i)w_{ij} = v_{k\ell}\mathbf{LM}(g_k)w_{k\ell}.$$

Among all such representations of f we can choose one such that

(3) u has the fewest occurrences on the right hand side of (1) and u is as small as possible.

To go further, let us simplify notation by writing $v = v_{ij}$, $g = g_i$, $w = w_{ij}$, $v' = v_{k\ell}$, $g' = g_k$, and $w' = w_{k\ell}$. Thus, (2) is turned into the form

(4) $$u = v\mathbf{LM}(g)w = v'\mathbf{LM}(g')w'.$$

Moreover, as usual we use $l(s)$ to denote the length of a monomial $s \in \mathcal{B}$. We show below, through a case by case study of (4), that the choice of f satisfying (3) is impossible.

Case A: $l(v) < l(v')$.

Case A.1: $l(w) < l(w')$.

This implies that $\mathbf{LM}(g)$ contains $\mathbf{LM}(g')$ as a submonomial, and hence, $\mathbf{LM}(g')|\mathbf{LM}(g)$, contradicting the hypothesis on $\mathcal{G}$.

Case A.2: $l(w) \geq l(w')$.

Then we have to deal with two possibilities.

Case A.2.1: $l(v') \geq l(v\mathbf{LM}(g))$.

This implies that the ordered pair $(\mathbf{LM}(g), \mathbf{LM}(g'))$ has no overlap relation in u. By the assumption on lengths, it follows that there is a segment w'' of v' such that $v' = v\mathbf{LM}(g)w''$ and $w = w''\mathbf{LM}(g')w'$, i.e.,

$$u = v\mathbf{LM}(g)w''\mathbf{LM}(g')w'.$$

Rewrite $g = \mathbf{LC}(g)\mathbf{LM}(g) + \sum \lambda_i u_i$, $g' = \mathbf{LC}(g')\mathbf{LM}(g') + \sum \mu_i u_i'$. Then

$$\begin{aligned} vgw &= vgw'' \cdot \frac{1}{\mathbf{LC}(g')}g'w' - vgw''\left(\frac{1}{\mathbf{LC}(g')}g' - \mathbf{LM}(g')\right)w' \\ &= \frac{\mathbf{LC}(g)}{\mathbf{LC}(g')}v\mathbf{LM}(g)w''g'w' + \sum \frac{\lambda_i}{\mathbf{LC}(g')}vu_iw''g'w' \\ &\qquad - \sum \frac{\mu_i}{\mathbf{LC}(g')}vgw''u_i'w' \\ &= \frac{\mathbf{LC}(g)}{\mathbf{LC}(g')}v'g'w' + \sum \frac{\lambda_i}{\mathbf{LC}(g')}vu_iw''g'w' - \sum \frac{\mu_i}{\mathbf{LC}(g')}vgw''u_i'w'. \end{aligned}$$

Thus, in writing vgw this way, the number of occurrences of u in the chosen representation of f satisfying the above (3) can be reduced, a contradiction.

Case A.2.2: $l(v') < l(v\mathbf{LM}(g))$.
This implies that the ordered pair $(\mathbf{LM}(g), \mathbf{LM}(g'))$ has an overlap relation in u, that is, there is a segment r of w and a segment s of v' such that $vs = v'$, $rw' = w$ and $\mathbf{LM}(g)r = s\mathbf{LM}(g')$. Hence,

$$(5) \qquad \begin{aligned} &u = v\mathbf{LM}(g)rw' = vs\mathbf{LM}(g')w' \text{ and} \\ &o(g,r;\ s,g') = \frac{1}{\mathbf{LC}(g)}(g\cdot r) - \frac{1}{\mathbf{LC}(g')}(s\cdot g'). \end{aligned}$$

Furthermore, it follows from $g\cdot r = \mathbf{LC}(g)\cdot o(g,r;\ s,g') + \frac{\mathbf{LC}(g)}{\mathbf{LC}(g')}(s\cdot g')$ that

$$(6) \qquad vgw = vgrw' = \mathbf{LC}(g)\cdot v\cdot o(g,r;\ s,g')\cdot w' + \frac{\mathbf{LC}(g)}{\mathbf{LC}(g')}(v'g'w').$$

By the assumption, $o(g,r;\ s,g')$ is reduced to 0 by the division by $\mathcal{G}$, namely,

$$(7) \qquad o(g,r;\ s,g') = \sum_{k,j} c_{kj}v_{kj}g_kw_{kj},\ v_{kj},w_{kj}\in\mathcal{B},\ g_k\in\mathcal{G},\ c_{kj}\in K^*,$$

satisfying

$$(8) \quad v_{kj}\mathbf{LM}(g_k)w_{kj} \neq 0 \text{ then } v_{kj}\mathbf{LM}(g_k)w_{kj} \prec \mathbf{LM}(g\cdot r) = \mathbf{LM}(s\cdot g').$$

Combining (5) – (8), once again we see that the number of occurrences of u in the chosen representation of f satisfying the foregoing (3) can be reduced, a contradiction.
Case B: $l(v) = l(v')$.
This implies $\mathbf{LM}(g)|\mathbf{LM}(g')$ or $\mathbf{LM}(g')|\mathbf{LM}(g)$, which contradicts the assumption that $\mathcal{G}$ is LM-reduced.
Case C: $l(v) > l(v')$.
This is similar to Case A. □

Remark Recalling the division procedure by a subset of $R-K$, the termination theorem tells us that for every ordered pair $(g_i,g_j)\in\mathcal{G}\times\mathcal{G}$, if each overlap element $o(g_i,u;\ v,g_j)$ of (g_i,g_j) has a finite representation

$$o(g_i,u;\ v,g_j) = \sum_{t,\ell}\lambda_{tl}s_{t\ell}g_tw_{t\ell},\ \lambda_{t\ell}\in K^*,\ s_{t\ell},w_{t\ell}\in\mathcal{B},\ g_t\in\mathcal{G},$$

satisfying $\mathbf{LM}(s_{t\ell}\mathbf{LM}(g_t)w_{t\ell}) \preceq \mathbf{LM}(o(g_i,u;\ v,g_j))$, then $\mathcal{G}$ is a Gröbner basis. This observation helps us

(1) to verify whether a given subset $\mathcal{G}$ is a Gröbner basis by using only the division algorithm without considering the order of taking elements from $\mathcal{G}$; and

(2) to construct a Gröbner basis computationally by using a given subset $\mathcal{G}$ in which elements are ordered in an arbitrary way, in particular, this is valid when $\mathcal{G} = \{g_1, \ldots, g_s\}$ is a finite set.

We illustrate (1) of the remark above by verifying several examples, among which, Gröbner bases provided by Examples 5 – 8 will run through the rest of this book.

Example 3. Consider again the ideal $I = \langle g \rangle$ in the free K-algebra $K\langle X \rangle = K\langle X_1, X_2 \rangle$, where $g = X_1^2 - X_1X_2$. It follows from the foregoing Example 1 that $I = \langle \mathcal{G} \rangle$ with $\mathcal{G} = \{g_n = X_1X_2^nX_1 - X_1X_2^{n+1} \mid n \in \mathbb{N}\}$, and if we use the $\mathbb{N}$-graded lexicographic ordering $X_2 \prec_{gr} X_1$ with respect to the natural $\mathbb{N}$-gradation of $K\langle X \rangle$, then each g_n has $\mathbf{LM}(g_n) = X_1X_2^nX_1$, $\mathcal{G}$ is LM-reduced, and for ordered pairs $(g_n, g_k), (g_k, g_n) \in \mathcal{G}$, including $n = k$, the overlap elements of (g_n, g_k) are given by

$$\begin{aligned} S_{n,k} = o(g_n, X_2^kX_1;\ X_1X_2^n, g_k) &= g_n \cdot X_2^kX_1 - X_1X_2^n \cdot g_k \\ &= X_1X_2^nX_1X_2^{k+1} - X_1X_2^{n+k+1}X_1. \end{aligned}$$

$$\begin{aligned} S_{k,n} = o(g_k, X_2^nX_1;\ X_1X_2^k, g_n) &= g_k \cdot X_2^nX_1 - X_1X_2^k \cdot g_n \\ &= X_1X_2^kX_1X_2^{n+1} - X_1X_2^{n+k+1}X_1. \end{aligned}$$

Note that $\mathbf{LM}(S_{n,k}) = X_1X_2^nX_1X_2^{k+1}$, $\mathbf{LM}(S_{k,n}) = X_1X_2^nX_1X_2^{n+1}$. After executing the division of $S_{n,k}$ by $\mathcal{G}$ we have $S_{n,k} = g_nX_2^{k+1} - g_{n+k+1}$, i.e., $\overline{S_{n,k}}^{\mathcal{G}} = 0$. Similarly, $S_{k,n} = g_kX_2^{n+1} - g_{n+k+1}$ implies $\overline{S_{k,n}}^{\mathcal{G}} = 0$. Hence $\mathcal{G}$ is a Gröbner basis for I. Furthermore, the division of monomials by $\mathbf{LM}(\mathcal{G})$ yields the set of normal monomials in $\mathcal{B}$ (mod I):

$$N(I) = \{X_1X_2^k,\ X_2^nX_1,\ X_2^n \mid k, n \in \mathbb{N}\}.$$

It follows from Theorem 3.3 and Proposition 4.8 that both the algebras $A = K\langle X \rangle / I$ and $K\langle X \rangle / \langle \mathbf{LM}(I) \rangle$ have the K-basis

$$\overline{N(I)} = \{\overline{X}_1\overline{X}_2^k,\ \overline{X}_2^n\overline{X}_1,\ \overline{X}_2^n \mid k, n \in \mathbb{N}\},$$

where each $\overline{X}_i$ is the canonical image of X_i in $K\langle X \rangle / I$ and $K\langle X \rangle / \langle \mathbf{LM}(I) \rangle$ respectively.

Note that the Gröbner basis $\mathcal{G}$ obtained above is a reduced infinite Gröbner basis for the principal ideal $I = \langle g \rangle$.

Example 4. For the ideal $I = \langle f, g \rangle \subset KQ$ considered in previous Example 2, it is easy to see that $f = abcdead - abc$, $g = deadea - dead$ are uniform, the generating set $\mathcal{G} = \{f, g\}$ is LM-reduced, and moreover,

$$\begin{aligned} &o(f, ea;\ ;\ abc, g) = abcdead = f + abc, \\ &o(f, eadea;\ abcdea, g) = abcdeadead = f \cdot ead, \\ &o(g, bcdead;\ deade, f) = deadeabc = g \cdot bc, \\ &o(g, dea;\ dea, g) = deadead = g \cdot d. \end{aligned}$$

Obviously $\overline{o(f, ea;\ ;\ abc, g)}^{\mathcal{G}} \neq 0$. Hence $\mathcal{G} = \{f, g\}$ is not a Gröbner basis for the ideal $I = \langle \mathcal{G} \rangle$.

Example 5. Consider in the free K-algebra $K\langle X \rangle = K\langle X_1, \ldots, X_n \rangle$ the subset

$$\mathcal{G} = \{g_{ji} = X_j X_i - \lambda_{ji} X_i X_j \mid 1 \leq i < j \leq n,\ \lambda_{ji} \in K\}.$$

This example provides a family of Gröbner bases that define algebras similar to the skew polynomial algebra $\mathcal{O}_n(\lambda_{ji})$ introduced in (Ch.1, Section 1), i.e., the algebra $K\langle X \rangle / \langle \mathcal{G} \rangle$ coincides with $\mathcal{O}_n(\lambda_{ji})$ provided all the $\lambda_{ji} \neq 0$.

If we use the $\mathbb{N}$-graded lexicographic ordering $X_1 \prec_{gr} X_2 \prec_{gr} \cdots \prec_{gr} X_n$ with respect to the natural $\mathbb{N}$-gradation of $K\langle X \rangle$, then $\mathbf{LM}(g_{ji}) = X_j X_i$, $1 \leq i < j \leq n$, and $\mathcal{G}$ is LM-reduced. It is not difficult to verify that all possible overlap elements of the ordered pairs in $\mathcal{G} \times \mathcal{G}$ are

$$S_{ji,ik} = o(g_{ji}, X_k;\ X_j, g_{ik}) = \lambda_{ik} X_j X_k X_i - \lambda_{ji} X_i X_j X_k,\ 1 \leq k < i < j \leq n.$$

Note that the parameters λ_{ji} are arbitrarily chosen. If $\lambda_{ji} = 0$ and $\lambda_{ik} \neq 0$, then the division of $S_{ji,ik}$ by $\mathcal{G}$ yields

$$S_{ji,ik} = \lambda_{ik} X_j X_k X_i = \lambda_{ik} g_{jk} X_i + \lambda_{ik} \lambda_{jk} X_k g_{ji},$$

showing $\overline{S_{ji,ik}}^{\mathcal{G}} = 0$. Similarly, if $\lambda_{ik} = 0$ and $\lambda_{ji} \neq 0$, then the division of $S_{ji,ik}$ by $\mathcal{G}$ yields

$$S_{ji,ik} = -\lambda_{ji} X_i X_j X_k = -\lambda_{ji} X_i g_{jk} - \lambda_{ji} \lambda_{jk} g_{ik} X_j,$$

showing $\overline{S_{ji,ik}}^{\mathcal{G}} = 0$. If $\lambda_{ik} \neq 0$ and $\lambda_{ji} \neq 0$, then $\mathbf{LM}(S_{ji,ik}) = X_j X_k X_i$. So the division of $S_{ji,ik}$ by $\mathcal{G}$ (always starting with the leading monomial of the remainder at each step) yields

$$\begin{aligned} S_{ji,ik} &= \lambda_{ik} g_{jk} X_i + \lambda_{ik} \lambda_{jk} X_k X_j X_i - \lambda_{ji} X_i X_j X_k \\ &= \lambda_{ik} g_{jk} X_i - \lambda_{ji} X_i g_{jk} - \lambda_{ji} \lambda_{jk} X_i X_k X_j + \lambda_{ik} \lambda_{jk} X_k X_j X_i \\ &= \lambda_{ik} g_{jk} X_i - \lambda_{ji} X_i g_{jk} - \lambda_{ji} \lambda_{jk} g_{ik} X_j - \lambda_{ji} \lambda_{jk} \lambda_{ik} X_k X_i X_j \\ &\quad + \lambda_{ik} \lambda_{jk} X_k X_j X_i \\ &= \lambda_{ik} g_{jk} X_i - \lambda_{ji} X_i g_{jk} - \lambda_{ji} \lambda_{jk} g_{ik} X_j + \lambda_{jk} \lambda_{ik} X_k g_{ji}, \end{aligned}$$

again showing $\overline{S_{ji,ik}}^{\mathcal{G}} = 0$. Therefore, $\mathcal{G}$ is a Gröbner basis for the ideal $I = \langle \mathcal{G} \rangle$.

Furthermore, the division of monomials by $\mathbf{LM}(\mathcal{G})$ yields

$$N(I) = \left\{ X_1^{\alpha_1} X_2^{\alpha_2} \cdots X_n^{\alpha_n} \;\middle|\; \alpha_i \in \mathbb{N} \right\}.$$

It follows from Theorem 3.3 and Proposition 4.8 that both the algebras $K\langle X \rangle / I$ and $K\langle X \rangle / \langle \mathbf{LM}(I) \rangle$ have the K-basis

$$\overline{N(I)} = \left\{ \overline{X}_1^{\alpha_1} \overline{X}_2^{\alpha_2} \cdots \overline{X}_n^{\alpha_n} \;\middle|\; \alpha_i \in \mathbb{N} \right\},$$

where each $\overline{X}_i$ is the canonical image of X_i in $K\langle X \rangle / I$ and $K\langle X \rangle / \langle \mathbf{LM}(I) \rangle$ respectively.

Example 6. Consider in the free K-algebra $K\langle X \rangle = K\langle X_1, \ldots, X_n \rangle$ the subset $\mathcal{G} = \Omega \cup \mathcal{R}$ consisting of

$$\begin{aligned} &\Omega \subseteq \{ g_i = X_i^p \mid 1 \le i \le n \} \text{ with } p \ge 2 \text{ a fixed integer,} \\ &\mathcal{R} = \{ g_{ji} = X_j X_i - \lambda_{ji} X_i X_j \mid 1 \le i < j \le n \} \text{ with } \lambda_{ji} \in K. \end{aligned}$$

This example provides a family of Gröbner bases that define algebras similar to the quantum grassmannian (or quantum exterior) algebra introduced in ([Man], Section3), i.e., if $\Omega = \{ g_i = x_i^2 \mid 1 \le i \le n \}$ and all the $\lambda_{ji} \ne 0$, then the algebra $K\langle X \rangle / \langle \mathcal{G} \rangle$ is just the quantum grassmannian algebra, and in this case $K\langle X \rangle / \langle \mathcal{G} \rangle$ is 2^n-dimensional over K.

If we use the $\mathbb{N}$-graded lexicographic ordering $X_1 \prec_{gr} X_2 \prec_{gr} \cdots \prec_{gr} X_n$ with respect to the natural $\mathbb{N}$-gradation of $K\langle X \rangle$, then

$$\mathbf{LM}(\mathcal{G}) = \{ X_s^p \mid g_s \in \Omega \} \cup \{ X_j X_i \mid 1 \le i < j \le n \},$$

and $\mathcal{G}$ is LM-reduced. It is not difficult to verify that all possible nonzero overlap elements of the ordered pairs in $\mathcal{G} \times \mathcal{G}$ are

$$\begin{aligned} &S_{i,i\ell} = o(g_i, X_\ell;\; X_i^{p-1}, g_{i\ell}) = \lambda_{i\ell} X_i^{p-1} X_\ell X_i, && i > \ell, \\ &S_{ji,i} = o(g_{ji}, X_i^{p-1};\; X_j, g_i) = -\lambda_{ji} X_i X_j X_i^{p-1}, && j > i, \\ &S_{j\ell,\ell t} = (g_{j\ell}, X_t;\; X_j, g_{\ell t}) = \lambda_{\ell t} X_j X_t X_\ell - \lambda_{j\ell} X_\ell X_j X_t, && j > \ell > t. \end{aligned}$$

The division of $S_{i,i\ell}$ and $S_{ji,i}$ by $\mathcal{G}$ (always starting with the leading monomial of the remainder at each step) yields the following results:

$$\begin{aligned} S_{i,i\ell} &= \sum_{k=2}^{p} \lambda_{i\ell}^{k-1} X_i^{p-k} g_{i\ell} X_i^{k-1} + \lambda_{i\ell}^p X_\ell g_i, \\ S_{ji,i} &= -\sum_{k=2}^{p} \lambda_{ji}^{k-1} X_i^{k-1} g_{ji} X_i^{p-k} - \lambda_{ji}^p g_i X_j, \end{aligned}$$

showing $\overline{S_{i,i\ell}}^{\mathcal{G}} = 0$ and $\overline{S_{ji,i}}^{\mathcal{G}} = 0$. Noticing that the parameters λ_{ji} are arbitrarily chosen, actually as in the last example, the division of $S_{j\ell,\ell t}$ by $\mathcal{G}$ yields

$$\begin{aligned} &\text{if } \lambda_{j\ell} = 0,\ \lambda_{\ell t} \neq 0, \text{ then } S_{j\ell,\ell t} = \lambda_{\ell t} g_{jt} X_\ell + \lambda_{\ell t}\lambda_{jt} X_t g_{j\ell}, \\ &\text{if } \lambda_{\ell t} = 0,\ \lambda_{j\ell} \neq 0, \text{ then } S_{j\ell,\ell t} = -\lambda_{j\ell} X_\ell g_{jt} - \lambda_{j\ell}\lambda_{jt} g_{\ell t} X_j, \\ &\text{if } \lambda_{\ell t} \neq 0,\ \lambda_{j\ell} \neq 0, \text{ then } S_{j\ell,\ell t} = \lambda_{\ell t} g_{jt} X_\ell - \lambda_{j\ell} X_\ell g_{jt} - \lambda_{j\ell}\lambda_{jt} g_{\ell t} X_j \\ &\qquad\qquad\qquad\qquad\qquad\qquad + \lambda_{jt}\lambda_{\ell t} X_t g_{j\ell}, \end{aligned}$$

showing $\overline{S_{j\ell,\ell k}}^{\mathcal{G}} = 0$. Hence, $\mathcal{G}$ is a Gröbner basis for the ideal $I = \langle \mathcal{G} \rangle$.

Furthermore, the division of monomials by $\mathbf{LM}(\mathcal{G})$ yields

$$N(I) = \left\{ X_1^{\alpha_1} X_2^{\alpha_2} \cdots X_n^{\alpha_n} \;\middle|\; \alpha_i \in \mathbb{N} \text{ and } 0 \le \alpha_s \le p-1 \text{ if } g_s \in \Omega \right\}.$$

It follows from Theorem 3.3 and Proposition 4.8 that both the algebras $K\langle X \rangle / I$ and $K\langle X \rangle / \langle \mathbf{LM}(I) \rangle$ have the K-basis

$$\overline{N(I)} = \left\{ \overline{X}_1^{\alpha_1} \overline{X}_2^{\alpha_2} \cdots \overline{X}_n^{\alpha_n} \;\middle|\; \alpha_i \in \mathbb{N} \text{ and } 0 \le \alpha_s \le p-1 \text{ if } g_s \in \Omega \right\},$$

where each $\overline{X}_i$ is the canonical image of X_i in $K\langle X \rangle / I$ and $K\langle X \rangle / \langle \mathbf{LM}(I) \rangle$ respectively.

Example 7. In this example we verify that the set of defining relations of a Clifford algebra C (an exterior algebra E) defined in (Ch.1, Section 1) forms a Gröbner basis.

Let $X = \{X_i\}_{i \in J}$ and $>$ be a well-ordering on J. Consider the subset $\mathcal{G}$ of the free algebra $K\langle X \rangle$ consisting of

$$\begin{aligned} g_i &= X_i^2 - q_i, && i \in J,\ q_i \in K, \\ g_{k\ell} &= X_k X_\ell + X_\ell X_k - q_{k\ell}, && k, \ell \in J,\ k > \ell,\ q_{k\ell} \in K. \end{aligned}$$

If we use the $\mathbb{N}$-graded lexicographic ordering

$$X_\ell \prec_{gr} X_k, \quad \ell, k \in J,\ \ell < k,$$

with respect to the natural $\mathbb{N}$-gradation of $K\langle X \rangle$, then

$$\begin{aligned} \mathbf{LM}(g_i) &= X_i^2, \quad i \in J, \\ \mathbf{LM}(g_{k\ell}) &= X_k X_\ell,\ k, \ell \in J,\ k > \ell, \end{aligned}$$

and $\mathcal{G}$ is LM-reduced. With a little patience, we can write down all possible nonzero overlap elements of the ordered pairs in $\mathcal{G} \times \mathcal{G}$ as follows.

$$\begin{aligned} S_{i,i\ell} &= o(g_i, X_\ell;\ X_i, g_{i\ell}) = -X_i X_\ell X_i - q_i X_\ell + q_{i\ell} X_i,\ i > \ell, \\ S_{ki,i} &= o(g_{ki}, X_i;\ X_k, g_i) = X_i X_k X_i - q_{ki} X_i + q_i X_k,\ k > i, \\ S_{k\ell,\ell t} &= o(g_{k\ell}, X_t;\ X_k, g_{\ell t}) = -X_k X_t X_\ell + X_\ell X_k X_t \\ &\qquad - q_{k\ell} X_t + q_{\ell t} X_k, \quad k > \ell > t. \end{aligned}$$

With a little more patience, the division of $S_{i,i\ell}$ by $\mathcal{G}$ (always starting with the leading monomial of the remainder at each step) turns out that

$$\begin{aligned} S_{i,i\ell} &= -(g_{i\ell}X_i - X_\ell g_i), \\ S_{ki,i} &= X_i g_{ki} - g_i X_k, \\ S_{k\ell,\ell t} &= -g_{kt}X_\ell + X_\ell g_{kt} - g_{\ell t}X_k + X_t g_{k\ell}, \end{aligned}$$

that is,

$$\overline{S_{i,i\ell}}^{\mathcal{G}} = \overline{S_{ki,i}}^{\mathcal{G}} = \overline{S_{k\ell,\ell t}}^{\mathcal{G}} = 0.$$

Consequently, $\mathcal{G}$ is a Gröbner basis for the ideal $I = \langle \mathcal{G} \rangle$. Furthermore, the division of monomials by $\mathbf{LM}(\mathcal{G})$ yields

$$N(I) = \left\{1,\ X_{i_1}X_{i_2}\cdots X_{i_s}\ \middle|\ X_{i_j} \in X,\ i_1 < i_2 < \cdots < i_s,\ s \geq 1\right\}.$$

By Theorem 3.3 and Proposition 4.8, both the algebras $K\langle X\rangle/I$ and $K\langle X\rangle/\langle \mathbf{LM}(I)\rangle$ have the K-basis

$$\overline{N(I)} = \left\{1,\ \overline{X}_{i_1}\overline{X}_{i_2}\cdots\overline{X}_{i_s}\ \middle|\ i_1 < i_2 < \cdots < i_s,\ s \geq 1\right\},$$

where each $\overline{X}_{i_j}$ is the canonical image of X_{i_j} in the respective quotient algebra.

Example 8. Consider in the free K-algebra $K\langle X\rangle = K\langle X_1, X_2\rangle$ the subset $\mathcal{G} = \{g_1, g_2\}$ with

$$\begin{aligned} g_1 &= X_1^2X_2 - \alpha X_1X_2X_1 - \beta X_2X_1^2 - F_1, \\ g_2 &= X_1X_2^2 - \alpha X_2X_1X_2 - \beta X_2^2X_1 - F_2, \end{aligned} \qquad \alpha, \beta \in K,$$

where

$$F_1, F_2 \in K\text{-span}\left\{1,\ X_2^3,\ X_1X_2,\ X_2X_1,\ X_1^2,\ X_2^2,\ X_1,\ X_2\right\}.$$

This example provides a family of Gröbner bases that define algebras similar to the down-up algebra introduced by G. Benkart and T. Roby ([Ben], [BR], 1998) in the study of algebras generated by the down and up operators on a differential or uniform partially ordered set (poset). That is, for any α, β, $\gamma \in K$, if $F_1 = \gamma X_1$ and $F_2 = \gamma X_2$, then the algebra $A(\alpha, \beta, \gamma) = K\langle X\rangle/\langle \mathcal{G}\rangle$ is a down-up algebra. The reader is referred to Example 6(c) of (Ch.4, Section 3) for another presentation of a down-up algebra.

If we use the $\mathbb{N}$-graded lexicographic ordering $X_2 \prec_{gr} X_1$ with respect to the natural $\mathbb{N}$-gradation of $K\langle X\rangle$, then $\mathbf{LM}(g_1) = X_1^2X_2$, $\mathbf{LM}(g_2) = X_1X_2^2$, and $\mathcal{G}$ is LM-reduced. It is easy to see that the only possible nonzero overlap element is

$$\begin{aligned} S_{12} &= o(g_1, X_2;\ X_1, g_2) \\ &= g_1X_2 - X_1g_2 \\ &= \beta(X_1X_2^2X_1 - X_2X_1^2X_2) - F_1X_2 + X_1F_2. \end{aligned}$$

Executing the division of S_{12} by g_2, g_1 successively, we obtain

$$S_{12} = \beta(g_2X_1 - X_2g_1) + \beta(X_2F_1 - F_2X_1) + (X_1F_2 - F_1X_2).$$

Put $G = \beta(X_2F_1 - F_2X_1) + (X_1F_2 - F_1X_2)$. Then for suitable choice of F_1 and F_2 such that $\overline{G}^{\mathcal{G}} = 0$ (for instance, $\beta = 0$, $F_1 = \lambda X_1^2 - \gamma X_1$ and $F_2 = \lambda X_1X_2 - \gamma X_2$ with $\lambda, \gamma \in K$; or for any α, β, $\gamma \in K$, $F_1 = \gamma X_1$ and $F_2 = \gamma X_2$), we have $\overline{S_{12}}^{\mathcal{G}} = 0$, and hence $\mathcal{G}$ forms a Gröbner basis for the ideal $I = \langle \mathcal{G} \rangle$. Furthermore, by the division of monomials by $\mathbf{LM}(\mathcal{G})$, we have

$$N(I) = \left\{ X_2^i(X_1X_2)^jX_1^k \;\middle|\; i, j, k \geq 0 \right\}.$$

Thus, by Theorem 3.3 and Proposition 4.8, both the algebras $K\langle X\rangle/I$ and $K\langle X\rangle/\langle \mathbf{LM}(I)\rangle$ have the K-basis

$$\overline{N(I)} = \left\{ \overline{X}_2^i(\overline{X}_1\overline{X}_2)^j\overline{X}_1^k \;\middle|\; i, j, k \geq 0 \right\},$$

where $\overline{X}_i$ is the canonical image of X_i in the respective algebra.

Remark Let R be a free K-algebra or a path algebra over K, with the admissible system $(\mathcal{B}, \prec)$, and let $\mathcal{G}$ be a nonempty subset of $R - K$. We should point out that whether $\mathcal{G}$ is a Gröbner basis depends on the choice of the monomial ordering $\prec$, though the division by $\mathcal{G}$ performed in the termination theorem does not depend on the order of elements in $\mathcal{G}$. For instance, by using $X_2 \prec_{gr} X_1$, we have shown in the foregoing Example 1 and Example 3 that $\mathcal{G} = \{g = X_1^2 - X_1X_2\}$ is not a Gröbner basis for the principal ideal $I = \langle g\rangle$, and indeed I has an infinite Gröbner basis. However, if the $\mathbb{N}$-graded monomial ordering $X_1 \prec_{gr} X_2$ is used, then it is straightforward that $\mathcal{G} = \{g = X_1^2 - X_1X_2\}$ is a Gröbner basis for I, and since in this case $\mathbf{LM}(g) = X_1X_2$, both the algebras $K\langle X_1, X_2\rangle/I$ and $K\langle X_1, X_2\rangle/\langle \mathbf{LM}(I)\rangle$ have the nice K-basis $\{\overline{X_2}^{\alpha}\overline{X_1}^{\beta} \mid \alpha, \beta \in \mathbb{N}\}$.

Construction of Gröbner bases

Let R be a free K-algebra or a path algebra defined by a finite directed graph, and let $(\mathcal{B}, \prec)$ be an admissible system of R, where $\mathcal{B}$ is the standard K-basis of R. It is now the time to implement part (2) of the remark made after Theorem 5.1, that is, to construct a Gröbner basis starting with a given finite subset.

Given an LM-reduced subset $G_0 = \{g_1, \ldots, g_s\}$ of uniform elements in $R - K$, we write $o(g_i, g_j)$ for *any overlap element* of the ordered pair $(g_i, g_j) \in G_0 \times G_0$, and let O_0 denote the set of all *nonzero* overlap elements.

By the division by G_0, every $o(g_i, g_j) \in O_0$ has a representation

$$(*) \qquad o(g_i, g_j) = \sum_i u_k g_k w_k + \overline{o(g_i, g_j)}^{G_0}, \quad u_k, w_k \in \mathcal{B}.$$

If $\overline{o(g_i, g_j)}^{G_0} = 0$ for all $o(g_i, g_j) \in O_0$, then by Theorem 4.1, G_0 is a Gröbner basis for the ideal $I = \langle G_0 \rangle$; otherwise, noticing $\overline{o(g_i, g_j)}^{G_0} \in I$, we let $g_{s+1}, g_{s+2}, \ldots, g_n$ denote all nonzero remainders $\overline{o(g_i, g_j)}^{G_0}$, and put

$$G_1 = \{g_1, \ldots, g_s, g_{s+1}, \ldots, g_n\}.$$

To test whether G_1 is a Gröbner basis for I, the above expression $(*)$ tells us that we need only to take the set O_1 consisting of all overlap elements

$$o(g_k, g_\ell) \text{ with } k > s \text{ or } \ell > s$$

into account. Continuing the same procedure for the data (G_n, O_n), $n \geq 1$, either, after a finite number of repetitions we obtain a finite Gröbner basis G_n, or, the repetition does not stop and nevertheless we obtain a countably infinite Gröbner basis $\mathcal{G} = \cup_{n \in \mathbb{N}} \mathcal{G}_n$ (as illustrated by Examples 9, 10 given later).

Before giving a better proof of the above conclusion, we first write the reduction procedure (in pseudo-code) into a (possibly non-stopping) algorithm which is usually referred to as the Buchberger-Bergman-Mora algorithm.

Algorithm-GB

Input: $G_0 = \{g_1, \ldots, g_s\}$
Output: $\mathcal{G} = \{g_1, \ldots, g_m, \ldots\}$, a Gröbner basis for $I = \langle G_0 \rangle$
Initialization: $\mathcal{G} := G_0$, $O := \{o(g_i, g_j) \mid g_i, g_j \in G_0\}$
While $O \neq \emptyset$ **do**
Choose any $o(g_i, g_j) \in O$
$O := O - \{o(g_i, g_j)\}$
$\overline{o(g_i, g_j)}^{\mathcal{G}} = r$
If $r \neq 0$ **then**
$\mathcal{G} := \mathcal{G} \cup \{r\}$
$O := O \cup \{o(g, r),\ o(r, g) \mid g \in \mathcal{G}\}$
⋮

Next, let us write $\mathcal{G}_{n+1}$ for the new set obtained after the n-th execution of the **While** loop in Algorithm-GB, that is,

$$\mathcal{G}_{n+1} = \mathcal{G}_n \cup \left\{ r_{ij} = \overline{o(g_i, g_j)}^{\mathcal{G}_n} \neq 0 \;\middle|\; g_i, g_j \in \mathcal{G}_n \right\}, \quad n \in \mathbb{N}.$$

Claim *$\mathcal{G} = \cup_{n \in \mathbb{N}} \mathcal{G}_n$ is a Gröbner basis for the ideal $I = \langle \mathcal{G}_0 \rangle$ of R, where $\mathcal{G}_0 = G_0$ in* **Algorithm GB**.

Proof. [Proof of Claim] That $\mathcal{G}$ is a set of generators of I is clear. It remains to prove that $\overline{o(g_i, g_j)}^{\mathcal{G}} = 0$ for all $g_i, g_j \in \mathcal{G}$.

Note that $\mathcal{G}_i = \mathcal{G}_{i+1}$ for some i means every overlap element of $\mathcal{G}_i$ reduces to zero after the i-th pass through the **While** loop in **Algorithm-GB**, and it turns out that $\mathcal{G}_i$ is a Gröbner basis of I. Suppose $\mathcal{G}_i \neq \mathcal{G}_j$ for all $i, j \in \mathbb{N}$. Let $g_i \in \mathcal{G}_i$, $g_j \in \mathcal{G}_j$, say $j > i$. Then it follows from **Algorithm-GB** that all overlap elements $o(g_i, g_j)$ reduces to zero after the $j+2$-nd execution of the **While** loop. This shows that $\overline{o(g_i, g_j)}^{\mathcal{G}} = 0$ for all $g_i, g_j \in \mathcal{G}$, as desired. □

One may check, indeed, that in **Algorithm-GB** the r may be replaced by the so-called *normal form* of $o(g_i, g_j)$, i.e., in dividing $o(g_i, g_j)$ by $\mathcal{G}$, if the remainder r_t obtained after the t-th pass through the **While** loop in **Algorithm-DIV** (Section 2) has the property that $\mathbf{LM}(g_k) \nmid \mathbf{LM}(r_t)$ for every $g_k \in \mathcal{G}$, then set $r := r_t$ in **Algorithm-GB**.

We give below two small examples to illustrate the constructive method formulated above.

Example 9. Consider the ideal $I = \langle g_1, g_2\rangle$ in the free K-algebra $K\langle X\rangle = K\langle X_1, X_2\rangle$, where $g_1 = X_1X_2 - X_1$, $g_2 = X_2X_1$. If we use the graded lexicographic ordering $X_2 \prec_{gr} X_1$ with respect to the natural $\mathbb{N}$-gradation of $K\langle X\rangle$, then $\mathbf{LM}(g_1) = X_1X_2$, $\mathbf{LM}(g_2) = X_2X_1$. To construct a Gröbner basis for I, start with the data (G_0, O_0), where $G_0 = \{g_1, g_2\}$ and O_0 consists of

$$\begin{aligned} S_{12} &= o(g_1, X_1;\ X_1, g_2) = g_1X_1 - X_1g_2 = -X_1^2, \\ S_{21} &= o(g_2, X_2;\ X_2, g_1) = g_2X_2 - X_2g_1 = X_2X_1. \end{aligned}$$

As $\overline{S_{12}}^{G_0} = -X_1^2$, $\overline{S_{21}}^{G_0} = 0$, we have to continue dealing with the new data (G_1, O_1), where $G_1 = \{g_1, g_2, g_3\}$ in which $g_3 = \overline{S_{12}}^{G_0} = -X_1^2$ with $\mathbf{LM}(g_3) = X_1^2$, and O_1 consists of

$$S_{31} = o(g_3, X_2;\ X_1, g_1) = -g_3X_2 - X_1g_1 = X_1^2.$$

Since $\overline{S_{31}}^{G_1} = 0$, G_1 is a Gröbner basis for the ideal $I = \langle g_1, g_2\rangle$.

Example 10. Consider the principle ideal $I = \langle g\rangle$ with $g = X_1X_2X_1 - X_1X_2$ in the free K-algebra $K\langle X\rangle = K\langle X_1, X_2\rangle$. Since X_1X_2 divides $X_1X_2X_1$ with respect to *any* monomial ordering $\prec$ on $K\langle X\rangle$, we have $\mathbf{LM}(g) = X_1X_2X_1$. Starting **Algorithm-GB** with the data (G_0, O_0) where $G_0 = \{g_1 = g\}$ and $O_0 = \{o(g_1, g_1) = X_1X_2X_1X_2 - X_1X_2^2X_1\}$, since $g_2 = \overline{o(g_1, g_1)}^{G_0} = X_1X_2^2 - X_1X_2^2X_1$ with $\mathbf{LM}(g_2) = X_1X_2^2X_1$, after the first pass through the **While** loop we obtain the new data (G_1, O_1) with $G_1 = \{g_1,\ g_2 = X_1X_2^2 - X_1X_2^2X_1\}$ and $O_1 = \{o(g_1, g_2) = X_1X_2X_1X_2^2 - X_1X_2^3X_1,\ o(g_2, g_1) = X_1X_2^2X_1X_2 - X_1X_2^3X_1,\ o(g_2, g_2) = X_1X_2^2X_1X_2^2 - X_1X_2^4X_1\}$. Executing the second pass through the **While** loop, we obtain the new data (G_2, O_2) with $G_2 = \{g_1,\ g_2,\ g_3 = X_1X_2^3 - X_1X_2^3X_1,\ g_4 = X_1X_2^4 - X_1X_2^4X_1\}$ and

$$O_2 = \left\{ \begin{matrix} o(g_1, g_3),\ o(g_3, g_1),\ o(g_2, g_3),\ o(g_3, g_2),\ o(g_3, g_3),\ o(g_1, g_4) \\ o(g_4, g_1),\ o(g_2, g_4),\ o(g_4, g_2),\ o(g_3, g_4),\ o(g_4, g_3),\ o(g_4, g_4) \end{matrix} \right\}.$$

Inductively we see, after executing the i-th pass through the **While** loop, that the output data is

$$G_i = \left\{ g_1,\ g_j = X_1X_2^j - X_1X_2^jX_1 \;\middle|\; 2 \le j \le 2^i \right\}.$$

It follows that **Algorithm-GB** does not stop, and that the ideal I has the countably infinite Gröbner basis $\mathcal{G} = \{g_j = X_1X_2^j - X_1X_2^jX_1 \mid j \ge 1\}$.

When does* Algorithm-GB *terminate?

We have seen from Example 10 above that **Algorithm-GB** may not terminate with respect to *any* monomial ordering, i.e., in certain cases it, in principle, produces a countably infinite Gröbner basis $\mathcal{G} = \cup_{n\in\mathbb{N}}\mathcal{G}_n$. Thus, the logical question to ask is what conditions do we need so that a finite Gröbner basis can be guaranteed to be the output data. Following [Mor2] and [Gr2], the next two propositions partially tell us when the algorithm may produce a finite Gröbner basis.

Proposition 5.2. *Let R be a finitely generated free K-algebra or a path algebra defined by a finite directed graph over K, and let $(\mathcal{B}, \prec)$ be an admissible system of R, where $\mathcal{B}$ is the standard K-basis of R. If I is an ideal of R generated by the subset $G_0 = \{g_1, \dots, g_s\}$ such that $\langle \mathbf{LM}(I)\rangle$, the monomial ideal generated by the set $\mathbf{LM}(I)$ of leading monomials of I, has a finite monomial generating set, then, starting with $\mathcal{G} = G_0$, the* **Algorithm-GB** *terminates in a finite number of steps and yields a finite Gröbner basis for I.*

Proof. With notation as before, let $\mathcal{G} = \cup_{n\in\mathbb{N}}\mathcal{G}_n$ be the Gröbner basis produced by **Algorithm-GB**. Then $\langle \mathbf{LM}(I)\rangle = \langle \mathbf{LM}(\mathcal{G})\rangle$. Suppose $\langle \mathbf{LM}(I)\rangle = \langle v_1, \dots, v_t\rangle$ with $v_i \in \mathcal{B}$, $1 \le i \le t$. Without loss of generality, we may assume that $\Omega = \{v_1, \dots, v_t\}$ is the unique reduced monomial generating set of $\langle \mathbf{LM}(I)\rangle$ (Theorem 4.4). Then since $\langle \mathbf{LM}(I)\rangle = \langle\Omega\rangle = \langle \mathbf{LM}(\mathcal{G})\rangle$, by Proposition 4.6 we have $\Omega \subseteq \{\mathbf{LM}(g_1), \dots, \mathbf{LM}(g_N)\}$ for sufficiently large N. Thus

$$\langle \mathbf{LM}(g_1), \dots, \mathbf{LM}(g_N)\rangle = \langle\Omega\rangle = \langle \mathbf{LM}(I)\rangle,$$

and it follows again from Proposition 4.6 that $\mathcal{G}_N = \{g_1, \dots, g_N\}$ is a Gröbner basis of I. Hence, $\overline{o(g_i, g_j)}^{\mathcal{G}_N} = 0$ for all $g_i, g_j \in \mathcal{G}_N$ and the algorithm terminates in a finite number of steps outptting a Gröbner basis. □

Proposition 5.3. *Let the ideal I of R be as in Proposition 5.2. If $dim_K(R/I) < \infty$, that is, the quotient algebra $A = R/I$ is finite dimensional over K, then the monomial ideal $\langle \mathbf{LM}(I)\rangle$ has a finite set of monomial generators.*

Proof. By the assumption, we may assume that the set of normal monomials (mod I) is $N(I) = \{u_1, \dots, u_s\}$. Let Ω be the unique reduced monomial generating set of $\langle \mathbf{LM}(I)\rangle$ (Theorem 4.4). We claim that Ω is a finite set. To see this, let the K-algebra R be generated by $\mathsf{X} = \{x_1, \ldots, x_p\}$, and let $v \in \Omega$, say $v = x_{i_1}x_{i_2}\cdots x_{i_m}$ with $x_{i_j} \in \mathsf{X}$. Then, by the minimality of elements in Ω, $u = x_{i_2}\cdots x_{i_m} \notin \mathbf{LM}(I)$. Hence, $u \in N(I)$ and it follows that $v = x_{i_1}u$ with $u \in N(I)$. This shows that

$$v \in \{x_i u \mid x_i \in \mathsf{X},\ u \in N(I)\} \cap \mathbf{LM}(I).$$

Therefore, Ω is a finite set as claimed. □

In Ch.8 we will see that some (finite) Gröbner bases in free algebras may be obtained by lifting finite Gröbner bases from certain algebras of solvable type.

Nowadays, there are many well-developed computer algebra systems implementing **Algorithm-GB** for ideals in free algebras and path algebras, such as BERGMAN, FELIX, GAP, MAGMA, MAPLE, MATHEMATICA, NCALGEBRA, OPAL, and SINGULAR.

The reader is also reminded that most of the well-known commutative and noncommutative computer algebra systems do a better job for producing *homogeneous Gröbner bases*, in particular, the recently developed programm LETTERPLACE, which is proposed by La Scala and Levandovskyy in [LL] and is implemented in SINGULAR, completely converts the calculation of homogeneous Gröbner bases for two-sided graded ideals in a free algebra to the calculation of Gröbner bases for graded ideals in a commutative polynomial algebra. This motivates a systematic study of (de)homogenized Gröbner bases from a structural-computational viewpoint in the next two sections; especially Proposition 6.14 given in the next section provides an effective way to obtain a Gröbner basis for a nonhomogeneous ideal in a free algebra via a homogeneous Gröbner basis that may be calculated by LETTERPLACE by passing to the commutative setting.

3.6 (De)homogenized Gröbner Bases

Let $R = \oplus_{p\in\mathbb{N}} R_p$ be an $\mathbb{N}$-graded K-algebra, and suppose $(\mathcal{B}, \prec_{gr})$ is an admissible system of R, where $\mathcal{B}$ is a *skew multiplicative K-basis of R consisting of $\mathbb{N}$-homogeneous elements*, and $\prec_{gr}$ is some $\mathbb{N}$-graded monomial ordering on $\mathcal{B}$. Then R has a Gröbner basis theory by the previous Section 4, in particular, every graded ideal of R has a *homogeneous Gröbner basis*, i.e., a Gröbner basis consisting of $\mathbb{N}$-homogeneous elements (Proposition

2.2(iii)). In this section, by employing the (de)homogenization techniques stemming from algebraic geometry, we study in detail the relation between Gröbner bases in R and homogeneous Gröbner bases in the polynomial ring $R[t]$, respectively the relation between Gröbner bases in the free algebra $K\langle X\rangle = K\langle X_1,\ldots,X_n\rangle$ and homogeneous Gröbner bases in the free algebra $K\langle X,T\rangle = K\langle X_1,\ldots,X_n,T\rangle$. As a consequence, this establishes a solid theoretical foundation for us to demonstrate, by passing to the graded ideal $\langle S^*\rangle$, how to obtain a Gröbner basis for the ideal $I = \langle S\rangle$ generated by a subset $S \subset R$ and hence a homogeneous Gröbner basis for the central homogenization ideal $\langle I^*\rangle$ of I in $R[t]$ with respect to t; respectively, this enables us to demonstrate, by passing to the graded ideal $\langle \widetilde{S}\rangle$, how to obtain a Gröbner basis for the ideal $I = \langle S\rangle$ generated by a subset $S \subset K\langle X\rangle$ and hence a homogeneous Gröbner basis for the non-central homogenization ideal $\langle \widetilde{I}\,\rangle$ of I in $K\langle X,T\rangle$ with respect to T.

Bearing in mind the foregoing **Algorithm-GB**, we start with a little discussion on the algorithmic construction of homogeneous Gröbner bases in free algebras and path algebras. This indicates roughly why common implementations of the Gröbner basis algorithm in the literature begin by homogenizing generators.

Construction of homogeneous Gröbner bases

Let R be a free K-algebra or a path algebra defined by a finite directed graph over K, which is equipped with a weight $\mathbb{N}$-gradation and an admissible system $(\mathcal{B},\prec)$, where $\prec$ is an arbitrarily fixed monomial ordering on the standard K-basis $\mathcal{B}$ of R. Let G be a subset of $\mathbb{N}$-*homogeneous elements* in R, $I = \langle G\rangle$, and $f \in R$. Then by the division by G (see Section 2) we have

$$f = \sum_{i,j} \lambda_{ij} v_{ij} g_i w_{ij} + \overline{f}^G, \quad \lambda_{ij} \in K^*,\ v_{ij}, w_{ij} \in \mathcal{B},\ g_i \in G,$$

where $v_{ij}\mathbf{LM}(g_i)w_{ij} \neq 0$, $\mathbf{LM}(v_{ij}g_iw_{ij}) \preceq_{gr} \mathbf{LM}(f)$ for all i,j, and $\mathbf{LM}(\overline{f}^G) \preceq_{gr} \mathbf{LM}(f)$ in case $\overline{f}^G \neq 0$. If furthermore $f \in R_n$ is a homogeneous element of degree n, then since $\mathcal{B}$ consists of $\mathbb{N}$-homogeneous elements, it follows that the above expression has the property that $d(g_i) \le n$ and either $\overline{f}^G = 0$ or $\overline{f}^G$ is an $\mathbb{N}$-homogeneous element of degree n. Also note that for $g_i, g_j \in G$, any nonzero overlap element of the ordered pair

(g_i, g_j), say

$$o(g_i, u;\ v, g_j) = g_i u - v g_j,$$

is an $\mathbb{N}$-homogeneous element of degree $\geq \max\{d(g_i), d(g_j)\}$. Hence, either $\overline{o(g_i, u;\ v, g_j)}^G = 0$ or $\overline{o(g_i, u;\ v, g_j)}^G$ is an $\mathbb{N}$-homogeneous element having the same degree as that of $o(g_i, u;\ v, g_j)$. In conclusion,

- starting with G in **Algorithm-GB**, a homogeneous Gröbner basis for the $\mathbb{N}$-graded ideal $I = \langle G \rangle$ may be constructed computationally.

n-Truncated Gröbner basis

Let R be a free K-algebra or a path algebra defined by a finite directed graph over K, which is equipped with a weight $\mathbb{N}$-gradation and an admissible system $(\mathcal{B}, \prec)$, where $\prec$ is an arbitrarily fixed monomial ordering on the standard K-basis $\mathcal{B}$ of R. In the light of the above discussion, **Algorithm-GB** can indeed be modified to produce a (finite or infinite) homogeneous Gröbner basis by computing successively the n-truncated homogeneous Gröbner basis (which is always finite) for each $n \geq 1$. More precisely, let $I = \langle G \rangle$ be the $\mathbb{N}$-graded ideal of R generated by a subset G of $\mathbb{N}$-homogeneous elements. Put

$$I_{\leq n} = \oplus_{p \leq n} I_p, \quad n \in \mathbb{N}.$$

An *n-truncated Gröbner basis* for I is a subset $\mathcal{G}_{\leq n} \subset I_{\leq n}$ consisting of homogeneous elements, such that

$$\langle \mathbf{LM}(\mathcal{G}_{\leq n}) \rangle_{\leq n} = \langle \mathbf{LM}(I) \rangle_{\leq n}.$$

As observed in the last part, if, in an execution of the **While** loop in **Algorithm-GB**, we consider only the overlap element $o(g_i, g_j)$ of degree n in O, then the only members of $\mathcal{G}$ used in the division of $o(g_i, g_j)$ by $\mathcal{G}$ are those of degree $\leq n$; and moreover, any nonzero remainder of $o(g_i, g_j)$ on division by $\mathcal{G}$ is also of degree n. Since for any fixed $m > 1$, $\mathcal{B} \cap R_m$ is a finite set, it follows that only a finite number of new elements of degree n can be created for the next execution of the **While** loop in **Algorithm-GB**, and that an n-truncated Gröbner basis $\mathcal{G}_{\leq n}$ may be obtained after a finite number of executions in running **Algorithm-GB**. We omit the technical details on this topic here.

With an n-truncated Gröbner basis $\mathcal{G}_{\leq n}$, it is easy to see that an element $f \in R$ of degree $\leq n$ is a member of I if and only if $\overline{f}^{\mathcal{G}_{\leq n}} = 0$, that is, the "degree-restricted membership problem" is solvable by computer. This solution may be viewed as another example of illustrating the mathematical principle of "mastering the infinity by the finite".

From the discussion of previous parts, though, we may see that an arbitrarily fixed monomial ordering $\prec$ on $\mathcal{B}$ works with no trouble for constructing a homogeneous Gröbner basis, very soon it will be clear that a suitable graded monomial ordering $\prec_{gr}$ on $\mathcal{B}$ is indispensable for achieving the goal of this section.

Central (De)homogenization

Now, let $R = \oplus_{p\in\mathbb{N}} R_p$ be an arbitrary $\mathbb{N}$-graded K-algebra, and let $R[t]$ be the polynomial ring in the commuting variable t over R. Then $R[t]$ has the mixed $\mathbb{N}$-gradation, that is, $R[t] = \oplus_{p\in\mathbb{Z}} R[t]_p$ is an $\mathbb{N}$-graded algebra with the degree-p homogeneous part

$$R[t]_p = \left\{ \sum_{i+j=p} f_i t^j \;\middle|\; f_i \in R_i,\ j \geq 0 \right\}, \quad p \in \mathbb{N}.$$

Considering the surjective ring homomorphism ϕ: $R[t] \to R$ defined by $\phi(t) = 1$, then for each $f \in R$, there exists a homogeneous element $F \in R[t]_p$, for some p, such that $\phi(F) = f$. More precisely, if $f = f_p + f_{p-1} + \cdots + f_{p-s}$ with $f_p \in R_p$, $f_{p-j} \in R_{p-j}$ and $f_p \neq 0$, then

$$f^* = f_p + tf_{p-1} + \cdots + t^s f_{p-s}$$

is a homogeneous element of degree p in $R[t]_p$ satisfying $\phi(f^*) = f$. We call the homogeneous element f^* obtained this way the *central homogenization* of f with respect to t (for the reason that t is in the center of $R[t]$). On the other hand, for an element $F \in R[t]$, we write

$$F_* = \phi(F)$$

and call it the *central dehomogenization* of F with respect to t (again for the reason that t is in the center of $R[t]$). Hence, if I is an ideal R, then we write $I^* = \{f^* \mid f \in I\}$ and call the $\mathbb{N}$-graded ideal $\langle I^* \rangle$ generated by I^* the *central homogenization ideal* of I in $R[t]$ with respect to t; and if J is an ideal of $R[t]$, then since ϕ is a ring epimorphism, $\phi(J)$ is an ideal of R, so we write J_* for $\phi(J) = \{H_* = \phi(H) \mid H \in J\}$ and call it the *central dehomogenization ideal* of J in R with respect to t. Consequently, henceforth we will also use the notation $(J_*)^* = \{(h_*)^* \mid h \in J\}$.

Lemma 6.1. *With the definitions and notation above, the following statements hold.*
(i) *For* $F, G \in R[t]$, $(F+G)_* = F_* + G_*$, $(FG)_* = F_*G_*$.
(ii) *For any* $f \in R$, $(f^*)_* = f$.
(iii) *If* $F \in R[t]_p$ *and if* $(F_*)^* \in R[t]_q$, *then* $p \geq q$ *and* $t^r(F_*)^* = F$ *with* $r = p - q$.
(iv) *If* $f, g \in R$ *are such that* fg *has nonzero* $\mathbf{LH}(fg) \in R_m$ *and* $f^*g^* \in R[t]_q$, *then* $f^*g^* = t^k(fg)^*$ *with* $k = q - m$, *where* $\mathbf{LH}(\ )$ *denotes taking the* $\mathbb{N}$*-leading homogeneous element of elements in both* R *and* $R[t]$ *(see Ch.2, Section 3).*
(v) *If* $f, g \in R$ *with nonzero* $\mathbf{LH}(f) \in R_p$ *and nonzero* $\mathbf{LH}(g) \in R_q$, *then* $(f+g)^* = f^* + g^*$ *in case* $p = q$, *and* $(f+g)^* = f^* + t^\ell g^*$ *in case* $p > q$, *where* $\ell = p - q$.
(vi) *If* I *is a two-sided ideal of* R, *then each homogeneous element* $F \in \langle I^* \rangle$ *is of the form* $t^r f^*$ *for some* $r \in \mathbb{N}$ *and* $f \in I$.
(vii) *If* J *is a graded ideal of* $R[t]$, *then for each* $h \in J_*$ *there is some homogeneous element* $F \in J$ *such that* $F_* = h$.

Proof. Exercise. □

Central (de)homogenized Gröbner bases

Suppose that the $\mathbb{N}$-graded K-algebra $R = \oplus_{p\in\mathbb{N}} R_p$ has an admissible system $(\mathcal{B}, \prec_{gr})$, where $\mathcal{B}$ is a skew multiplicative K-basis of R consisting of $\mathbb{N}$-homogeneous elements, and $\prec_{gr}$ is an $\mathbb{N}$-graded monomial ordering on $\mathcal{B}$, i.e., R has a Gröbner basis theory. Consider the mixed $\mathbb{N}$-gradation of $R[t]$ as defined in the previous part and the K-basis

$$\mathcal{B}^* = \{t^r w \mid w \in \mathcal{B},\ r \in \mathbb{N}\}$$

of $R[t]$. Since $\mathcal{B}^*$ is obviously a skew multiplicative K-basis for $R[t]$, the $\mathbb{N}$-graded monomial ordering $\prec_{gr}$ on $\mathcal{B}$ extends to a monomial ordering on $\mathcal{B}^*$, denoted $\prec_{t\text{-}gr}$, as follows:

$$t^{r_1}w_1 \prec_{t\text{-}gr} t^{r_2}w_2 \Longleftrightarrow w_1 \prec_{gr} w_2, \text{ or } w_1 = w_2 \text{ and } r_1 < r_2.$$

It is clear that $\prec_{t\text{-}gr}$ *is not a graded monomial ordering* on $\mathcal{B}^*$. Nevertheless, $R[t]$ has a Gröbner basis theory with respect to the admissible system $(\mathcal{B}^*, \prec_{t\text{-}gr})$, in particular, since $\mathcal{B}^*$ consists of $\mathbb{N}$-homogeneous elements, every $\mathbb{N}$-graded ideal of $R[t]$ has a homogeneous Gröbner basis by Proposition 2.2(iii).

Before dealing with the relation between Gröbner basis in R and homogeneous Gröbner bases in $R[t]$, let us see how leading monomials behave with respect to both $\prec_{gr}$ and $\prec_{t\text{-}gr}$. The properties listed in the next lemma indeed highlight the *key* advantages of using a graded monomial ordering $\prec_{gr}$ on R (see also Lemma 6.9 later), that will be essential not only for the present discussion but also for the subsequent chapters.

Lemma 6.2. *With notation given above, the following statements hold.*
(i) *If $f \in R$, then*

$$\mathbf{LM}(f) = \mathbf{LM}(\mathbf{LH}(f)) \text{ w.r.t. } \prec_{gr} \text{ on } \mathcal{B}.$$

(ii) *If $f \in R$, then*

$$\mathbf{LM}(f^*) = \mathbf{LM}(f) \text{ w.r.t. } \prec_{t\text{-}gr} \text{ on } \mathcal{B}^*.$$

(iii) *If F is a nonzero homogeneous element of $R[t]$, then*

$$\mathbf{LM}(F_*) = \mathbf{LM}(F)_* \text{ w.r.t. } \prec_{gr} \text{ on } \mathcal{B}.$$

Proof. The proof of (i) and (ii) is an easy exercise. To prove (iii), let $F \in R[t]_p$ be a nonzero homogeneous element of degree p, say

$$F = \lambda t^r w + \lambda_1 t^{r_1} w_1 + \cdots + \lambda_s t^{r_s} w_s,$$

where $\lambda, \lambda_i, \in K^*$, $r, r_i \in \mathbb{N}$, $w, w_i \in \mathcal{B}$, such that $\mathbf{LM}(F) = t^r w$. Since $\mathcal{B}$ consists of $\mathbb{N}$-homogeneous elements and $R[t]$ has the mixed $\mathbb{N}$-gradation by the previously fixed assumption, we have $d(t^r w) = d(t^{r_i} w_i) = p$, $1 \leq i \leq s$. Thus $w = w_i$ will imply $r = r_i$ and thereby $t^r w = t^{r_i} w_i$. So we may assume that $w \neq w_i$, $1 \leq i \leq s$. Then it follows from the definition of $\prec_{t\text{-}gr}$ that $w_i \prec_{gr} w$ and $r \leq r_i$, $1 \leq i \leq s$. Therefore $\mathbf{LM}(F_*) = w = \mathbf{LM}(F)_*$, as desired. □

The next result is a generalization of ([LWZ] Theorem 2.3.2).

Theorem 6.3. *With notions and notations as fixed before, let $I = \langle \mathcal{G} \rangle$ be the ideal of R generated by a subset $\mathcal{G}$, and $\langle I^* \rangle$ the central homogeneization ideal of I in $R[t]$ with respect to t. The following two statements are equivalent.*
(i) *$\mathcal{G}$ is a Gröbner basis for I in R with respect to the admissible system $(\mathcal{B}, \prec_{gr})$.*
(ii) *$\mathcal{G}^* = \{g^* \mid g \in \mathcal{G}\}$ is a Gröbner basis for $\langle I^* \rangle$ in $R[t]$ with respect to the admissible system $(\mathcal{B}^*, \prec_{t\text{-}gr})$.*

Proof. In proving the equivalence below, without specific indication we shall use Lemma 6.2(ii) wherever it is needed.

(i) $\Rightarrow$ (ii). First note that $\mathbf{LM}(\mathcal{G}^*) \subset \mathbf{LM}(I^*)$. We have to prove that $\mathbf{LM}(\mathcal{G}^*)$ generates $\langle \mathbf{LM}(I^*)\rangle$ in order to see that $\mathcal{G}^*$ is a Gröbner basis for $\langle I^*\rangle$. If $F \in \langle I^*\rangle$, then since $\langle I^*\rangle$ is a graded ideal, we may assume, without loss of generality, that F is a homogeneous element. So, by Lemma 6.1(vi) we have $F = t^r f^*$ for some $f \in I$. It follows from $\mathbf{LM}(f^*) = \mathbf{LM}(f)$ that

$$\mathbf{LM}(F) = t^r\mathbf{LM}(f^*) = t^r\mathbf{LM}(f). \tag{1}$$

Since $\mathcal{G}$ is a Gröbner basis for I, $\mathbf{LM}(f) = \lambda v\mathbf{LM}(g_i)w$ for some $\lambda \in K^*$, $g_i \in \mathcal{G}$, and $v, w \in \mathcal{B}$. Thus, it follows from $\mathbf{LM}(g) = \mathbf{LM}(g^*)$ and (1) that

$$\mathbf{LM}(F) = t^r\mathbf{LM}(f) = \lambda t^r v\mathbf{LM}(g_i^*)w \in \langle \mathbf{LM}(\mathcal{G}^*)\rangle.$$

This shows that $\langle \mathbf{LM}(\langle I^*\rangle)\rangle = \langle \mathbf{LM}(\mathcal{G}^*)\rangle$, as desired.

(ii) $\Rightarrow$ (i). Suppose $\mathcal{G}^*$ is a Gröbner basis for the homogenization ideal $\langle I^*\rangle$ of I in $R[t]$. Let $f \in I$. Then $\mathbf{LM}(f^*) = \lambda t^{r_1}v\mathbf{LM}(g_i^*)t^{r_2}w$ for some $\lambda \in K^*$, $r_1, r_2 \in \mathbb{N}$, $v, w \in \mathcal{B}$ and $g_i^* \in \mathcal{G}^*$. Since $\mathbf{LM}(f) = \mathbf{LM}(f^*)$ and $\mathbf{LM}(g_i) = \mathbf{LM}(g_i^*)$, after taking dehomogenization we get

$$\mathbf{LM}(f) = \lambda v\mathbf{LM}(g_i)w \in \langle \mathbf{LM}(\mathcal{G})\rangle.$$

This shows that $\langle \mathbf{LM}(I)\rangle = \langle \mathbf{LM}(\mathcal{G})\rangle$, i.e., $\mathcal{G}$ is a Gröbner basis for I in R. □

We call the Gröbner basis $\mathcal{G}^*$ obtained in Theorem 6.3 the *central homogenization of* $\mathcal{G}$ in $R[t]$ with respect to t, or $\mathcal{G}^*$ is a *central homogenized Gröbner basis* with respect to t.

By Lemma 6.2 and Theorem 6.3, we have immediately the following corollary.

Corollary 6.4. *Let I be an arbitrary ideal of R. With notation as before, if $\mathcal{G}$ is a Gröbner basis of I with respect to the data $(\mathcal{B}, \prec_{gr})$, then, with respect to the data $(\mathcal{B}^*, \prec_{t\text{-}gr})$ we have*

$$\mathcal{B}^* - \langle \mathbf{LM}(\mathcal{G}^*)\rangle = \{t^r w \mid w \in \mathcal{B} - \langle \mathbf{LM}(\mathcal{G})\rangle,\ r \in \mathbb{N}\},$$

that is, the set $N(\langle I^*\rangle)$ of normal monomials (mod $\langle I^*\rangle$) in $\mathcal{B}^*$ is determined by the set $N(I)$ of normal monomials (mod I) in $\mathcal{B}$. Hence, the algebra $R[t]/\langle I^*\rangle = R[t]/\langle \mathcal{G}^*\rangle$ has the K-basis

$$\overline{N(\langle I^*\rangle)} = \{\overline{t^r w} \mid w \in N(I),\ r \in \mathbb{N}\}.$$

□

Theoretically we may also obtain a Gröbner basis for an ideal I of R by dehomogenizing a homogeneous Gröbner basis of the ideal $\langle I^*\rangle \subset R[t]$. We give below a more general approach to this assertion.

Theorem 6.5. *Let J be a graded ideal of $R[t]$. If $\mathscr{G}$ is a homogeneous Gröbner basis of J with respect to the data $(\mathcal{B}^*, \prec_{t\text{-}gr})$, then $\mathscr{G}_* = \{G_* \mid G \in \mathscr{G}\}$ is a Gröbner basis for the ideal J_* in R with respect to the data $(\mathcal{B}, \prec_{gr})$.*

Proof. If $\mathscr{G}$ is a Gröbner basis of J, then $\mathscr{G}$ generates J and hence $\mathscr{G}_* = \phi(\mathscr{G})$ generates $J_* = \phi(J)$. For a nonzero $f \in J_*$, by Lemma 6.1(vii), there exists a homogeneous element $H \in J$ such that $H_* = f$. It follows from Lemma 6.2 that

(1) $$\mathbf{LM}(f) = \mathbf{LM}(f^*) = \mathbf{LM}((H_*)^*).$$

On the other hand, there exists some $G \in \mathscr{G}$ such that $\mathbf{LM}(G)|\mathbf{LM}(H)$, i.e.,

(2) $$\mathbf{LM}(H) = \lambda t^{r_1} w \mathbf{LM}(G) t^{r_2} v$$

for some $\lambda \in K^*$, $r_1, r_2 \in \mathbb{N}$, $w, v \in \mathcal{B}$. But by Lemma 6.1(iii) we also have $t^r(H_*)^* = H$ for some $r \in \mathbb{N}$, and hence

(3) $$\mathbf{LM}(H) = \mathbf{LM}(t^r(H_*)^*) = t^r\mathbf{LM}(H_*)^*).$$

So, (1), (2) and (3) yield

$$\begin{aligned} \lambda t^{r_1+r_2} w \mathbf{LM}(G) v &= \mathbf{LM}(H) \\ &= t^r \mathbf{LM}((H_*)^*) \\ &= t^r \mathbf{LM}(f). \end{aligned}$$

Taking the central dehomogenization for the above equality, by Lemma 6.2(iii) we obtain

$$\lambda w \mathbf{LM}(G_*) v = \lambda w \mathbf{LM}(G)_* v = \mathbf{LM}(f).$$

This shows that $\mathbf{LM}(G_*)|\mathbf{LM}(f)$. Therefore, $\mathscr{G}_*$ is a Gröbner basis for J_*. □

We call the Gröbner basis $\mathscr{G}_*$ obtained in Theorem 6.5 the *central dehomogenization of $\mathscr{G}$* in R with respect to t, or $\mathscr{G}_*$ is a *central dehomogenized Gröbner basis* with respect to t.

Corollary 6.6. *Let I be an ideal of R. If $\mathscr{G}$ is a homogeneous Gröbner basis of $\langle I^*\rangle$ in $R[t]$ with respect to the data $(\mathcal{B}^*, \prec_{t\text{-}gr})$, then $\mathscr{G}_* = \{g_* \mid g \in \mathscr{G}\}$*

is a Gröbner basis for I in R with respect to the data $(B, \prec_{gr})$. Moreover, if I is generated by the subset F and $F^* \subset \mathscr{G}$, then $F \subset \mathscr{G}_*$.

Proof. Put $J = \langle I^* \rangle$. Then since $J_* = I$, it follows from Theorem 6.5 that if $\mathscr{G}$ is a homogeneous Gröbner basis of J then $\mathscr{G}_*$ is a Gröbner basis for I. The second assertion of the theorem is clear by Lemma 6.1(ii). □

Let S be a nonempty subset of R and $I = \langle S \rangle$ the ideal generated by S. Then, with $S^* = \{f^* \mid f \in S\}$, in general $\langle S^* \rangle \subsetneq \langle I^* \rangle$ in $R[t]$ (for instance, consider $S = \{y^3 - x - y,\ y^2 + 1\}$ in the commutative polynomial ring $K[x, y]$ and the homogenization in $K[x, y, t]$ with respect to t). So, from a practical and computational viewpoint (as argued in the first two parts of this section), it is the right place to set up the procedure of getting a Gröbner basis for I and hence a Gröbner basis for $\langle I^* \rangle$ by producing a homogeneous Gröbner basis of the graded ideal $\langle S^* \rangle$.

Proposition 6.7. *Let $I = \langle S \rangle$ be the ideal of R as fixed above. Suppose that Gröbner bases are algorithmically computable in R and hence in $R[t]$. Then a Gröbner basis for I and a homogeneous Gröbner basis for $\langle I^* \rangle$ may be obtained by implementing the following procedure:*

Step 1. *Starting with the initial subset $S^* = \{f^* \mid f \in S\}$, compute a homogeneous Gröbner basis $\mathscr{G}$ for the graded ideal $\langle S^* \rangle$ of $R[t]$.*

Step 2. *Noticing $\langle S^* \rangle_* = I$, use Theorem 6.5 and dehomogenize $\mathscr{G}$ with respect to t in order to obtain the Gröbner basis $\mathscr{G}_*$ for I.*

Step 3. *Use Theorem 6.3 and homogenize $\mathscr{G}_*$ with respect to t in order to obtain the homogeneous Gröbner basis $(\mathscr{G}_*)^*$ for the graded ideal $\langle I^* \rangle$.*

□

In a similar way, we proceed now to consider the free algebra $K\langle X \rangle = K\langle X_1, \ldots, X_n \rangle$ of n generators as well as the free algebra $K\langle X, T \rangle = K\langle X_1, \ldots, X_n, T \rangle$ of $n+1$ generators, and demonstrate how Gröbner bases in $K\langle X \rangle$ are related with homogeneous Gröbner bases in $K\langle X, T \rangle$ if the non-central (de)homogenization with respect to T is employed.

Non-central (de)homogenization

Let $K\langle X \rangle$ be equipped with a fixed *weight* $\mathbb{N}$-*gradation*, say each X_i has degree $n_i > 0$, $1 \leq i \leq n$. Assigning T to the degree 1 in $K\langle X, T \rangle$ and using the same weight n_i for each X_i as in $K\langle X \rangle$, we get the weight $\mathbb{N}$-

gradation of $K\langle X,T\rangle$ which extends the weight $\mathbb{N}$-gradation of $K\langle X\rangle$. Let $\mathcal{B}$ and $\widetilde{\mathcal{B}}$ denote the standard K-bases of $K\langle X\rangle$ and $K\langle X,T\rangle$ respectively. For convenience, we use lowercase letters $w,u,v,\dots$ to denote monomials in $\mathcal{B}$ as before, but use capitals $W,U,V,\dots$ to denote monomials in $\widetilde{\mathcal{B}}$.

In what follows, we fix an admissible system $(\mathcal{B},\prec_{gr})$ for $K\langle X\rangle$, where $\prec_{gr}$ is an $\mathbb{N}$-*graded lexicographic ordering* on $\mathcal{B}$ with respect to the fixed weight $\mathbb{N}$-gradation of $K\langle X\rangle$, such that

$$X_{i_1}\prec_{gr}X_{i_2}\prec_{gr}\cdots\prec_{gr}X_{i_n}.$$

Then it is not difficult to see that $\prec_{gr}$ can be extended to an $\mathbb{N}$-*graded lexicographic ordering* $\prec_{_{T\text{-}gr}}$ on $\widetilde{\mathcal{B}}$ with respect to the fixed weight $\mathbb{N}$-gradation of $K\langle X,T\rangle$, such that

$$T\prec_{_{T\text{-}gr}}X_{i_1}\prec_{_{T\text{-}gr}}X_{i_2}\prec_{_{T\text{-}gr}}\cdots\prec_{_{T\text{-}gr}}X_{i_n},$$

and thus we get the admissible system $(\widetilde{\mathcal{B}},\prec_{_{T\text{-}gr}})$ for $K\langle X,T\rangle$.

Consider the fixed $\mathbb{N}$-graded structures $K\langle X\rangle=\oplus_{p\in\mathbb{N}}K\langle X\rangle_p$, $K\langle X,T\rangle=\oplus_{p\in\mathbb{N}}K\langle X,T\rangle_p$, and the ring epimorphism

$$\psi:\ K\langle X,T\rangle\ \longrightarrow\ K\langle X\rangle$$

defined by $\psi(X_i)=X_i$ and $\psi(T)=1$. Then each $f\in K\langle X\rangle$ is the image of some homogeneous element in $K\langle X,T\rangle$. More precisely, if $f=f_p+f_{p-1}+\cdots+f_{p-s}$ with $f_p\in K\langle X\rangle_p$, $f_{p-j}\in K\langle X\rangle_{p-j}$ and $f_p\neq 0$, then

$$\widetilde{f}=f_p+Tf_{p-1}+\cdots+T^sf_{p-s}$$

is a homogeneous element of degree p in $K\langle X,T\rangle_p$ such that $\psi(\widetilde{f})=f$. We call the homogeneous element $\widetilde{f}$ obtained this way the *non-central homogenization* of f with respect to T (for the reason that T is not a commuting variable). On the other hand, for $F\in K\langle X,T\rangle$, we write

$$F_{\sim}=\psi(F)$$

and call $F_{\sim}$ the *non-central dehomogenization* of F with respect to T (again for the reason that T is not a commuting variable). Furthermore, if $I=\langle S\rangle$ is the ideal of $K\langle X\rangle$ generated by a subset S, then we define

$$\widetilde{S}=\{\widetilde{f}\mid f\in S\}\cup\{X_iT-TX_i\mid 1\le i\le n\},$$
$$\widetilde{I}=\{\widetilde{f}\mid f\in I\}\cup\{X_iT-TX_i\mid 1\le i\le n\},$$

and call the graded ideal $\langle\widetilde{I}\rangle$ generated by $\widetilde{I}$ the *non-central homogenization ideal* of I in $K\langle X,T\rangle$ with respect to T; while if J is an ideal of $K\langle X,T\rangle$, then since ψ is a surjective ring homomorphism, $\psi(J)$ is an ideal of $K\langle X\rangle$, so we write $J_\sim$ for $\psi(J) = \{H_\sim | H \in J\}$ and call it the *non-central dehomogenization ideal* of J in $K\langle X\rangle$ with respect to T. Consequently, henceforth we will also use the notation

$$(J_\sim)^\sim = \{(h_\sim)^\sim \mid h \in J\} \cup \{X_iT - TX_i \mid 1 \le i \le n\}.$$

It is straightforward to check that with resspect to the data $(\widetilde{\mathcal{B}}, \prec_{T\text{-}gr})$, the subset $\{X_iT - TX_i \mid 1 \le i \le n\}$ of $K\langle X, T\rangle$ forms a homogeneous Gröbner basis with $\mathbf{LM}(X_iT - TX_i) = X_iT$, $1 \le i \le n$. In the latter discussion we will freely use this fact without extra indication.

Lemma 6.8. *With the definitions and notation above, the following properties hold.*

(i) *If* $F, G \in K\langle X, T\rangle$, *then* $(F + G)_\sim = F_\sim + G_\sim$, $(FG)_\sim = F_\sim G_\sim$.

(ii) *For each nonzero* $f \in K\langle X\rangle$, $(\widetilde{f})_\sim = f$.

(iii) *Let* $\mathscr{C}$ *be the graded ideal of* $K\langle X, T\rangle$ *generated by* $\{X_iT - TX_i \mid 1 \le i \le n\}$. *If* $F \in K\langle X, T\rangle_p$, *then there exists an* $L \in \mathscr{C}$ *and a unique homogeneous element of the form* $H = \sum_i \lambda_i T^{r_i} w_i$, *where* $\lambda_i \in K^*$, $w_i \in \mathcal{B}$, *such that* $F = L + H$; *moreover there is some* $r \in \mathbb{N}$ *such that* $T^r(H_\sim)^\sim = H$, *and hence* $F = L + T^r(F_\sim)^\sim$.

(iv) *Let* $\mathscr{C}$ *be as in* (iii) *above. If* I *is an ideal of* $K\langle X\rangle$, $F \in \langle \widetilde{I}\rangle$ *is a homogeneous element, then there exist some* $L \in \mathscr{C}$, $f \in I$ *and* $r \in \mathbb{N}$ *such that* $F = L + T^r\widetilde{f}$.

(v) *If* J *is a graded ideal of* $K\langle X, T\rangle$ *and* $\{X_iT - TX_i \mid 1 \le i \le n\} \subset J$, *then for each nonzero* $h \in J_\sim$, *there exists a homogeneous element* $H = \sum_i \lambda_i T^{r_i} w_i \in J$, *where* $\lambda_i \in K^*$, $r_i \in \mathbb{N}$, *and* $w_i \in \mathcal{B}$, *such that for some* $r \in \mathbb{N}$, $T^r(H_\sim)^\sim = H$ *and* $H_\sim = h$.

Proof. (i) and (ii) follow from the definitions of non-central homogenization and non-central dehomogenization directly.

(iii). Since the subset $\{X_iT - TX_i \mid 1 \le i \le n\}$ is a Gröbner basis in $K\langle X, T\rangle$ with respect to $(\widetilde{\mathcal{B}}, \prec_{T\text{-}gr})$, such that $\mathbf{LM}(X_iT - TX_i) = X_iT$, $1 \le i \le n$, if $F \in K\langle X, T\rangle_p$, then the division of F by this subset yields $F = L + H$, where $L \in \mathscr{C}$, and $H = \sum_i \lambda_i T^{r_i} w_i$ is the unique remainder with $\lambda_i \in K^*$, $w_i \in \mathcal{B}$, in which each monomial $T^r w_i$ is of degree p. By the definitions of $\prec_{T\text{-}gr}$, non-central homogenization and non-central dehomogenization, it is not difficult to see that H has the desired property.

(iv). By (iii), $F = L + T^r(F_\sim)^\sim$ with $L \in \mathscr{C}$ and $r \in \mathbb{N}$. Since by (ii) we have $F_\sim \in \langle \widetilde{I}\rangle_\sim = I$, thus $f = F_\sim$ is the desired element.

(v). Using basic properties of homogeneous element and graded ideal in a graded ring, this follows from the preceding (iii). □

Non-central (de)homogenized Gröbner bases

As in the case of using the central (de)homogenization, before turning to deal with the relation between Gröbner bases in $K\langle X\rangle$ and homogeneous Gröbner bases in $K\langle X,T\rangle$, we are also concerned about the behavior of leading monomials under taking the $\mathbb{N}$-leading homogeneous element and non-central (de)homogenization with respect to $\prec_{gr}$ and $\prec_{_{T\text{-}gr}}$.

Lemma 6.9. *With the assumptions and notations as fixed before, the following statements hold.*
(i) *If $f \in K\langle X\rangle$, then*

$$\mathbf{LM}(f) = \mathbf{LM}(\mathbf{LH}(f)) \text{ w.r.t. } \prec_{gr} \text{ on } \mathcal{B}.$$

If $F \in K\langle X,T\rangle$, then

$$\mathbf{LM}(F) = \mathbf{LM}(\mathbf{LH}(F)) \text{ w.r.t. } \prec_{_{T\text{-}gr}} \text{ on } \widetilde{\mathcal{B}},$$

where $\mathbf{LH}(\ )$ *denotes taking the $\mathbb{N}$-leading homogeneous element of elements in both $K\langle X\rangle$ and $K\langle X,T\rangle$.*
(ii) *For each nonzero $f \in K\langle X\rangle$, we have*

$$\mathbf{LM}(f) = \mathbf{LM}(\widetilde{f}) \text{ w.r.t. } \prec_{_{T\text{-}gr}} \text{ on } \widetilde{\mathcal{B}}.$$

(iii) *If F is a homogeneous element in $K\langle X,T\rangle$ such that $X_iT \nmid \mathbf{LM}(F)$ with respect to $\prec_{_{T\text{-}gr}}$ for all $1 \le i \le n$, then $\mathbf{LM}(F) = T^rw$ for some $r \in \mathbb{N}$ and $w \in \mathcal{B}$, such that*

$$\mathbf{LM}(F_\sim) = w = \mathbf{LM}(F)_\sim \text{ w.r.t. } \prec_{gr} \text{ on } \mathcal{B}.$$

Proof. The proof of (i) and (ii) is an easy exercise. To prove (iii), let $F \in K\langle X,T\rangle_p$ be a nonzero homogeneous element of degree p. Then by the assumption F may be written as

$$F = \lambda T^rw + \lambda_1T^{r_1}X_{j_1}W_1 + \lambda_2T^{r_2}X_{j_2}W_2 \cdots + \lambda_sT^{r_s}X_{j_s}W_s,$$

where $\lambda, \lambda_i, \in K^*$, $r, r_i \in \mathbb{N}$, $w \in \mathcal{B}$ and $W_i \in \widetilde{\mathcal{B}}$, such that $\mathbf{LM}(F) = T^rw$. Since $\mathcal{B}$ consists of $\mathbb{N}$-homogeneous elements and the $\mathbb{N}$-gradation of $K\langle X\rangle$ extends to give the $\mathbb{N}$-gradation of $K\langle X,T\rangle$, we have $d(T^rw) = d(T^{r_i}X_{j_i}W_i) = p$, $1 \le i \le s$. Also note that T has degree 1. Thus $w = X_{j_i}W_i$ will imply $r = r_i$ and thereby $T^rw = T^{r_i}X_{j_i}W_i$. So we may assume, without loss of generality, that $w \neq X_{j_i}W_i$, $1 \le i \le s$. Then it follows from the definition of $\prec_{T\text{-}gr}$ that $r \le r_i$, $1 \le i \le n$. Hence $X_{j_i}W_i \prec_{T\text{-}gr} w$, $1 \le i \le s$. Therefore $(X_{j_i}W_i)_\sim \prec_{gr} w$, $1 \le i \le n$, and consequently $\mathbf{LM}(F_\sim) = w = \mathbf{LM}(F)_\sim$, as desired. □

The next result strengthens ([LWZ], Theorem 2.3.1), ([Li1], Theorem 3.8 without proof), and ([Li3], Theorem 8.2), in particular, the proof of (i) ⇒ (ii) given below improves the proof of the same implication given in [Li3].

Theorem 6.10. *With the notions and notations as given before, let* $I = \langle \mathcal{G} \rangle$ *be the ideal of* $K\langle X \rangle$ *generated by a subset* $\mathcal{G}$*, and* $\langle \widetilde{I} \rangle$ *the non-central homogenization ideal of* I *in* $K\langle X, T\rangle$ *with respect to* T*. The following two statements are equivalent.*
(i) $\mathcal{G}$ *is a Gröbner basis of* I *with respect to the admissible system* $(\mathcal{B}, \prec_{gr})$ *of* $K\langle X \rangle$;
(ii) $\widetilde{\mathcal{G}} = \{\widetilde{g} \mid g \in \mathcal{G}\} \cup \{X_iT - TX_i \mid 1 \le i \le n\}$ *is a homogeneous Gröbner basis for* $\langle \widetilde{I} \rangle$ *with respect to the admissible system* $(\widetilde{\mathcal{B}}, \prec_{T\text{-}gr})$ *of* $K\langle X, T\rangle$.

Proof. In proving the equivalence below, without specific indication we shall use Lemma 6.9(ii) wherever it is needed.

(i) ⇒ (ii). Suppose that $\mathcal{G}$ is a Gröbner basis for I with respect to the data $(\mathcal{B}, \prec_{gr})$. Let $F \in \langle \widetilde{I} \rangle$. Then since $\langle \widetilde{I} \rangle$ is a graded ideal, we may assume, without loss of generality, that F is a nonzero homogeneous element. We want to show that there is some $D \in \widetilde{\mathcal{G}}$ such that $\mathbf{LM}(D)|\mathbf{LM}(F)$, and hence $\widetilde{\mathcal{G}}$ is a Gröbner basis.

Note that $\{X_iT - TX_i \mid 1 \le i \le n\} \subset \widetilde{\mathcal{G}}$ with $\mathbf{LM}(X_iT - TX_i) = X_iT$. If $X_iT|\mathbf{LM}(F)$ for some X_iT, then we are done. Otherwise, $X_iT \nmid \mathbf{LM}(F)$ for all $1 \le i \le n$. Thus, by Lemma 6.9(iii), $\mathbf{LM}(F) = T^r w$ for some $r \in \mathbb{N}$ and $w \in \mathcal{B}$ such that

$$\text{(1)} \qquad \mathbf{LM}(F_\sim) = w = \mathbf{LM}(F)_\sim .$$

On the other hand, by Lemma 6.8(iv) we have $F = L + T^q\widetilde{f}$, where L is an element in the ideal $\mathscr{C}$ generated by $\{X_iT - TX_i \mid 1 \le i \le n\}$ in $K\langle X, T\rangle$, $q \in \mathbb{N}$, and $f \in I$. It turns out that

$$\text{(2)} \qquad F_\sim = (\widetilde{f})_\sim = f \text{ and hence } \mathbf{LM}(F_\sim) = \mathbf{LM}(f).$$

But since $\mathcal{G}$ is a Gröbner basis for I, there is some $g \in \mathcal{G}$ such that $\mathbf{LM}(g)|\mathbf{LM}(f)$, i.e., there are $u, v \in \mathcal{B}$ such that

$$\text{(3)} \qquad \mathbf{LM}(f) = u\mathbf{LM}(g)v = u\mathbf{LM}(\widetilde{g})v.$$

Combining (1), (2) and (3) above, we have

$$w = \mathbf{LM}(F_{\sim}) = \mathbf{LM}(f) = u\mathbf{LM}(\widetilde{g})v.$$

Therefore, $\mathbf{LM}(\widetilde{g})|T^r w$, i.e., $\mathbf{LM}(\widetilde{g})|\mathbf{LM}(F)$, as desired.

(ii) $\Rightarrow$ (i). Suppose that $\widetilde{\mathcal{G}}$ is a Gröbner basis of the graded ideal $\langle \widetilde{I} \rangle$ in $K\langle X, T\rangle$. If $f \in I$, then since $\widetilde{f} \in \widetilde{I}$, there is some $D \in \widetilde{\mathcal{G}}$ such that $\mathbf{LM}(D)|\mathbf{LM}(\widetilde{f})$. Note that $\mathbf{LM}(\widetilde{f}) = \mathbf{LM}(f)$ and thus $T \nmid \mathbf{LM}(\widetilde{f})$. Hence $D = \widetilde{g}$ for some $g \in \mathcal{G}$, and there are $w, v \in \mathcal{B}$ such that

$$\mathbf{LM}(f) = \mathbf{LM}(\widetilde{f}) = w\mathbf{LM}(\widetilde{g})v = w\mathbf{LM}(g)v,$$

i.e., $\mathbf{LM}(g)|\mathbf{LM}(f)$. This shows that $\mathcal{G}$ is a Gröbner basis for I in R. □

We call the Gröbner basis $\widetilde{\mathcal{G}}$ obtained in Theorem 6.10 the *non-central homogenization of* $\mathcal{G}$ in $K\langle X, T\rangle$ with respect to T, or $\widetilde{\mathcal{G}}$ is a *non-central homogenized Gröbner basis* with respect to T.

By Lemma 6.8 and Theorem 6.10, the following Corollary is straightforward.

Corollary 6.11. *Let I be an arbitrary ideal of $K\langle X\rangle$. With notation as before, if $\mathcal{G}$ is a Gröbner basis of I with respect to the data $(\mathcal{B}, \prec_{gr})$, then, with respect to the data $(\widetilde{\mathcal{B}}, \prec_{_{T\text{-}gr}})$ we have*

$$\widetilde{\mathcal{B}} - \langle \mathbf{LM}(\widetilde{\mathcal{G}})\rangle = \{T^r w \mid w \in \mathcal{B} - \langle \mathbf{LM}(\mathcal{G})\rangle,\ r \in \mathbb{N}\},$$

that is, the set $N(\langle \widetilde{I}\rangle)$ of normal monomials (mod $\langle \widetilde{I}\rangle$) in $\widetilde{\mathcal{B}}$ is determined by the set $N(I)$ of normal monomials (mod I) in $\mathcal{B}$. Hence, the algebra $K\langle X, T\rangle/\langle \widetilde{I}\rangle = K\langle X, T\rangle/\langle \widetilde{\mathcal{G}}\rangle$ has the K-basis

$$\overline{N(\langle \widetilde{I}\rangle)} = \left\{\overline{T^r w} \mid w \in N(I),\ r \in \mathbb{N}\right\}.$$

□

As with the central (de)homogenization with respect to the commuting variable t in previous discussion, theoretically we may also obtain a Gröbner basis for an ideal I of $K\langle X\rangle$ by dehomogenizing a homogeneous Gröbner basis of the ideal $\langle \widetilde{I}\rangle \subset K\langle X, T\rangle$. We give below a more general approach to this assertion, which guarantees the completion and consistency of subsequent results (Corollary 6.13, Proposition 6.14, and Theorem 7.6 – Theorem 7.10).

Theorem 6.12. *Let J be a graded ideal of $K\langle X,T\rangle$, and suppose that $\{X_iT - TX_i \mid 1 \le i \le n\} \subset J$. If $\mathscr{G}$ is a homogeneous Gröbner basis of J with respect to the data $(\widetilde{\mathcal{B}}, \prec_{_{T\text{-}gr}})$, then $\mathscr{G}_\sim = \{G_\sim \mid G \in \mathscr{G}\}$ is a Gröbner basis for the ideal $J_\sim$ in $K\langle X\rangle$ with respect to the data $(\mathcal{B}, \prec_{gr})$.*

Proof. If $\mathscr{G}$ is a Gröbner basis of J, then $\mathscr{G}$ generates J and hence $\mathscr{G}_\sim = \phi(\mathscr{G})$ generates $J_\sim = \phi(J)$. We show next that for each nonzero $h \in J_\sim$, there is some $G_\sim \in \mathscr{G}_\sim$ such that $\mathbf{LM}(G_\sim)|\mathbf{LM}(h)$, and hence $\mathscr{G}_\sim$ is a Gröbner basis for $J_\sim$.

Since $\{X_iT - TX_i \mid 1 \le i \le n\} \subset J$, by Lemma 6.8(v) there exists a homogeneous element $H \in J$ and some $r \in \mathbb{N}$ such that $T^r(H_\sim)^\sim = H$ and $H_\sim = h$. It follows that

(1) $$\mathbf{LM}(H) = T^r\mathbf{LM}((H_\sim)^\sim) = T^r\mathbf{LM}(\widetilde{h}) = T^r\mathbf{LM}(h).$$

On the other hand, there is some $G \in \mathscr{G}$ such that $\mathbf{LM}(G)|\mathbf{LM}(H)$, i.e., there are $W, V \in \widetilde{\mathcal{B}}$ such that

(2) $$\mathbf{LM}(H) = W\mathbf{LM}(G)V.$$

But by the above (1) we must have $\mathbf{LM}(G) = T^q w$ for some $q \in \mathbb{N}$ and $w \in \mathcal{B}$. Thus, by Lemma 6.9(iii),

(3) $$\mathbf{LM}(G_\sim) = w = \mathbf{LM}(G)_\sim \text{ w.r.t. } \prec_{gr} \text{ on } \mathcal{B}.$$

Combining (1), (2), and (3) above, we then obtain

$$\begin{aligned} \mathbf{LM}(h) &= \mathbf{LM}(H)_\sim \\ &= (W\mathbf{LM}(G)V)_\sim \\ &= W_\sim\mathbf{LM}(G)_\sim V_\sim \\ &= W_\sim\mathbf{LM}(G_\sim)V_\sim. \end{aligned}$$

This shows that $\mathbf{LM}(G_\sim)|\mathbf{LM}(h)$, as asserted. □

We call the Gröbner basis $\mathscr{G}_\sim$ obtained in Theorem 6.12 the *non-central dehomogenization of* $\mathscr{G}$ in $K\langle X\rangle$ with respect to T, or $\mathscr{G}_\sim$ is a *non-central dehomogenized Gröbner basis* with respect to T.

Remark At this point, let us point out that Theorem 6.12 generalizes ([Nor1], Theorem 5) in which $\mathscr{G}$ is taken to be a *reduced Gröbner basis* and the proof given there depends essentially on the reducibility of $\mathscr{G}$.

Corollary 6.13. *Let I be an ideal of $K\langle X\rangle$. If $\mathscr{G}$ is a homogeneous Gröbner basis of $\langle\widetilde{I}\rangle$ in $K\langle X,T\rangle$ with respect to the data $(\widetilde{\mathcal{B}}, \prec_{_{T\text{-}gr}})$, then*

$\mathscr{G}_\sim = \{g_\sim \mid g \in \mathscr{G}\}$ is a Gröbner basis for I in $K\langle X\rangle$ with respect to the data $(B, \prec_{gr})$. Moreover, if I is generated by the subset F and $\widetilde{F} \subset \mathscr{G}$, then $F \subset \mathscr{G}_\sim$.

Proof. Put $J = \langle \widetilde{I}\rangle$. Then since $J_\sim = I$, it follows from Theorem 6.12 that if $\mathscr{G}$ is a homogeneous Gröbner basis of J then $\mathscr{G}_\sim$ is a Gröbner basis for I. The second assertion of the theorem is clear by Lemma 6.8(ii). □

Let S be a nonempty subset of $K\langle X\rangle$ and $I = \langle S\rangle$ the ideal generated by S. Then, with $\widetilde{S} = \{\widetilde{f} \mid f \in S\} \cup \{X_iT - TX_i \mid 1 \le i \le n\}$, in general $\langle \widetilde{S}\rangle \subsetneq \langle \widetilde{I}\rangle$ in $K\langle X, T\rangle$ (for instance, consider $S = \{Y^3 - XY - X - Y,\ Y^2 - X + 3\}$ in the free algebra $K\langle X, Y\rangle$ and the homogenization in $K\langle X, Y, T\rangle$ with respect to T). Again, as we did in the case dealing with (de)homogenized Gröbner bases with respect to the commuting variable t, we take this place to set up the procedure of getting a Gröbner basis for I and hence a Gröbner basis for $\langle \widetilde{I}\rangle$ by producing a homogeneous Gröbner basis of the graded ideal $\langle \widetilde{S}\rangle$.

Proposition 6.14. *Let $I = \langle S\rangle$ be the ideal of $K\langle X\rangle$ as fixed above. Suppose the ground field K is computable. Then a Gröbner basis for I and a homogeneous Gröbner basis for $\langle \widetilde{I}\rangle$ may be obtained by implementing the following procedure:*

Step 1. *Starting with the initial subset*

$$\widetilde{S} = \{\widetilde{f} \mid f \in S\} \cup \{X_iT - TX_i \mid 1 \le i \le n\},$$

compute a homogeneous Gröbner basis $\mathscr{G}$ for the graded ideal $\langle \widetilde{S}\rangle$ of $K\langle X, T\rangle$.

Step 2. *Noticing $\langle \widetilde{S}\rangle_\sim = I$, use Theorem 6.12 and dehomogenize $\mathscr{G}$ with respect to T in order to obtain the Gröbner basis $\mathscr{G}_\sim$ for I.*

Step 3. *Use Theorem 6.10 and homogenize $\mathscr{G}_\sim$ with respect to T in order to obtain the homogeneous Gröbner basis $(\mathscr{G}_\sim)^\sim$ for the graded ideal $\langle \widetilde{I}\rangle$.*

□

Remark At this point, the reader is especially reminded that if, in the above proposition, S is finite, then the programm LETTERPLACE developed by La Scala and Levandovskyy in [LL] may be used to calculate the homogeneous Gröbner basis $\mathscr{G}$ quite effectively.

3.7 dh-Closed Homogeneous Gröbner Bases

Based on Theorem 6.3, Theorem 6.5, Theorem 6.10 and Theorem 6.12, in this section we introduce the notion of a dh-closed homogeneous Gröbner basis in the polynomial ring $R[t]$, respectively in the free algebra $K\langle X, T\rangle$, and establish a one-to-one correspondence

$$\{\text{Gröbner bases in } R\} \longleftrightarrow \left\{\begin{matrix}\text{dh-closed homogeneous}\\ \text{Gröbner bases in } R[t]\end{matrix}\right\}$$

as well as a one-to-one correspondence

$$\{\text{Gröbner bases in } K\langle X\rangle\} \longleftrightarrow \left\{\begin{matrix}\text{dh-closed homogeneous}\\ \text{Gröbner bases in } K\langle X, T\rangle\end{matrix}\right\}.$$

Moreover, we characterize graded ideals which are generated by dh-closed homogeneous Gröbner bases in the polynomial algebra $R[t]$ and the free algebra $K\langle X, T\rangle$, respectively.

All notions and notations used in the last section remain valid throughout the present section.

dh-Closed homogeneous Gröbner bases in $R[t]$

Let the $\mathbb{N}$-graded K-algebra $R = \oplus_{p\in\mathbb{N}} R_p$ and the admissible system $(\mathcal{B}, \prec_{gr})$ be as in the last section, i.e., where $\mathcal{B}$ is a skew multiplicative K-basis of R consisting of $\mathbb{N}$-homogeneous elements, and $\prec_{gr}$ is an $\mathbb{N}$-graded monomial ordering on $\mathcal{B}$. Thus we have the corresponding admissible system $(\mathcal{B}^*, \prec_{t\text{-}gr})$ for the polynomial ring $R[t] = \oplus_{p\in\mathbb{N}} R[t]_p$ which has the mixed $\mathbb{N}$-gradation (see Section 6). The goal of this part is to find those homogeneous Gröbner bases in $R[t]$ that correspond bijectively to Gröbner bases in R.

Considering the (de)homogenization with respect to t, a homogeneous element $F \in R[t]$ is called *dh-closed* if $(F_*)^* = F$; a subset S of $R[t]$ consisting of dh-closed homogeneous elements is called a *dh-closed homogeneous set*; if a dh-closed homogeneous set $\mathcal{G}$ in $R[t]$ forms a Gröbner basis with respect to $\prec_{t\text{-}gr}$, then it is called a *dh-closed homogeneous Gröbner basis.*

To better understand the dh-closed property introduced above, we characterize a dh-closed homogeneous element as follows.

Lemma 7.1. *With notation as before, for a homogeneous element* $F \in$

$R[t]$, with respect to $(\mathcal{B}, \prec_{gr})$ and $(\mathcal{B}^*, \prec_{t\text{-}gr})$ the following statements are equivalent:
(i) *F is dh-closed, i.e., $(F_*)^* = F$;*
(ii) $\mathbf{LM}(F_*) = \mathbf{LM}(F)$;
(iii) *F cannot be written as $F = t^r H$ with H a homogeneous element of $R[t]$ and $r \geq 1$;*
(iv) $t \not| \mathbf{LM}(F)$.

Proof. By the definitions of $\prec_{t\text{-}gr}$ and $(F_*)^*$, the verification of (i) $\Rightarrow$ (ii) $\Rightarrow$ (iii) $\Rightarrow$ (iv) $\Rightarrow$ (i) is straightforward. □

Theorem 7.2. *With respect to the system $(\mathcal{B}, \prec_{gr})$ and the system $(\mathcal{B}^*, \prec_{t\text{-}gr})$, there is a one-to-one correspondence between the set of all Gröbner bases in R and the set of all dh-closed homogeneous Gröbner bases in $R[t]$:*

$$\begin{array}{ccc} \{\text{Gröbner bases } \mathcal{G} \text{ in } R\} & \longleftrightarrow & \left\{\begin{array}{c}\text{dh-closed homogeneous} \\ \text{Gröbner bases } \mathscr{G} \text{ in } R[t]\end{array}\right\} \\ \mathcal{G} & \longrightarrow & \mathcal{G}^* \\ \mathscr{G}_* & \longleftarrow & \mathscr{G} \end{array}$$

and this correspondence also gives rise to a bijective map between the set of all minimal Gröbner bases in R and the set of all dh-closed minimal homogeneous Gröbner bases in $R[t]$.

Proof. Bearing in mind the definitions of homogenization and dehomogenization with respect to t, by Theorem 6.3 and Theorem 6.5 it can be verified directly that the given rule of correspondence defines a one-to-one map. By the definition of a minimal Gröbner basis, the second assertion follows from Lemma 6.2(ii) and Lemma 7.1(ii). □

dh-Closed graded ideals in $R[t]$

Let $R = \oplus_{p \in \mathbb{N}} R_p$, $(\mathcal{B}, \prec_{gr})$, $R[t]$, and $(\mathcal{B}^*, \prec_{t\text{-}gr})$ be as before. Considering the central (de)homogenization with respect to t, a graded ideal J of $R[t]$ is called a *dh-closed graded ideal* if $\langle (J_*)^* \rangle = J$. This definition generalizes the notion of a $(\phi_*)^*$-closed graded ideal introduced in [LVO5].

It is easy to see that under central (de)homogenization there is a one-to-one correspondence between the set of all ideals in R and the set of all dh-closed graded ideals in $R[t]$. We characterize below the relation between dh-closed homogeneous Gröbner bases and dh-closed graded ideals.

Theorem 7.3. *With notation as before, let J be a graded ideal of $R[t]$ and $\mathscr{G}$ a minimal homogeneous Gröbner basis of J. Under $(\mathcal{B}, \prec_{gr})$ and $(\mathcal{B}^*, \prec_{t\text{-}gr})$, the following statements are equivalent:*
(i) $\mathscr{G}$ is a dh-closed homogeneous Gröbner basis;
(ii) J is a dh-closed graded ideal of $R[t]$;
(iii) The $R[t]$-module $R[t]/J$ is t-torsionfree, i.e., if $\overline{f} = f + J \in R[t]/J$ and $\overline{f} \neq 0$, then $t\overline{f} \neq 0$, or equivalently, $tf \notin J$;
(iv) $tR[t] \cap J = tJ$.

Proof. (i) $\Rightarrow$ (ii). By Theorem 6.5, $\mathscr{G}_*$ is a Gröbner basis for J_* in R with respect to $(\mathcal{B}, \prec_{gr})$. Since $\mathscr{G}$ is dh-closed, it follows from Theorem 6.3 that $\mathscr{G} = (\mathscr{G}_*)^*$ is a Gröbner basis for $\langle (J_*)^* \rangle$. This shows that $J = \langle \mathscr{G} \rangle = \langle (J_*)^* \rangle$, that is, J is dh-closed in $R[t]$.

(ii) $\Rightarrow$ (iii). Note that $R[t]/J$ is an $\mathbb{N}$-graded $R[t]$-module and t is a homogeneous element in $R[t]$. It is sufficient to prove that t does not annihilate any nonzero homogeneous element of $R[t]/J$. Thus, assuming $F \in R[t]_p$ and $tF \in J$, then since J is dh-closed, we have

$$(F_*)^* = ((tF)_*)^* \in \langle (J_*)^* \rangle = J.$$

It follows from Lemma 6.1(iii) that there exists some $r \in \mathbb{N}$ such that $F = t^r (F_*)^* \in J$, as desired.

(iii) $\Leftrightarrow$ (iv). Obvious.

(iii) $\Rightarrow$ (i). Note that $\mathscr{G}$ is a homogeneous Gröbner basis by the assumption. For each $g \in \mathscr{G}$, by Lemma 6.1(iii) there is some $r \in \mathbb{N}$ such that $t^r (g_*)^* = g$. It follows that with respect to $\prec_{t\text{-}gr}$ we have $\mathbf{LM}(g) = t^r \mathbf{LM}((g_*)^*)$. Since $R[t]/J$ is t-torsionfree, if $r > 0$, then $(g_*)^* \in J$. Thus, there is some $g' \in \mathscr{G}$ such that $g' \neq g$ and $\mathbf{LM}(g') | \mathbf{LM}((g_*)^*)$. Hence $\mathbf{LM}(g') | \mathbf{LM}(g)$, contradicting the assumption that $\mathscr{G}$ is a minimal Gröbner basis. Therefore, we must have $r = 0$, i.e., $(g_*)^* = g$. This shows that $\mathscr{G}$ is dh-closed. □

Corollary 7.4. *With notation as before, let J be a graded ideal of $R[t]$. If, with respect to $(\mathcal{B}, \prec_{gr})$ and $(\mathcal{B}^*, \prec_{t\text{-}gr})$, J has a dh-closed minimal homogeneous Gröbner basis, then every minimal homogeneous Gröbner basis of J is dh-closed.*

Proof. This follows from the fact that each of the properties (ii) – (iv) in Theorem 7.3 does not depend on the choice of the generating set for J. □

By the foregoing argument, to know wether a given graded ideal J of $R[t]$ is dh-closed, it is sufficient to compute a minimal homogeneous Gröbner basis $\mathscr{G}$ for J (if Gröbner basis is computable in R), and then use the definition of a dh-closed homogeneous set or Lemma 7.1 to check whether $\mathscr{G}$ is dh-closed. This procedure may be realized, for instance, in commutative polynomial K-algebras, noncommutative free K-algebras, path algebras over K, the skew polynomial algebra $\mathcal{O}_n(\lambda_{ji})$ (or the coordinate ring of a quantum affine n-space over K defined in Ch.1, Section 1), and exterior K-algebras, because Gröbner bases are computable in these algebras and their polynomial extensions.

dh-Closed homogeneous Gröbner bases in $K\langle X,T\rangle$

Let the free K-algebras $K\langle X\rangle = K\langle X_1,\dots,X_n\rangle$, the admissible system $(\mathcal{B},\prec_{gr})$ for $K\langle X\rangle$, the free K-algebra $K\langle X,T\rangle = K\langle X_1,\dots,X_n,T\rangle$, and the admissible system $(\widetilde{\mathcal{B}},\prec_{_{T\text{-}gr}})$ for $K\langle X,T\rangle$ be as fixed in Section 6. In this part we find those homogeneous Gröbner bases in $K\langle X,T\rangle$ that correspond bijectively to Gröbner bases in $K\langle X\rangle$.

A homogeneous element $F\in K\langle X,T\rangle$ is called *dh-closed* if $(F_\sim)^\sim = F$; a subset S of $K\langle X,T\rangle$ consisting of dh-closed homogeneous elements is called a *dh-closed homogeneous set.*

To better understand the dh-closed property introduced above for homogeneous elements in $K\langle X,T\rangle$, we characterize a dh-closed homogeneous element as follows.

Lemma 7.5. *With notation as before, for a nonzero homogeneous element $F\in K\langle X,T\rangle$, the following two properties are equivalent with respect to $(\mathcal{B},\prec_{gr})$ and $(\widetilde{\mathcal{B}},\prec_{_{T\text{-}gr}})$:*
(i) *F is dh-closed;*
(ii) *$F=\sum_i\lambda_iT^{r_i}w_i$ satisfying $\mathbf{LM}(F_\sim)=\mathbf{LM}(F)$, where $\lambda_i\in K^*$, $r_i\in\mathbb{N}$, $w_i\in\mathcal{B}$.*

Proof. This follows easily from Lemma 6.8(iii). □

Recall from Section 6 that we have defined

$$\widetilde{S} = \{\widetilde{f} \mid f \in S\} \cup \{X_iT - TX_i \mid 1 \le i \le n\}$$

if $I = \langle S \rangle$ is an ideal of $K\langle X \rangle$ generated by the subset S. If $\mathscr{H}$ is a nonempty dh-closed homogeneous set in $K\langle X, T \rangle$ such that the subset

$$\mathscr{G} = \mathscr{H} \cup \{X_iT - TX_i \mid 1 \le i \le n\}$$

forms a Gröbner basis for the graded ideal $J = \langle \mathscr{G} \rangle$ with respect to $(\widetilde{\mathcal{B}}, \prec_{T\text{-}gr})$, then we call $\mathscr{G}$ a *dh-closed homogeneous Gröbner basis.*

Theorem 7.6. *With respect to the system $(\mathcal{B}, \prec_{gr})$ and the system $(\widetilde{\mathcal{B}}, \prec_{T\text{-}gr})$, there is a one-to-one correspondence between the set of all Gröbner bases in $K\langle X \rangle$ and the set of all dh-closed homogeneous Gröbner bases in $K\langle X, T \rangle$:*

$$\begin{array}{ccc} \{\textit{Gröbner bases } \mathcal{G} \textit{ in } K\langle X \rangle\} & \longleftrightarrow & \left\{ \begin{array}{l} \textit{dh-closed homogeneous} \\ \textit{Gröbner bases } \mathscr{G} \textit{ in } K\langle X, T \rangle \end{array} \right\} \\ \mathcal{G} & \longrightarrow & \widetilde{\mathcal{G}} \\ \mathscr{G}_{\sim} & \longleftarrow & \mathscr{G} \end{array}$$

and this correspondence also gives rise to a bijective map between the set of all minimal Gröbner bases in $K\langle X \rangle$ and the set of all dh-closed minimal homogeneous Gröbner bases in $K\langle X, T \rangle$.

Proof. By the definitions of homogenization and dehomogenization with respect to T, Theorem 6.10, and Theorem 6.12, it can be verified directly that the given rule of correspondence defines a one-to-one map. The second assertion follows from Lemma 6.9(ii) and Lemma 7.5(ii). □

dh-Closed graded ideals in $K\langle X, T \rangle$

Considering the non-central (de)homogenization with respect to T as before, a graded ideal J of $K\langle X, T \rangle$ is called a *dh-closed graded ideal* if $\langle (J_{\sim})^{\sim} \rangle = J$.

It is easy to see that under non-central (de)homogenization there is a one-to-one correspondence between the set of all ideals in $K\langle X \rangle$ and the set of all dh-closed graded ideals in $K\langle X, T \rangle$. We characterize below the relation between dh-closed homogeneous Gröbner bases and dh-closed graded ideals in $K\langle X, T \rangle$.

For convenience, we let $\mathscr{C}$ denote the ideal of $K\langle X,T\rangle$ generated by the Gröbner basis $\{X_iT - TX_i \mid 1 \le i \le n\}$, and let K-span$N(\mathscr{C})$ denote the K-space spanned by the set $N(\mathscr{C})$ of normal monomials in $\widetilde{\mathcal{B}}$ (mod $\mathscr{C}$). Noticing that with respect to the monomial ordering $\prec_{_{T\text{-}gr}}$ on $\widetilde{\mathcal{B}}$, $\mathbf{LM}(X_iT - TX_i) = X_iT$ for all $1 \le i \le n$, so each element $F \in K$-span$N(\mathscr{C})$ is of the form $F = \sum_i \lambda_i T^{r_i} w_i$ with $\lambda_i \in K^*$, $r_i \in \mathbb{N}$, and $w_i \in \mathcal{B}$.

Theorem 7.7. *With the convention made above, let $\mathscr{H} \subset K$-span$N(\mathscr{C})$ be a subset consisting of nonzero homogeneous elements. Suppose that the subset $\mathscr{G} = \mathscr{H} \cup \{X_iT - TX_i \mid 1 \le i \le n\}$ forms a minimal Gröbner basis for the graded ideal $J = \langle \mathscr{G}\rangle$ in $K\langle X,T\rangle$ with respect to the data $(\widetilde{\mathcal{B}}, \prec_{_{T\text{-}gr}})$. Then, the following statements are equivalent:*
(i) $\mathscr{G}$ is a dh-closed homogeneous Gröbner basis, i.e., $\mathscr{H}$ is a dh-closed homogeneous set;
(ii) J is a dh-closed graded ideal of $K\langle X,T\rangle$, i.e., $\langle (J_\sim)^\sim\rangle = J$;
(iii) $K\langle X,T\rangle/J$ is a T-torsionfree (left) $K\langle X,T\rangle$-module, i.e., if $\overline{f} = f+J \in K\langle X,T\rangle/J$ and $\overline{f} \ne 0$, then $T\overline{f} \ne 0$, or equivalently, $Tf \notin J$;
(iv) $TR[t] \cap J = TJ$.

Proof. (i) $\Rightarrow$ (ii). If $\mathscr{G}$ is dh-closed, then by Theorem 6.12, $\mathscr{G}_\sim$ is a Gröbner basis for the ideal $J_\sim$ with respect to $(\mathcal{B}, \prec_{gr})$. Furthermore, it follows from Theorem 6.10 that $\mathscr{G}$ is a Gröbner basis for $\langle (J_\sim)^\sim\rangle$ with respect to $(\widetilde{\mathcal{B}}, \prec_{_{T\text{-}gr}})$. Hence, $J = \langle \mathscr{G}\rangle = \langle (J_\sim)^\sim\rangle$, i.e., J is dh-closed in $K\langle X,T\rangle$.

(ii) $\Rightarrow$ (iii). Noticing that $K\langle X,T\rangle/J$ is an $\mathbb{N}$-graded $K\langle X,T\rangle$-module and T is a homogeneous element in $K\langle X,T\rangle$, it is sufficient to show that T does not annihilate any nonzero homogeneous element of $K\langle X,T\rangle/J$. Suppose $F \in K\langle X,T\rangle_p$ and $TF \in J$. Then since J is dh-closed, we have

$$(F_\sim)^\sim = ((TF)_\sim)^\sim \in \langle (J_\sim)^\sim\rangle = J. \tag{1}$$

Moreover, by Lemma 6.8(iii) there exist $L \in J$ and $r \in \mathbb{N}$ such that

$$F = L + T^r (F_\sim)^\sim. \tag{2}$$

Hence (1) and (2) yield $F \in J$, as desired.

(iii) $\Leftrightarrow$ (iv). Obvious.

(iii) $\Rightarrow$ (i). Let $H \in \mathscr{H} - \{X_iT - TX_i \mid 1 \le i \le n\}$. Then $H = \sum_i \lambda_i T^{r_i} w_i \in K$-span$N(\mathscr{C})$ such that $H = T^r (H_\sim)^\sim$ for some $r \in \mathbb{N}$. If $r \ge 1$, then since $K\langle X,T\rangle/J$ is T-torsionfree we must have $(H_\sim)^\sim \in J$, and $\mathbf{LM}(H) \not| \mathbf{LM}((H_\sim)^\sim)$. Hence, there exists some $H' \in \mathscr{H} - \{H\}$ such that $\mathbf{LM}(H')|\mathbf{LM}((H_{\sim})^{\sim})$ and consequently $\mathbf{LM}(H')|\mathbf{LM}(H)$, contradicting the minimality of $\mathscr{G}$. Therefore, $r = 0$, i.e., $H = (H_{\sim})^{\sim}$. This shows that $\mathscr{H}$ is dh-closed. □

Corollary 7.8. *With notation as before, let J be a graded ideal of $K\langle X, T\rangle$. If, with respect to $(\mathcal{B}, \prec_{gr})$ and $(\widetilde{\mathcal{B}}, \prec_{_{T\text{-}gr}})$, J has a dh-closed minimal homogeneous Gröbner basis, then every minimal homogeneous Gröbner basis of J is dh-closed.*

Proof. This follows from the fact that each of the properties (ii) – (iv) in Theorem 7.7 does not depend on the choice of the generating set for J. □

Note that Gröbner bases are computable in $K\langle X, T\rangle$ if the ground field K is computable. By the foregoing argument, to know wether a given graded ideal J of $K\langle X, T\rangle$ is dh-closed, it is sufficient to check if J contains a minimal homogeneous Gröbner basis of the form $\mathscr{G} = \mathscr{H} \cup \{X_iT - TX_i \mid 1 \le i \le n\}$ in which $\mathscr{H} \subset K\text{-span}N(\mathscr{C})$ is a dh-closed homogeneous set.

An application to $K[x_1, \ldots, x_n]$ and $K\langle X_1, \ldots, X_n\rangle$

Finally, let us point out that if we replace $R[t]$ by the commutative polynomial K-algebra $K[x_1, \ldots, x_n]$ with $R = K[x_1, \ldots, x_{i-1}, x_{i+1}, \ldots, x_n]$, $t = x_i$, $1 \le i \le n$, respectively, if we replace $K\langle X, T\rangle$ by the free K-algebra $K\langle X_1, \ldots, X_n\rangle$ with $K\langle X\rangle = K\langle X_1, \ldots, X_{i-1}, X_{i+1}, \ldots, X_n\rangle$, $T = X_i$, $1 \le i \le n$, then everything we have done in this and the last section can be done with respect to each $x_i = t$ and $X_i = T$ respectively, $1 \le i \le n$. Instead of mentioning a version of each result obtained before with respect to $x_i = t$ and $X_i = T$ respectively, we highlight the analogues of Theorem 6.5, Theorem 6.12, Theorem 7.1 and Theorem 7.6 as follows.

Consider the commutative polynomial K-algebra $K[x_1, \ldots, x_n]$. For each fixed $x_i = t$, $1 \le i \le n$, let $R = K[x_1, \ldots, x_{i-1}, x_{i+1}, \ldots, x_n]$, and $K[x_1, \ldots, x_n] = R[t]$. Moreover, let $(\mathcal{B}, \prec_{gr})$ be any fixed admissible system for R, where $\prec_{gr}$ is a graded monomial ordering on the standard K-basis $\mathcal{B}$ of R with respect to a fixed (weight) $\mathbb{N}$-gradation. Then, as in Section 6, $R[t]$ has the mixed $\mathbb{N}$-gradation and the corresponding admissible system $(\mathcal{B}^*, \prec_{t\text{-}gr})$, where $\prec_{t\text{-}gr}$ is the monomial ordering obtained by extending

$\prec_{gr}$ on the standard K-basis $\mathcal{B}^*$ of $R[t]$ such that $t^r \prec_{t\text{-}gr} w$ for all $r \in \mathbb{N}$ and $w \in \mathcal{B}$.

Theorem 7.9. *With the convention made above, the following statements hold.*
(i) *For each fixed $x_i = t$, $1 \le i \le n$, if $\mathscr{G}$ is a homogeneous Gröbner basis of the graded ideal $J = \langle \mathscr{G} \rangle$ in $R[t] = K[x_1, \dots, x_n]$ with respect to $(\mathcal{B}^*, \prec_{t\text{-}gr})$, then $\mathscr{G}_* = \{g_* \mid g \in \mathscr{G}\}$ is a Gröbner basis for the ideal J_* in $R = K[x_1, \dots, x_{i-1}, x_{i+1}, \dots, x_n]$ with respect to $(\mathcal{B}, \prec_{gr})$.*
(ii) *For each fixed $x_i = t$, $1 \le i \le n$, there is a one-to-one correspondence between the set of all dh-closed homogeneous Gröbner bases in $R[t] = K[x_1, \dots, x_n]$ and the set of all Gröbner bases in $R = K[x_1, \dots, x_{i-1}, x_{i+1}, \dots, x_n]$, under which minimal Gröbner bases correspond to minimal Gröbner bases.*

□

Geometrically, Theorem 7.9 may be viewed as a Gröbner basis realization of the correspondence between algebraic sets in the projective space $\mathbb{P}_K^{n-1}$ and algebraic sets in the affine space $\mathbb{A}_K^{n-1}$, where $n \ge 2$.

Next, we consider the noncommutative free K-algebra $K\langle X_1, \dots, X_n \rangle$. For each fixed $X_i = T$, $1 \le i \le n$, let $K\langle X_1, \dots, X_{i-1}, X_{i+1}, \dots, X_n \rangle = K\langle X \rangle$, and $K\langle X_1, \dots, X_n \rangle = K\langle X, T \rangle$. Moreover, let $(\mathcal{B}, \prec_{gr})$ be any fixed admissible system for $K\langle X \rangle$, where $\prec_{gr}$ is an $\mathbb{N}$-graded lexicographic ordering on the standard K-basis $\mathcal{B}$ of $K\langle X \rangle$ with respect to a fixed (weight) $\mathbb{N}$-gradation. Then, as in Section 6, $K\langle X, T \rangle$ has the extended (weight) $\mathbb{N}$-gradation and the corresponding admissible system $(\widetilde{\mathcal{B}}, \prec_{T\text{-}gr})$, where $\prec_{T\text{-}gr}$ is the $\mathbb{N}$-graded lexicographic ordering obtained by extending $\prec_{gr}$ on the standard K-basis $\widetilde{\mathcal{B}}$ of $K\langle X, T \rangle$ such that $T \prec_{T\text{-}gr} X_i$, $1 \le i \le n$.

Theorem 7.10. *With the convention made above, the following statements hold.*
(i) *For each fixed $X_i = T$, $1 \le i \le n$, if J is a graded ideal of $K\langle X, T \rangle = K\langle X_1, \dots, X_n \rangle$ that contains the subset $\{X_jT - TX_j \mid j \ne i\}$ and if $\mathscr{G}$ is a homogeneous Gröbner basis of J with respect to $(\widetilde{\mathcal{B}}, \prec_{T\text{-}gr})$, then $\mathscr{G}_\sim = \{g_\sim \mid g \in \mathscr{G}\}$ is a Gröbner basis for the ideal $J_\sim$ in $K\langle X \rangle = K\langle X_1, \dots, X_{i-1}, X_{i+1}, \dots, X_n \rangle$ with respect to $(\mathcal{B}, \prec_{gr})$.*
(ii) *For each fixed $X_i = T$, $1 \le i \le n$, there is a one-to-one corre-*

spondence between the set of all dh-closed homogeneous Gröbner bases in $K\langle X,T\rangle = K\langle X_1,\ldots,X_n\rangle$ *and the set of all Gröbner bases in* $K\langle X\rangle = K\langle X_1,\ldots,X_{i-1},X_{i+1},\ldots,X_n\rangle$*, under which minimal Gröbner bases correspond to minimal Gröbner bases.*

□

According to the treatment of Rees algebras demonstrated in [LWZ], ([Li1], Ch.3), [Li3], and [Li5], the results obtained in this and the last section indeed tell us that algebras defined by dh-closed homogenous Gröbner bases may be studied as Rees algebras (defined by grading filtration) via studying algebras with simpler Gröbner defining relations (see Ch.7 for more detals).

Final remark on Ch.3

Let $K\langle X\rangle = K\langle X_1,\ldots,X_n\rangle$ be the free algebra of n generators over a field K, and let $R\langle X\rangle = R\langle X_1,\ldots,X_n\rangle$ be the free R-algebra of n generators over an arbitrary commutative ring R. In [Li7], by examining the feasibility of a version of the termination theorem (Thoerem 5.1) for verifying monic Gröbner bases in $R\langle X\rangle$ (i.e., every element g in such Gröbner bases has $\mathbf{LC}(g) = 1$), it was clarified that any monic Gröbner basis in $K\langle X\rangle$ may give rise to a monic Gröbner basis of the same type in $R\langle X\rangle$, and vice versa. It turns out that numerous important R-algebras have defining relations which form a monic Gröbner basis, and consequently,

(a) based on a constructive PBW theory over rings established in [Li7], which is a (nontrivial) analogue of our next chapter, many global structural properties of such R-algebras may be studied via their Gröbner defining relations as that demonstrated for studying quotient algebras of $K\langle X\rangle$ in the forthcoming Ch.5 – Ch.8;

(b) such monic Gröbner bases over rings may be effectively used to study certain topics such as valuation extensions and many reduction properties (in terms of algebras over residue fields) as shown in ([Li8], [LVO6]).

Moreover, the principle we mentioned above works as well for a commutative polynomial ring $R[x_1,\ldots,x_n]$ over an arbitrary commutative ring R, where overlap elements are replaced by S-polynomials. Detailed discussion on monic Gröbner bases over rings is referred to [Li7].

Chapter 4

Gröbner Basis Theory Meets PBW Theory

In this chapter we develop a constructive PBW theory by means of Gröbner bases in a quite extensive context. More precisely, after establishing in Section 1 an equivalence between the existence of a Γ-standard basis and the existence of a Γ-PBW isomorphism proposed by (Ch.2, Theorem 3.2), we first realize, by means of Gröbner bases, the Γ-PBW isomorphism with respect to different filtered-graded structures (Section 2); and then in Section 3, for a quotient algebra $A = K\langle X\rangle/I$ of the free K-algebra $K\langle X\rangle = K\langle X_1, \ldots, X_n\rangle$, we consider, in light of (Ch.3, Theorem 3.3, Proposition 4.8), the naturally raised twofold question: What kind of a Gröbner basis $\mathcal{G}$ for I should we have in order to obtain a classical PBW K-basis for A, and if, without using the Gröbner basis technique, a classical PBW K-basis of A with respect to a given generating set $\mathcal{G}$ of I is obtained in a different way, under what condition can $\mathcal{G}$ be a Gröbner basis? As the first dividend of a solution to the above question (Theorem 3.1), checkable conditions are derived (Proposition 3.2) for recognizing and constructing interesting algebras that have a classical PBW K-basis and a nice associated $\mathbb{N}$-graded algebra. In Section 4 we revisit the solvable polynomial algebras (in the sense of [K-RW]) in terms of certain results obtained in Section 3.

The PBW theory we formulated in this chapter covers not only the classical PBW theory for enveloping algebras of Lie algebras (cf. [Hu], [Jac]), but also some other types of PBW property studied in (e.g., [BG1], [BG2], [FV], [Pr], [Pos]), as well as the so called PBW-deformation of a finitely presented $\mathbb{N}$-graded algebra studied in (e.g., [C-BH], [CS2], [EG], [FV], [GG]).

As before we maintain previously used conventions and notations.

4.1 Γ-Standard Basis and Γ-PBW Isomorphism

Let $R = \oplus_{\gamma\in\Gamma}R_\gamma$ be a Γ-graded K-algebra, where Γ is a totally ordered semigroup with the total ordering $\prec$, and let I be an arbitrary ideal of R. Consider the quotient algebra $A = R/I$ and its Γ-filtration FA induced by the Γ-grading filtration FR of R as in (Ch.2, Section 3). In this section we establish an equivalence between the existence of a Γ-standard basis F (see Definition 1.1 below) of I and the existence of a Γ-PBW isomorphism

$$R/\langle \mathbf{LH}(F)\rangle \underset{\cong}{\xrightarrow{\rho}} G^\Gamma(A)$$

with respect to the Γ-filtration FA (see Definition 1.2 below).

First, note that the Γ-grading filtration FR of R has two basic properties:

(1) If $f \in R$ with the Γ-leading homogeneous element $\mathbf{LH}(f) = r_\gamma \in R_\gamma$, then

$$f \in F_\gamma R - F^*_\gamma R,\ \text{ where } F^*_\gamma R = \oplus_{\gamma'\prec\gamma}R_{\gamma'}.$$

It turns out that $\cap_{\gamma\in\Gamma}F_\gamma R = \{0\}$, namely, the Γ-grading filtration FR is separated.

(2) $G^\Gamma(R) \cong R$ as Γ-graded algebras.

So, for $f \in R$ with $\mathbf{LH}(f) = r_\gamma \neq 0$, we may denote the (nonzero) image of f in $G^\Gamma(R)_\gamma = F_\gamma R/F^*_\gamma R \cong R_\gamma$ by $\sigma(f)$ as in (Ch.2, Section 1), and it is thereby clear that $\sigma(f) = \mathbf{LH}(f) = r_\gamma$. Applying this fact to an ideal I of R and considering the Γ-filtration FI induced by FR, we have

$$\begin{aligned}&\sigma(I) = \{\sigma(f) \mid f \in I\} = \{\mathbf{LH}(f) \mid f \in I\} = \mathbf{LH}(I),\\ &G^\Gamma(I) = \langle\sigma(I)\rangle = \langle \mathbf{LH}(I)\rangle.\end{aligned}$$

Consequently, we may adopt the notion as given in [Gol] to specify a standard basis for I as follows.

Definition 1.1. Let F be a set of generators of the ideal I. If $\langle\mathbf{LH}(I)\rangle = \langle\mathbf{LH}(F)\rangle$, then we call F a Γ-*standard basis* of I.

Remark At this point, let us point out, that it was the work of Golod ([Gol], 1986) that defined, for the first time, a standard basis (and hence a Gröbner basis) for an ideal I in a Γ-filtered algebra R by using clearly the associated Γ-graded ideal $G^\Gamma(I)$ of I in the associated Γ-graded algebra

$G^{\Gamma}(R)$ of R. With such a filtered-graded structural definition, a homological characterization of standard bases (and hence Gröbner bases) was given in [Gol] in terms of the first homology module of the Shafarevich complex [GS] (respectively in terms of the Koszul complex in the commutative case); as applications of this characterization, three celebrated algorithmic results were reproved in the same paper [Gol], namely the confluence or diamond lemma ([Sh2], [Ber3]), the Buchberger algorithm [Bu2], and the algorithm for the construction of Gröbner bases of ideals in enveloping algebras of Lie algebras [Lat].

On the other hand, considering the quotient algebra $A = R/I$, from the proof of (Ch.2, Theorem 3.2) we may see that if the ideal I is generated by the subset F, then since $\mathbf{LH}(F) \subseteq \mathbf{LH}(I) = \mathrm{Ker}\phi$, there is the naturally induced commutative diagram of K-algebra homomorphisms

$$\begin{array}{ccccc} 0 \longrightarrow \langle \mathbf{LH}(F)\rangle & \longrightarrow & R & \longrightarrow & R/\langle \mathbf{LH}(F)\rangle \\ & & {\scriptstyle\phi}\big\downarrow & \swarrow{\scriptstyle\rho} & \\ & & G^{\Gamma}(A) & & \end{array}$$

in which $\mathrm{Ker}\rho = \langle \mathbf{LH}(I)\rangle/\langle \mathbf{LH}(F)\rangle$. It turns out that $\langle \mathbf{LH}(I)\rangle = \langle \mathbf{LH}(F)\rangle$ if and only if ρ is an isomorphism. Comparing with the classical PBW theorem, actually, having such an isomorphism amounts to having a version of PBW theorem for Γ-filtered algebras of the form A, that is, the following definition and theorem make sense.

Definition 1.2. The algebra $A = R/I$ is said to have a Γ-*PBW isomorphism* if I has a set of generators F such that

$$R/\langle \mathbf{LH}(F)\rangle \xrightarrow[\cong]{\rho} G^{\Gamma}(A).$$

Theorem 1.3. *The algebra $A = R/I$ has a Γ-PBW isomorphism if and only I has a Γ-standard basis.*

□

Obviously, Definition 1.2 covers the classical $\mathbb{N}$-PBW isomorphism $G^{\mathbb{N}}(U(\mathrm{g})) \cong K[x_1, \ldots, x_n]$ for an n-dimensional K-Lie algebra g. We explain below that Definition 1.2 also covers the important p-type PBW property defined for $\mathbb{N}$-filtered algebras in connection with the non-homogeneous

Koszulity ([BG1], [BG2], [FV], [Pr], [Pos]), as well as the so called PBW-deformation of a finitely presented $\mathbb{N}$-graded algebra ([C-BH], [CS2], [EG], [FV], [GG]). The reader is referred to later (Ch.6, Section 1) for the definition of a non-homogeneous p-Koszul algebra.

p-type PBW property

Let the free K-algebra $K\langle X\rangle = K\langle X_1, \ldots, X_n\rangle$ of n generators be equipped with the natural $\mathbb{N}$-filtration $FK\langle X\rangle$. Consider a K-subspace $V_p \subset F_pK\langle X\rangle - F_{p-1}K\langle X\rangle$, $p \geq 2$, and let $\mathbf{LH}(V_p)$ be the set of $\mathbb{N}$-leading homogeneous elements of V_p taken with respect to the natural $\mathbb{N}$-gradation of $K\langle X\rangle$. The K-algebra $A = K\langle X\rangle/\langle V_p\rangle$ is said to have the *p-type PBW property*, if, in the language we used above, the $\mathbb{N}$-graded K-algebra homomorphism ρ appearing in the commutative diagram

$$\begin{array}{ccccccc} 0 \longrightarrow & \langle\mathbf{LH}(V_p)\rangle & \longrightarrow & K\langle X\rangle & \longrightarrow & K\langle X\rangle/\langle\mathbf{LH}(V_p)\rangle \\ & & & \phi\downarrow & \swarrow\rho & \\ & & & G^{\mathbb{N}}(A) & & \end{array}$$

is an isomorphism, where, $G^{\mathbb{N}}(A)$ is the associated $\mathbb{N}$-graded algebra of A with respect to the natural $\mathbb{N}$-filtration FA, or equivalently, V_p is an $\mathbb{N}$-standard basis for the ideal $\langle V_p\rangle$. Note that all elements in $\mathbf{LH}(V_p)$ are $\mathbb{N}$-homogeneous elements of degree p. For $p = 2$, A. Braverman and D. Gaitsgory [BG1] studied this isomorphism problem posed by Joseph Bernstein. Applying graded deformations to both the graded hochschild cohomology and Koszul algebras, they obtained a PBW theorem as follows.

Theorem A ([BG1], Theorem 0.5). *Suppose that V_2 satisfies*
(I) $V_2 \cap F_1K\langle X\rangle = \{0\}$ *and*
(J) $(F_1K\langle X\rangle \cdot V_2 \cdot F_1K\langle X\rangle) \cap F_2K\langle X\rangle = V_2$.
If the quadratic algebra $\overline{A} = K\langle X\rangle/\langle\mathbf{LH}(V_2)\rangle$ is classical 2-Koszul, then ρ is an isomorphism.

□

Generally for $p \geq 2$, the p-type PBW property was studied in [FV] and [BG2] respectively. In [FV], G. Floystad and J. E. Vatne dealt with the p-type PBW property for deformations of p-Koszul K-algebras and obtained the following p-type PBW theorem.

Theorem B ([FV], Theorem 4.1). *Suppose that the* $\mathbb{N}$*-graded algebra* $\overline{A} = K\langle X\rangle/\langle \mathbf{LH}(V_p)\rangle$ *is a* p*-Koszul algebra in the sense of* [Ber]. *Then* ρ *is an isomorphism if and only if*

$$\langle V_p\rangle \cap F_pK\langle X\rangle = V_p.$$

□

While R. Berger and V. Ginzburg dealt with the p-type PBW property in [BG2] for ungraded quotients of the tensor algebra over a von Neumann regular ring K, and a p-type PBW theorem was obtained as well.

Theorem C ([BG2], Theorem 3.4). *Suppose that* V_p *satisfies*
(a) $V_p \cap F_{p-1}K\langle X\rangle = \{0\}$ *and*
(b) $(V_p \cdot K\langle X\rangle_1 + K\langle X\rangle_1 \cdot V_p) \cap F_pK\langle X\rangle = V_p$.
If the graded left K*-module* $Tor_3^A(K_A,\ {}_AK)$ *is concentrated in degree* $p+1$, *then* ρ *is an isomorphism.*

□

As a consequence of Theorem C, an extension of the p-Koszulity [Ber] for $\mathbb{N}$-graded algebras to ungraded algebras was realized through the p-type PBW property in [BG2].

Note that we have stated Theorems A–C above in the language we used in our context. For instance, in [FV], the algebra $\overline{A}$ is a given p-Koszul algebra defined by $\mathbb{N}$-homogeneous elements of degree p, and the algebra A is a deformation of $\overline{A}$ such that its associated $\mathbb{N}$-graded algebra with respect to the natural $\mathbb{N}$-filtration FA is exactly $\overline{A}$ (see the definition below).

PBW-deformation

Let the free K-algebra $K\langle X\rangle = K\langle X_1, \ldots, X_n\rangle$ of n generators be equipped with the natural $\mathbb{N}$-gradation, $G = \{g_1, \ldots, g_s\}$ a finite set of $\mathbb{N}$-homogeneous elements of $K\langle X\rangle$, and $R = K\langle X\rangle/\langle G\rangle$ the $\mathbb{N}$-graded quotient algebra defined by the graded ideal $\langle G\rangle$. Recall from the literature (e.g., [CS2]. [FV]) that if $f_1, \ldots, f_s \in K\langle X\rangle$ satisfy $d(f_i) < d(g_i)$, $1 \le i \le s$, where $d(\ )$ denotes the degree function on elements of $K\langle X\rangle$ with respect to the natural $\mathbb{N}$-gradation, and if $I = \langle \mathcal{G}\rangle$ with $\mathcal{G} = \{g_i + f_i \mid 1 \le i \le s\}$, then the quotient algebra $A = K\langle X\rangle/I$ is called a *deformation* of R. Furthermore, considering $G^{\mathbb{N}}(A)$, the associated $\mathbb{N}$-graded algebra of A with respect to the natural $\mathbb{N}$-filtration FA (which is induced by the natural $\mathbb{N}$-grading filtration $FK\langle X\rangle$), if there is an $\mathbb{N}$-graded K-algebra isomorphism $G^{\mathbb{N}}(A) \cong R$, then A is said to be a *PBW-deformation* of R. Taking the $\mathbb{N}$-leading homogeneous elements into account, it is clear that $G = \mathbf{LH}(\mathcal{G}) \subset \mathbf{LH}(I)$, which yields the following commutative diagram of $\mathbb{N}$-graded algebra homomorphisms (as described before Theorem A):

$$\begin{array}{ccccc} 0 \longrightarrow \langle \mathbf{LH}(\mathcal{G})\rangle & \longrightarrow & K\langle X\rangle & \longrightarrow & K\langle X\rangle/\langle \mathbf{LH}(\mathcal{G})\rangle \\ & & \phi\big\downarrow & \swarrow\rho & \\ & & G^{\mathbb{N}}(A) & & \end{array}$$

So, it follows from Theorem 1.3 that if $\mathcal{G}$ is an $\mathbb{N}$-standard basis, then ρ is an isomorphism, or equivalently, A is a PBW-deformation of R.

In (Ch.7, Section 4), we will see that the study of the deformation A of R is closely related to the study of the so-called regular central extension B of R, and that B may be realized as the usual Rees algebra $\widetilde{A}$ of A in case $\mathcal{G}$ is a Gröbner basis.

We finish this section by a proposition which is a modification of the equivalent characterization of a standard basis in [Gol]. One may see that this proposition is a more general version of previous (Ch.3, Theorem 3.4(i)). It is this result that enables us to gain insights into the possibility of realizing the PBW isomorphism of different types by means of Gröbner bases.

Proposition 1.4. *Let $R = \oplus_{\gamma\in\Gamma}R_\gamma$, I, and $A = R/I$ be as fixed at the beginning of this section, and let F be a subset of the ideal I. The following two statements hold.*

(i) *Suppose that F is a generating set of I having the property that each nonzero $f \in I$ has a finite representation*

$$f = \sum_{i,j} v_{ij} f_j w_{ij},$$

where the v_{ij}, w_{ij} are Γ-homogeneous elements, $f_j \in F$, satisfying $v_{ij}\mathbf{LH}(f_j)w_{ij} \neq 0$, $d(v_{ij}\mathbf{LH}(f_j)w_{ij}) \preceq d(f)$, $d(\ )$ denotes the degree function on elements of R (Ch.2, Section 3). Then $\langle \mathbf{LH}(I)\rangle = \langle \mathbf{LH}(F)\rangle$, i.e., F is a Γ-standard basis for I.

(ii) *If $\prec$ is a well-ordering on Γ and $\langle \mathbf{LH}(I)\rangle = \langle \mathbf{LH}(F)\rangle$, then F is a Γ-standard basis of I having the property mentioned in* (i) *above.*

Proof. (i). By the definition, if $f \in R$, $f \neq 0$ and $\mathbf{LH}(f) \in R_\gamma$, then $d(f) = d(\mathbf{LH}(f)) = \gamma$. Since Γ is a totally ordered semigroup with the total ordering $\prec$, by the assumption on the representation of f, the Γ-leading homogeneous element $\mathbf{LH}(f)$ of f must have the form $\mathbf{LH}(f) = \sum v'_{ij}\mathbf{LH}(f'_j)w'_{ij}$ with $f'_j \in F$, i.e., $\mathbf{LH}(f) \in \langle \mathbf{LH}(F)\rangle$. Hence $\langle \mathbf{LH}(I)\rangle = \langle \mathbf{LH}(F)\rangle$.

(ii). For $f \in I$, $f \neq 0$, suppose $\mathbf{LH}(f) \in R_\gamma$. By the assumption, the Γ-leading homogeneous element $\mathbf{LH}(f)$ can be written as

$$\mathbf{LH}(f) = \sum_{i,j} v_{ij}\mathbf{LH}(f_j)w_{ij},$$

in which v_{ij}, w_{ij} are Γ-homogeneous elements of R and $f_j \in F$, satisfying $v_{ij}\mathbf{LH}(f_j)w_{ij} \neq 0$ and $d(v_{ij}\mathbf{LH}(f_j)w_{ij}) = d(\mathbf{LH}(f)) = d(f) = \gamma$, $j = 1, \ldots, s$. Thus, rewriting each f_j as $f_j = \mathbf{LH}(f_j) + f'_j$ such that $d(f'_j) \prec d(f)$, we have

$$\mathbf{LH}(f) = \sum_{i,j} v_{ij}f_jw_{ij} - \sum_{i,j} v_{ij}f'_jw_{ij},$$

in which $d(\sum_{ij} v_{ij}f_jw_{ij}) = d(\mathbf{LH}(f)) = d(f) = \gamma$, and $d(\sum_{i,j} v_{ij}f'_jw_{ij}) \prec \gamma$. It turns out that the element

$$f' = f - \sum_{i,j} v_{ij}f_jw_{ij} \in I$$

has $d(f') \prec d(f) = \gamma$. Let $d(f') = \gamma'$. We may repeat the same procedure for f' and get some element

$$f'' = f' - \sum_{\ell,j} v_{\ell j}f_\ell w_{\ell j} \in I$$

with $d(f'') \prec \gamma'$, where $v_{\ell j}$, $w_{\ell j}$ are Γ-homogeneous elements of R, $f_\ell \in F$, satisfying each $v_{\ell j}\mathbf{LH}(f_\ell)w_{\ell j} \neq 0$ and $d(\sum_{\ell,j} v_{\ell j}f_\ell w_{\ell j}) = \gamma'$. Since the Γ-grading filtration FR is separated by the foregoing (1), after a finite number of repetitions, such reduction procedure of decreasing degrees must stop to give us an expression $f = \sum_{i,j} v_{ij}f_jw_{ij}$ with the desired property. □

Corresponding to (Ch.2, Theorem 3.3), similar statements of Proposition 1.4 hold for a left ideal by referring to (Ch.3, Theorem 3.4(ii)).

Proposition 1.5. *Let F be a subset of a left ideal L of R. The following two statements hold.*

(i) *Suppose that F is a generating set of L having the property that each nonzero $f \in L$ has a finite representation*

$$f = \sum_{i,j} v_{ij} f_j,$$

where the v_{ij} are Γ-homogeneous elements, $f_j \in F$, satisfying $v_{ij}\mathbf{LH}(f_j) \neq 0$ and $d(v_{ij}\mathbf{LH}(f_j)) \preceq d(f)$. Then $\langle \mathbf{LH}(L)] = \langle \mathbf{LH}(F)]$.

(ii) *In the case that $\prec$ is a well-ordering on Γ, if $\langle \mathbf{LH}(L)] = \langle \mathbf{LH}(F)]$, then F is a generating set of L having the property mentioned in* (i) *above.* □

4.2 Realizing Γ-PBW Isomorphism by Gröbner Basis

By means of Gröbner basis, in this section we realize the PBW isomorphism (as described in Section 1) with respect to different filtered-graded structures.

Throughout this section R denotes a K-algebra with an admissible system $(\mathcal{B}, \prec)$ in which $\mathcal{B}$ is a *skew multiplicative K-basis* of R (in the sense of Ch.3, Section 4) and $\prec$ is a monomial ordering on $\mathcal{B}$. Hence R has a Gröbner basis theory in the sense of (Ch.3, Definition 2.3).

Realization of the $\mathcal{B}$-PBW isomorphism

Let R and $(\mathcal{B}, \prec)$ be as fixed above. In this part we assume that R *does not have divisors of zero* (thus, for instance path algebras and exterior algebras are excluded).

Since $\mathcal{B}$ is a skew multiplicative K-basis, R is turned into a $\mathcal{B}$-graded algebra (though $\mathcal{B}$ is generally not a semigroup), namely, $R = \oplus_{u\in\mathcal{B}} R_u$ with $R_u = Ku$ satisfying $R_u R_v \subseteq R_{uv}$. In this case, for each $f \in R$ we see that the $\mathcal{B}$-leading homogeneous element $\mathbf{LH}(f)$ of f defined in (Ch.2, Section 3) is the same as the leading term $\mathbf{LT}(f)$ of f defined in (Ch.3, Section 1), that is,

$$\mathbf{LH}(f) = \mathbf{LT}(f) = \mathbf{LC}(f)\mathbf{LM}(f).$$

Thus, for an ideal I of R we have $\langle \mathbf{LH}(I)\rangle = \langle \mathbf{LM}(I)\rangle$. It turns out that the quotient algebra $A = R/I$ has its $\mathcal{B}$-leading homogeneous algebra

$$A^{\mathcal{B}}_{\mathrm{LH}} = R/\langle \mathbf{LH}(I)\rangle = R/\langle \mathbf{LM}(I)\rangle.$$

Remark The algebra $R/\langle \mathbf{LM}(I)\rangle$ is usually called the associated monomial algebra of A in the literature (e.g., see [An2], [G-IL], [G-I2], [Gr2]). To avoid confusing notions in the general Γ-context of this book, we still use $A^{\mathcal{B}}_{\rm LH}$ to denote $R/\langle \mathbf{LM}(I)\rangle$ and call it the $\mathcal{B}$-leading homogeneous algebra of A as before, for, here we have $\Gamma = \mathcal{B}$.

Furthermore, since R has no divisors of zero, it has the $\mathcal{B}$-grading filtration $FR = \{F_uR\}_{u\in\mathcal{B}}$ with $F_uR = \oplus_{v\preceq u}R_v$, and hence the algebra $A = R/I$ has the induced $\mathcal{B}$-filtration FA. It follows that we may talk about the *$\mathcal{B}$-PBW isomorphism* of A determined by a $\mathcal{B}$-standard basis of I as described in Definition 1.2.

Theorem 2.1. *Let I be an ideal of R and $A = R/I$. With notation as fixed above, consider the $\mathcal{B}$-filtration FA of A induced by FR and its associated $\mathcal{B}$-graded algebra $G^{\mathcal{B}}(A)$. For a subset $\mathcal{G}$ of the ideal I, the following two statements are equivalent.*
(i) *$\mathcal{G}$ is a Gröbner basis for I with respect to $(\mathcal{B},\prec)$.*
(ii) *$\mathcal{G}$ is a $\mathcal{B}$-standard basis of I and consequently $\mathcal{G}$ determines the $\mathcal{B}$-PBW isomorphism*

$$A^{\mathcal{B}}_{\rm LH} = R/\langle \mathbf{LM}(I)\rangle = R/\langle \mathbf{LM}(\mathcal{G})\rangle \xrightarrow[\cong]{\rho} G^{\mathcal{B}}(A).$$

Proof. By the discussion on the relation between $\mathcal{B}$-leading homogeneous elements and $\prec$-leading monomials made above, a Gröbner basis in R is obviously a $\mathcal{B}$-standard basis in R. Also note that the proof of (Ch.2, Theorem 3.2) works completely for $\Gamma = \mathcal{B}$, which is the skew multiplicative K-basis of R. So, the equivalence mentioned in the theorem is clear now. □

Realization of the Γ-PBW isomorphism

In this part we assume that R is a Γ-graded K-algebra, i.e., $R = \oplus_{\gamma\in\Gamma}R_\gamma$, where Γ is a totally ordered semigroup with the total ordering $<$, and that $(\mathcal{B},\prec_{gr})$ is an admissible system of R in which $\mathcal{B}$ is a skew multiplicative K-basis of R *consisting of Γ- homogeneous elements*, and $\prec_{gr}$ is a *Γ-graded monomial ordering* on R as defined in (Ch.3, Section 1), that is, there is a well-ordering $\prec$ on $\mathcal{B}$ such that the new ordering $\prec_{gr}$ defined for u, $v\in\mathcal{B}$ by the rule

$$\begin{aligned} u \prec_{gr} v \Leftrightarrow\ & d(u) < d(v) \\ & \text{or } d(u) = d(v) \text{ and } u \prec v \end{aligned}$$

is a monomial ordering on $\mathcal{B}$, where $d(\ )$ is the degree function on elements of R with respect to the Γ-gradation. Moreover, noticing that in the last part the condition that R has no divisors of zero is only used to guarantee the existence of $\mathcal{B}$-grading filtration, now in the Γ-graded case it is clear that the Γ-grading filtration FR does exist, and so we may allow R to *have divisors of zero* (thus, for instance path algebras and exterior algebras are included).

Let I be an ideal of R and $A = R/I$. Considering the Γ-filtration FA of A induced by the Γ-grading filtration FR of R, our aim is to show that if $\mathcal{G}$ is a Gröbner basis of I with respect to $(\mathcal{B}, \prec_{gr})$, then $\mathcal{G}$ is a Γ-standard basis of I and consequently $\mathcal{G}$ determines a Γ-PBW isomorphism of A in the sense of Definition 1.2. To see this, we first prove a useful result concerning the relation between a Gröbner basis $\mathcal{G}$ and the associated set of Γ-leading homogeneous elements $\mathbf{LH}(\mathcal{G})$ in R (see Ch.2, Section 3 for the definition).

Proposition 2.2. *Let I be an ideal of R and $\mathbf{LH}(I)$ the set of Γ-leading homogeneous elements of I. Put $J = \langle \mathbf{LH}(I)\rangle$. The following two statements hold.*

(i) $\mathbf{LM}(J) = \mathbf{LM}(I)$.

(ii) *A subset $\mathcal{G}$ of I is a Gröbner basis for I with respect to $(\mathcal{B}, \prec_{gr})$ if and only if $\mathbf{LH}(\mathcal{G})$, the set of Γ-leading homogeneous elements of $\mathcal{G}$, is a Gröbner basis for the Γ-graded ideal J with respect to $(\mathcal{B}, \prec_{gr})$.*

Proof. (i). First, note that $\prec_{gr}$ is a graded monomial ordering on $\mathcal{B}$ and every element of $\mathcal{B}$ is a Γ-homogeneous element. For $f \in R$, it follows from a Γ-analogue of (Ch.3, Lemma 6.2(i)) that

$$(*) \qquad \mathbf{LM}(f) = \mathbf{LM}(\mathbf{LH}(f)),$$

and this turns out $\mathbf{LM}(I) = \mathbf{LM}(\mathbf{LH}(I))$. Hence $\mathbf{LM}(I) \subset \mathbf{LM}(J)$. It remains to prove the inverse inclusion. Since J is a Γ-graded ideal of R, noticing the formula $(*)$ above we need only to consider the $\mathcal{B}$-leading monomials of Γ-homogeneous elements. Let $F \in J$ be a Γ-homogeneous element of degree γ. Then $F = \sum_{i,j} G_{ij}\mathbf{LH}(f_i)H_{ij}$, where G_{ij}, H_{ij} are Γ-homogeneous elements of R and $f_i \in I$, such that $d(G_{ij})d(f_i)d(H_{ij}) = \gamma$ whenever $G_{ij}\mathbf{LH}(f_i)H_{ij} \neq 0$. Write $f_i = \mathbf{LH}(f_i) + f_i'$ such that $d(f_i') < d(f_i)$. Then

$$\sum_{i,j} G_{ij}f_iH_{ij} = F + \sum_{i,j} G_{ij}f_i'H_{ij},$$

in which $d(\sum_{i,j} G_{ij}f_i'H_{ij}) < \gamma = d(F)$. Hence

$$\mathbf{LM}(F) = \mathbf{LM}\left(\sum_{i,j} G_{ij}f_iH_{ij}\right) \in \mathbf{LM}(I).$$

This shows that $\mathbf{LM}(J) \subset \mathbf{LM}(I)$, and consequently, the desired equality follows.

(ii). Note that the formula $(*)$ given in the proof of (i) yields $\langle \mathbf{LM}(\mathcal{G})\rangle = \langle \mathbf{LM}(\mathbf{LH}(\mathcal{G}))\rangle$. By the equality obtained in (i) we have

$$\langle \mathbf{LM}(I)\rangle = \langle \mathbf{LM}(\mathcal{G})\rangle \text{ if and only if } \langle \mathbf{LM}(J)\rangle = \langle \mathbf{LM}(\mathbf{LH}(\mathcal{G}))\rangle.$$

It follows that $\mathcal{G}$ is a Gröbner basis of I with respect to $(\mathcal{B}, \prec_{gr})$ if and only if $\mathbf{LH}(\mathcal{G})$ is a Gröbner basis for the Γ-graded ideal J with respect to $(\mathcal{B}, \prec_{gr})$, as desired. □

Theorem 2.3. *Let I and $A = R/I$ be as fixed above, and let $G^{\Gamma}(A)$ be the associated Γ-graded algebra of A determined by the Γ-filtration FA. Then any Gröbner basis $\mathcal{G}$ of I with respect to $(\mathcal{B}, \prec_{gr})$ is a Γ-standard basis of I, and thereby $\mathcal{G}$ determines the Γ-PBW isomorphism*

$$A_{\mathrm{LH}}^{\Gamma} = R/\langle \mathbf{LH}(I)\rangle = R/\langle \mathbf{LH}(\mathcal{G})\rangle \underset{\cong}{\xrightarrow{\rho}} G^{\Gamma}(A).$$

Proof. First note that by (Ch.2, Theorem 3.2), $A_{\mathrm{LH}}^{\Gamma} = R/\langle \mathbf{LH}(I)\rangle \cong G^{\Gamma}(A)$. If $\mathcal{G}$ is a Gröbner basis of I, then by the above Proposition 2.2, the set $\mathbf{LH}(\mathcal{G})$ of Γ-leading homogeneous elements of $\mathcal{G}$ is a Gröbner basis for the ideal $\langle \mathbf{LH}(I)\rangle$. It follows that $\langle \mathbf{LH}(I)\rangle = \langle \mathbf{LH}(\mathcal{G})\rangle$. Therefore, $\mathbf{LH}(\mathcal{G})$ is a Γ-standard basis by Definition 1.1. This proves the theorem. □

It is a good exercise to check directly that if $\mathcal{G}$ is a Gröbner basis of I with respect to the fixed data $(\mathcal{B}, \prec_{gr})$, then $\mathcal{G}$ satisfies the condition of Proposition 1.4(i), and hence $\mathcal{G}$ is a Γ-standard basis for I.

Realization of the $\mathbb{N}$-PBW isomorphism

The foregoing Theorem 2.3 applies immediately to quotient algebras of $\mathbb{N}$-graded algebras (with weight $\mathbb{N}$-gradation), including free algebras, commutative polynomial algebras, algebras of type $\mathcal{O}_n(\lambda_{ji})$ (see Ch.1, Section 1), path algebras and exterior algebras, etc., in particular, the classical

PBW isomorphism concerning enveloping algebras of Lie algebras, the p-type PBW property, and the PBW-deformation of a finitely presented $\mathbb{N}$-graded algebra (see Section 1) may be realized by means of Gröbner bases in free algebras.

Theorem 2.4. *Let $R = \oplus_{p\in\mathbb{N}} R_p$ be an $\mathbb{N}$-graded K-algebra, I an ideal of R and $A = R/I$. If $(\mathcal{B}, \prec_{gr})$ is an admissible system of R, where $\mathcal{B}$ is a skew multiplicative K-basis of R consisting of $\mathbb{N}$-homogeneous elements and $\prec_{gr}$ is an $\mathbb{N}$-graded monomial ordering on $\mathcal{B}$, then any Gröbner basis of I with respect to $(\mathcal{B}, \prec_{gr})$ is an $\mathbb{N}$-standard basis of I, and thereby $\mathcal{G}$ determines the $\mathbb{N}$-PBW isomorphism*

$$A^{\mathbb{N}}_{\mathrm{LH}} = R/\langle \mathbf{LH}(I)\rangle = R/\langle \mathbf{LH}(\mathcal{G})\rangle \xrightarrow[\cong]{\rho} G^{\mathbb{N}}(A),$$

where $G^{\mathbb{N}}(A)$ is the associate $\mathbb{N}$-graded K-algebra of A with respect to the $\mathbb{N}$-filtration FA induced by the $\mathbb{N}$-grading filtration FR of R. □

In Example 1 (or more generally Example 3) of the next section we will see that the defining relations of the enveloping algebra $U(\mathsf{g})$ of a Lie algebra g form a Gröbner basis. Thereby Theorem 2.4 may be applied to recapture the classical PBW isomorphism theorem. We revisit below the class of Clifford algebras and the class of algebras similar to the down-up algebra provided by Example 8 of (Ch.3, Section 5).

Example 1. Let $\mathsf{C} = K[T]$ be the Clifford algebra generated by $T = \{x_i\}_{i\in J}$ over the field K, where J is ordered by a well-ordering $<$. Then, by referring to Example 7 of (Ch.3, Section 5), Theorem 2.4 entails that with respect to the natural $\mathbb{N}$-filtration $F\mathsf{C}$ of C we have

$$K\langle X\rangle/\langle S\rangle \xrightarrow[\cong]{\rho} G^{\mathbb{N}}(\mathsf{C}),$$

where $K\langle X\rangle$ is the free K-algebra generated by $X = \{X_i\}_{i\in J}$ and S consists of

$$\begin{aligned} \mathbf{LH}(g_i) &= X_i^2, & i \in J,\\ \mathbf{LH}(g_{k\ell}) &= X_kX_\ell + X_\ell X_k, & k,\ell \in J,\ k > \ell. \end{aligned}$$

It is clear that in the above formula the algebra $K\langle X\rangle/\langle S\rangle$ is nothing but the exterior algebra E generated by $\{x_i\}_{i\in J}$. Hence $G^{\mathbb{N}}(\mathsf{C}) \cong \mathsf{E}$, as we claimed in Example (3) of (Ch.2, Section 3).

Example 2. Let $K\langle X\rangle = K\langle X_1, X_2\rangle$ be the free K-algebra generated by $X = \{X_1, X_2\}$, and let $A = K\langle X\rangle/I$ be the quotient algebra as defined in

Example 8 of (Ch.3, Section 5). If in the defining relations of A the F_1, F_2 have degree less than or equal to 2 (with respect to the natural $\mathbb{N}$-gradation of $K\langle X\rangle$), then with respect to the natural $\mathbb{N}$-filtration FA of A we have

$$R/\langle S\rangle \underset{\cong}{\xrightarrow{\rho}} G^{\mathbb{N}}(A),$$

where S consists of

$$\begin{array}{l}\mathbf{LH}(g_1) = X_1^2X_2 - \alpha X_1X_2X_1 - \beta X_2X_1^2 \\ \mathbf{LH}(g_2) = X_1X_2^2 - \alpha X_2X_1X_2 - \beta X_2^2X_1\end{array} \quad \alpha, \beta \in K.$$

In particular, this is true for every down-up algebra $A = A(\alpha, \beta, \gamma)$ in the sense of ([Ben], [BR]).

More algebras with nice $G^{\mathbb{N}}(A)$ defined by quadratic Gröbner bases may be seen in the next two sections.

4.3 Classical PBW K-Bases vs Gröbner Bases

Let $A = K[a_1, \ldots, a_n]$ be a finitely generated K-algebra with generators $a_1, \ldots, a_n$. If, with respect to a fixed permutation $i_1i_2\cdots i_n$ of the sequence $12\cdots n$, the set $\mathscr{B} = \{a_{i_1}^{\alpha_1}a_{i_2}^{\alpha_2}\cdots a_{i_n}^{\alpha_n} \mid \alpha_j \in \mathbb{N}\}$ forms a K-basis of A, then, in honor of the classical PBW (Poincaré-Birkhoff-Witt) theorem for enveloping algebras of Lie algebras over a ground field K, the set $\mathscr{B}$ is usually referred to as a (classical) *PBW K-basis* of A. Presenting A as a quotient algebra of the free K-algebra $K\langle X\rangle = K\langle X_1, \ldots, X_n\rangle$, i.e., $A = K\langle X\rangle/I$ with I an ideal of $K\langle X\rangle$, we know from the last section that the $\mathbb{N}$-PBW isomorphism problem for A can be solved by means of a Gröbner basis of I. So the next work on the agenda of this chapter is to characterize a PBW K-basis of A in terms of Gröbner basis of I. Indeed, the main result (Theorem 3.1) obtained in this section tells us more, namely, on one hand it enables us to obtain PBW K-bases by means of specific Gröbner bases; and on the other hand, since it is well known that in practice there are different ways to find a PBW K-basis for a given algebra (if it exists), this result also enables us to obtain Gröbner bases via already known PBW K-bases.

By the construction of the free algebra $K\langle X\rangle = K\langle X_1, \ldots, X_n\rangle$ and the definition of a PBW K-basis, it is clear that in our discussion on PBW K-bases we can always assume that $i_1 = 1, i_2 = 2, \ldots, i_n = n$.

Classical PBW K-bases vs Gröbner bases

Let $(\mathcal{B}, \prec)$ be an admissible system of $K\langle X\rangle$, where $\mathcal{B}$ is the standard K-basis of $K\langle X\rangle$ and $\prec$ is a monomial ordering on $K\langle X\rangle$. Consider an ideal I of $K\langle X\rangle$ and the quotient algebra $A = K\langle X\rangle/I$. Suppose that A has the PBW R-basis $\mathscr{B} = \left\{\overline{X}_1^{\alpha_1}\overline{X}_2^{\alpha_2}\cdots\overline{X}_n^{\alpha_n} \mid \alpha_i \in \mathbb{N}\right\}$, where each $\overline{X}_i$ is the canonical image of X_i in A. Then I necessarily contains a subset G consisting of the $\frac{n(n-1)}{2}$ elements of the form

$$g_{ji} = X_jX_i - \sum_{\alpha} \lambda_\alpha w_\alpha,$$

where $1 \le i < j \le n$, $\lambda_\alpha \in K$, $w_\alpha = X_1^{\alpha_1}X_2^{\alpha_2}\cdots X_n^{\alpha_n}$. In view of (Ch.3, Theorem 3.3, Proposition 4.8(ii)), this observation inspires the following characterization of a PBW K-basis in terms of Gröbner basis of specific type.

Theorem 3.1. *Let I be an ideal of $K\langle X\rangle$, $A = K\langle X\rangle/I$. Suppose that I contains a subset of $\frac{n(n-1)}{2}$ elements $G = \{g_{ji} \mid 1 \le i < j \le n\}$ such that with respect to some monomial ordering $\prec$ on the standard K-basis $\mathcal{B}$ of $K\langle X\rangle$, $\mathbf{LM}(g_{ji}) = X_jX_i$ for all $1 \le i < j \le n$. The following two statements are equivalent:*
(i) *The K-algebra A has the PBW K-basis $\mathscr{B} = \{\overline{X}_1^{\alpha_1}\overline{X}_2^{\alpha_2}\cdots\overline{X}_n^{\alpha_n} \mid \alpha_j \in \mathbb{N}\}$ where each $\overline{X}_i$ is the canonical image of X_i in A;*
(ii) *Any subset $\mathcal{G}$ of I containing G is a Gröbner basis for I with respect to the data $(\mathcal{B}, \prec)$.*

Proof. (i) $\Rightarrow$ (ii). Let $\mathcal{G}$ be a subset of I containing G, and let

$$N(\mathcal{G}) = \{u \in \mathcal{B} \mid \mathbf{LM}(g) \nmid u,\ g \in \mathcal{G}\}.$$

If $f \in I$ and $f \ne 0$, then, after implementing the division of f by $\mathcal{G}$ (with respect to the given monomial ordering $\prec$) we have

$$f = \sum_{i,j} \lambda_{ij}u_{ij}g_iv_{ij} + r_f,$$

where $\lambda_{ij} \in K$, $u_{ij}, v_{ij} \in \mathcal{B}$, $g_i \in \mathcal{G}$, satisfying $\mathbf{LM}(u_{ij}g_iv_{ij}) \preceq \mathbf{LM}(f)$ whenever $\lambda_{ij} \ne 0$, and either $r_f = 0$ or $r_f = \sum_p \lambda_pw_p$ with $\lambda_p \in K$ and $w_p \in N(\mathcal{G})$. Note that $g_{ji} \in G \subseteq \mathcal{G}$ and $\mathbf{LM}(g_{ji}) = X_jX_i$ by the assumption. It follows that $N(\mathcal{G}) \subseteq \{X_1^{\alpha_1}X_2^{\alpha_2}\cdots X_n^{\alpha_n} \mid \alpha_j \in \mathbb{N}\}$. Thus, since $\mathscr{B}$ is a K-basis of A, $r_f = \sum_p \lambda_pw_p = f - \sum_{i,j} u_{ij}g_iv_{ij} \in I$ implies $\lambda_p = 0$ for all p. Consequently $r_f = 0$. This shows that every nonzero element of I has a Gröbner representation by the elements of $\mathcal{G}$. Hence $\mathcal{G}$ is a Gröbner basis for I by (Ch.3, Theorem 3.4).

(ii) $\Rightarrow$ (i). By (ii), the subset G itself is a Gröbner basis of I with respect to $\prec$. Noticing that $\mathbf{LM}(g_{ji}) = X_jX_i$, $1 \le i < j \le n$, it follows from (Ch.3, Proposition 4.8(ii)) that the set $N(I)$ of normal monomials in $\mathcal{B}$ (mod I) is given by

$$N(I) = \mathcal{B} - \langle \mathbf{LM}(G) \rangle = \left\{ X_1^{\alpha_1} X_2^{\alpha_2} \cdots X_n^{\alpha_n} \;\middle|\; \alpha_j \in \mathbb{N} \right\}.$$

Hence the algebra A has the desired PBW R-basis $\mathscr{B}$ by (Ch.3, Theorem 3.3). □

Note that Theorem 3.1 generalizes ([Gr3], Proposition 2.14) and ([Li1] Ch.3, Theorem 1.5). We illustrate Theorem 3.1 by several examples. The first four examples given below serve to obtain Gröbner bases by means of already known PBW K-bases which are obtained in the literature without using the theory of Gröbner basis.

Example 1. Let $\mathbf{g} = K[V]$ be the K-Lie algebra defined by the K-space $V = \oplus_{i=1}^n Kx_i$ and the bracket product $[x_j, x_i] = \sum_{\ell=1}^n \lambda_{ji}^\ell x_\ell$, $1 \le i < j \le n$, $\lambda_{ji}^\ell \in K$. By the classical PBW theorem (e.g., [Jac], [Hu]), the universal enveloping algebra $U(\mathbf{g})$ of $\mathbf{g}$ has the PBW K-basis $\mathscr{B} = \{x_1^{\alpha_1} x_2^{\alpha_2} \cdots x_n^{\alpha_n} \mid \alpha_j \in \mathbb{N}\}$. If we use the $\mathbb{N}$-graded lexicographic ordering $X_1 \prec_{gr} X_2 \prec_{gr} \cdots \prec_{gr} X_n$ on the standard K-basis $\mathcal{B}$ of $K\langle X \rangle = K\langle X_1, \ldots, X_n \rangle$ with respect to the natural $\mathbb{N}$-gradation of $K\langle X \rangle$ (i.e., $\deg X_i = 1$, $1 \le i \le n$), then the set of defining relations

$$\mathcal{G} = \left\{ g_{ji} = X_jX_i - X_iX_j - \sum_{\ell=1}^n \lambda_{ji}^\ell X_\ell \;\middle|\; 1 \le i < j \le n \right\}$$

of $U(\mathbf{g})$ satisfies $\mathbf{LM}(g_{ji}) = X_jX_i$ for all $1 \le i < j \le n$. Hence, by Theorem 3.1, $\mathcal{G}$ is a Gröbner basis for the ideal $I = \langle \mathcal{G} \rangle$ in $K\langle X \rangle$. Moreover, by Theorem 2.4, with respect to the natural $\mathbb{N}$-filtration induced by the natural $\mathbb{N}$-grading filtration of $K\langle X \rangle$, $U(\mathbf{g})$ has the associated graded algebra $G^{\mathbb{N}}(U(\mathbf{g})) \cong K\langle X \rangle / \langle \mathbf{LH}(\mathcal{G}) \rangle$ with

$$\mathbf{LH}(\mathcal{G}) = \{ \mathbf{LH}(g_{ji}) = X_jX_i - X_iX_j \mid 1 \le i < j \le n \},$$

which is clearly the commutative polynomial algebra in n variables.

Example 2. Let $U_q^+(A_N)$ be the (+)-part of the Drinfeld-Jimbo quantum group of type A_N over a field K. In [Ros] and [Yam] it was proved that

$U_q^+(A_N)$ has a PBW K-basis with respect to the defining relations (Jimbo relations) of $U_q^+(A_N)$; later in [BM] such a PBW basis was recaptured by showing that the Jimbo relations form a Gröbner basis ([BM] Theorem 4.1), where the proof was sketched to check that all compositions (overlaps) of Jimbo relations reduce to zero on the base argument of [Yam]. Furthermore, a very detailed elementary verification of the fact that all compositions (overlaps) of Jimbo relations reduce to zero and hence the Jimbo relations form a Gröbner basis (namely Theorem 4.1 of [BM]) was carried out by [CSS]. Now, by using Theorem 3.1 we will see that it is very easy to conclude: the Jimbo relations form a Gröbner basis.

Recall that the Jimbo relations (as described in [Yam]) are given by

$$\begin{array}{ll} x_{mn}x_{ij} - q^{-2}x_{ij}x_{mn}, & ((i,j),(m,n)) \in C_1 \cup C_3, \\ x_{mn}x_{ij} - x_{ij}x_{mn}, & ((i,j),(m,n)) \in C_2 \cup C_6, \\ x_{mn}x_{ij} - x_{ij}x_{mn} + (q^2 - q^{-2})x_{in}x_{mj}, & ((i,j),(m,n)) \in C_4, \\ x_{mn}x_{ij} - q^2x_{ij}x_{mn} + qx_{in}, & ((i,j),(m,n)) \in C_1 \cup C_3, \end{array}$$

where with $\Lambda_N = \{(i,j) \in \mathbb{N} \times \mathbb{N} \mid 1 \le i < j \le N+1\}$, and

$$\begin{aligned} C_1 &= \{((i,j),(m,n)) \mid i = m < j < n\}, \\ C_2 &= \{((i,j),(m,n)) \mid i < m < n < j\}, \\ C_3 &= \{((i,j),(m,n)) \mid i < m < j = n\}, \\ C_4 &= \{((i,j),(m,n)) \mid i < m < j < n\}, \\ C_5 &= \{((i,j),(m,n)) \mid i < j = m < n\}, \\ C_6 &= \{((i,j),(m,n)) \mid i < j < m < n\}. \end{aligned}$$

By [Yam], for $q^8 \ne 1$, $U_q^+(A_N)$ has the PBW basis consisting of elements

$$x_{i_1j_1}x_{i_2j_2}\cdots x_{i_kj_k} \text{ with } (i_1,j_1) \le (i_2,j_2) \le \cdots \le (i_k,j_k),\ k \ge 0,$$

where $(i_\ell, j_\ell) \in \Lambda_N$ and $<$ is the lexicographic ordering on Λ_N. If we use the $\mathbb{N}$-graded monomial ordering $\prec_{gr}$ (on the standard K-basis $\mathcal{B}$ of the corresponding free algebra with the natural $\mathbb{N}$-gradation) subject to

$$x_{ij} \prec_{gr} x_{mn} \Longleftrightarrow (i,j) < (m,n),$$

then it is clear that for each pair $((i,j),(m,n)) \in C_i$ with $(i,j) < (m,n)$, the leading monomial of the corresponding relation is of the form $x_{mn}x_{ij}$ as desired by Theorem 3.1. Moreover, by Theorem 2.4, with respect to the natural $\mathbb{N}$-filtration, $U_q^+(A_N)$ has the associated graded algebra $G^{\mathbb{N}}(U_q^+(A_N))$ which has the quadratic defining relations:

$$\begin{array}{ll} x_{mn}x_{ij} - q^{-2}x_{ij}x_{mn}, & ((i,j),(m,n)) \in C_1 \cup C_3, \\ x_{mn}x_{ij} - x_{ij}x_{mn}, & ((i,j),(m,n)) \in C_2 \cup C_6, \\ x_{mn}x_{ij} - x_{ij}x_{mn} + (q^2 - q^{-2})x_{in}x_{mj}, & ((i,j),(m,n)) \in C_4, \\ x_{mn}x_{ij} - q^2x_{ij}x_{mn}, & ((i,j),(m,n)) \in C_1 \cup C_3. \end{array}$$

Example 3. With the free K-algebra $K\langle X\rangle = K\langle X_1, \ldots, X_n\rangle$ as before, recall from [Ber1] that a q-algebra $A = K\langle X\rangle/\langle\mathcal{G}\rangle$ over K is defined by the set $\mathcal{G}$ of quadric relations

$$g_{ji} = X_jX_i - q_{ji}X_iX_j - \{X_j, X_i\},\ 1 \le i < j \le n,$$

where $q_{ji} \in K$, and

$$\{X_j, X_i\} = \sum \alpha_{ji}^{k\ell}X_kX_\ell + \sum \alpha_hX_h + c_{ji},\ \alpha_{ji}^{k\ell}, \alpha_h, c_{ji} \in K,$$

satisfying if $\alpha_{ji}^{kl} \ne 0$, then $i < k \le \ell < j$, and $k - i = j - \ell$. Consider two K-subspaces of $K\langle X\rangle$:

$$\mathcal{E}_1 = K\text{-span}\left\{g_{ji} \;\middle|\; 1 \le i < j \le n\right\},$$

$$\mathcal{E}_2 = K\text{-span}\left\{X_ig_{ji},\ g_{ji}X_i,\ X_jg_{ji},\ g_{ji}X_j \;\middle|\; 1 \le i < j \le n\right\}.$$

If, for all $1 \le i < j < k \le n$, every Jacobi sum

$$\begin{aligned} J(X_k, X_j, X_i) = &\{X_k, X_j\}X_i - \lambda_{ki}\lambda_{ji}X_i\{X_k, X_j\} \\ &-\lambda_{ji}\{X_k, X_i\}X_j + \lambda_{kj}X_j\{X_k, X_i\} \\ &+\lambda_{kj}\lambda_{ki}\{X_j, X_i\}X_k - X_k\{X_j, X_i\} \end{aligned}$$

is contained in $\mathcal{E}_1 + \mathcal{E}_2$, then A is called a *q-enveloping algebra*. Clearly, enveloping algebras of K-Lie algebras are special q-enveloping algebras with $q = 1$. In [Ber1], a q-PBW theorem for q-enveloping algebras (over a commutative ring K) was obtained along the line similar to the classical argument on enveloping algebras of Lie algebras as given in [Jac], that is, if A is a q-enveloping K-algebra then A has the PBW K-basis $\mathscr{B} = \{\overline{X}_1^{\alpha_1}\overline{X}_2^{\alpha_2}\cdots\overline{X}_n^{\alpha_n} \mid \alpha_j \in \mathbb{N}\}$.

Now, if we use the $\mathbb{N}$-graded lexicographic ordering $X_1 \prec_{gr} X_2 \prec_{gr} \cdots \prec_{gr} X_n$ on the standard K-basis $\mathcal{B}$ of $K\langle X\rangle$ with respect to the natural $\mathbb{N}$-gradation of $K\langle X\rangle$ (i.e., $\deg X_i = 1$, $1 \le i \le n$), then $\mathcal{G}$ satisfies $\mathbf{LM}(g_{ji}) = X_jX_i$ for all $1 \le i < j \le n$. Hence, by Theorem 3.1, the set $\mathcal{G}$ of the defining relations of a q-enveloping K-algebra forms a Gröbner basis for the ideal $I = \langle\mathcal{G}\rangle$ in $K\langle X\rangle$. Moreover, by Theorem 2.4, with respect to the natural $\mathbb{N}$-filtration induced by the natural $\mathbb{N}$-grading filtration of $K\langle X\rangle$, A has the associated graded algebra $G^{\mathbb{N}}(A) \cong K\langle X\rangle/\langle\mathbf{LH}(\mathcal{G})\rangle$ with

$$\mathbf{LH}(\mathcal{G}) = \left\{\mathbf{LH}(g_{ji}) = X_jX_i - q_{ji}X_iX_j - \sum \alpha_{ji}^{k\ell}X_kX_\ell \;\middle|\; 1 \le i < j \le n\right\}.$$

Example 4. This example generalizes the previous three examples but uses an ad hoc monomial ordering. As an application we show that the

PBW generators of the quantum algebra $U_q^+(A_N)$ derived in [Rin] provides another Gröbner defining set for $U_q^+(A_N)$.

With the free K-algebra $K\langle X\rangle = K\langle X_1,\dots,X_n\rangle$ as before, consider the K-algebra $A = K\langle X\rangle/\langle\mathcal{G}\rangle$ defined by the subset $\mathcal{G}$ consisting of the $\frac{n(n-1)}{2}$ elements

$$g_{ji} = X_jX_i - q_{ji}X_iX_j - \sum_{\alpha}\lambda_\alpha X_{i_1}^{\alpha_1}X_{i_2}^{\alpha_2}\cdots X_{i_s}^{\alpha_s} + \lambda_{ji},\ 1\le i<j\le n,$$

where $q_{ji},\lambda_\alpha,\lambda_{ji}\in K$, $\alpha_k\in\mathbb{N}$, $i<i_1\le i_2\le\cdots\le i_s<j$. It is well known that numerous iterated skew polynomial algebras over K are defined subject to such relations, and consequently they have the PBW K-basis $\mathscr{B} = \{\overline{X}_1^{\alpha_1}\overline{X}_2^{\alpha_2}\cdots\overline{X}_n^{\alpha_n} \mid \alpha_j\in\mathbb{N}\}$. Under the assumption that A has the PBW K-basis as described we aim to show that $\mathcal{G}$ is a Gröbner basis of $\langle\mathcal{G}\rangle$. By Theorem 3.1, it is sufficient to introduce a monomial ordering on the standard K-basis $\mathcal{B}$ of $K\langle X\rangle$ so that $\mathbf{LM}(g_{ji}) = X_jX_i$ for all $1\le i<j\le n$. To this end, let $K[\mathrm{t}] = K[t_1,\dots,t_n]$ be the commutative polynomial K-algebra of n variables. Consider the canonical algebra epimorphism π: $K\langle X\rangle\to K[\mathrm{t}]$ with $\pi(X_i)=t_i$. If we fix the lexicographic ordering $X_1 <_{lex} X_2 <_{lex}\cdots<_{lex} X_n$ on $\mathcal{B}$ (note that $<_{lex}$ is not a monomial ordering on $\mathcal{B}$) and fix an arbitrarily chosen monomial ordering $\prec$ on the standard K-basis $\mathbb{B} = \{t_1^{\alpha_1}t_2^{\alpha_2}\cdots t_n^{\alpha_n}\mid\alpha_j\in\mathbb{N}\}$ of $K[\mathrm{t}]$, respectively, then, as in [EPS], a monomial ordering $\prec_{et}$ on $\mathcal{B}$, which is called the *lexicographic extension* of the given monomial ordering $\prec$ on $\mathbb{B}$, may be obtained as follows: for $u,v\in\mathcal{B}$,

$$u\prec_{et} v \text{ if } \begin{cases}\pi(u)\prec\pi(v),\\ \text{or}\\ \pi(u)=\pi(v)\text{ and } u<_{lex} v \text{ in } \mathcal{B}_R.\end{cases}$$

In particular, with respect to the monomial ordering $\prec_{et}$ obtained by using the lexicographic ordering $t_n\prec_{lex} t_{n-1}\prec_{lex}\cdots\prec_{lex} t_1$ on $\mathbb{B}$, we see that $\mathbf{LM}(g_{ji}) = X_jX_i$ for all $1\le i<j\le n$, as desired by Theorem 3.1.

In [Rin] it was proved that $U_q^+(A_N)$ has $m = \frac{N(N+1)}{2}$ generators $x_1,\dots,x_m$ satisfying the relations:

$$x_jx_i = q^{v_{ji}}x_ix_j - r_{ji},\ 1\le i<j\le m,$$

where $v_{ji} = (wt(x_i),wt(x_j))$, and r_{ji} is a linear combination of monomials of the form $x_{i+1}^{\alpha_{i+1}}x_{i+2}^{\alpha_{i+2}}\cdots x_{j-1}^{\alpha_{j-1}}$, and that $U_q^+(A_N)$ is an iterated skew polynomial algebra generated by $x_1,\dots,x_m$ subject to the above relations. Thus $U_q^+(A_N)$ has the PBW basis $\{x_1^{\alpha_1}x_2^{\alpha_2}\cdots x_m^{\alpha_m}\mid\alpha_j\in\mathbb{N}\}$, and consequently $\mathcal{G} = \{g_{ji} = x_jx_i - q^{v_{ij}}x_ix_j - r_{ji}\mid 1\le i<j\le m\}$ forms a

Gröbner defining set of $U_q^+(A_N)$ with respect to the monomial ordering $\prec_{et}$ as described before.

Remark (i) If, in the defining relations given in the last example, the condition $i < i_1 \le i_2 \le \cdots \le i_s < j$ is replaced by $1 \le i_1 \le i_2 \le \cdots \le i_s \le i-1$, then a similar result holds.
(ii) Considering $U_q^+(A_N)$ over a commutative ring R, in [Li7] it was shown, by using an analogue of Theorem 3.1 over R, that the Jimbo relations, respectively the relations satisfied by the PBW generators (as described in Example 2 and Example 4), form a Gröbner basis over R.
(iii) In the case where K is a field, the fact that the defining relations of a q-enveloping K-algebra A form a Gröbner basis was verified in [LWZ] and ([Li1], Ch.3) directly by using the termination theorem through the division algorithm. The reader is referred to [Li7] for a general assertion that defining relations of a q-enveloping algebra over an arbitrary commutative ring R form a monic Gröbner basis, in particular, that all quantum algebras over the ring $R = \mathbb{C}[[h]]$ which are q-enveloping algebras appeared in [Ber1] are defined by monic Gröbner bases.

The next three examples provide Gröbner bases which are not necessarily the type as described in previous Examples 3 – 4, but they all give rise to PBW R-bases.

Example 5. Let I be the ideal of the free K-algebra $K\langle X\rangle = K\langle X_1, X_2\rangle$ generated by the single element

$$g_{21} = X_2X_1 - qX_1X_2 - \alpha X_2 - f(X_1),$$

where $q, \alpha \in K$, and $f(X_1)$ is a polynomial in the variable X_1. Assigning X_1 the degree 1, then in either of the following two cases:
(a) $\deg f(X_1) \le 2$, and X_2 is assigned the degree 1;
(b) $\deg f(X_1) = n \ge 3$, and X_2 is assigned the degree n,
$\mathcal{G} = \{g_{21}\}$ forms a Gröbner basis for I. For, in both cases we may use the $\mathbb{N}$-graded lexicographic ordering $X_1 \prec_{gr} X_2$ with respect to the natural $\mathbb{N}$-gradation of $K\langle X\rangle$, respectively the weight $\mathbb{N}$-gradation of $K\langle X\rangle$ with weight $\{1, n\}$, such that $\mathbf{LM}(g_{21}) = X_2X_1$, and then we see that the ordered pair (g_{21}, g_{21}) has the only overlap element $o(g_{21}, 1;\ 1, g_{21}) = 0$. By (Ch.3, Theorem 5.1), $\mathcal{G}$ is a Gröbner basis for the ideal I. Thus, by Theorem 3.1, in both cases the algebra $A = K\langle X\rangle/I$ has the PBW R-basis $\mathscr{B} = \{\overline{X}_1^{\alpha}\overline{X}_2^{\beta} \mid \alpha, \beta \in \mathbb{N}\}$. Moreover, by Theorem 2.4, with respect to both the

natural $\mathbb{N}$-filtration and the weight $\mathbb{N}$-filtration induced by the weight $\mathbb{N}$-grading filtration of $K\langle X\rangle$, A has the associated graded algebra $G^{\mathbb{N}}(A) \cong K\langle X\rangle/\langle X_2X_1 - qX_1X_2\rangle$, which, in the case that $q \neq 0$, is the coordinate ring of the quantum plane.

Example 6. Let $K\langle X\rangle = K\langle X_1, X_2, X_3\rangle$ be the free K-algebra generated by $X = \{X_1, X_2, X_3\}$. This example provides a family of algebras similar to the enveloping algebra $U(\mathsf{sl}(2,K))$ of the K-Lie algebra $\mathsf{sl}(2,K)$ (see Ch.1, Section 1), that is, we consider the algebra $A = K\langle X\rangle/\langle\mathcal{G}\rangle$ with $\mathcal{G}$ consisting of

$$\begin{aligned} g_{31} &= X_3X_1 - \lambda X_1X_3 + \gamma X_3, \\ g_{12} &= X_1X_2 - \lambda X_2X_1 + \gamma X_2, \\ g_{32} &= X_3X_2 - \omega X_2X_3 + f(X_1), \end{aligned}$$

where $\lambda, \gamma, \omega \in K$, and $f(X_1)$ is a polynomial in the variable X_1. It is clear that $A = U(\mathsf{sl}(2,K))$ provided $\lambda = \omega = 1$, $\gamma = 2$ and $f(X_1) = -X_1$.

Suppose $f(X_1)$ has degree $n \geq 1$. Then, according to $n \leq 2$ or $n > 2$, we can always equip $K\langle X\rangle$ with a weight $\mathbb{N}$-gradation by assigning X_1, X_2 and X_3 to the positive degree n_1, n_2, n_3 respectively (for instance, $(1,1,1)$ if $\deg f(X_1) = n \leq 2$; $(1,n,n)$ if $\deg f(X_1) = n > 2$), such that $\mathbf{LM}(\mathcal{G}) = \{X_3X_1,\ X_1X_2,\ X_3X_2\}$ with respect to the $\mathbb{N}$-graded lexicographic ordering $X_2 \prec_{gr} X_1 \prec_{gr} X_3$ on $\mathcal{B}_R$, and in this case only the ordered pair (g_{31}, g_{12}) has a unique overlap element

$$S_{312} = o(g_{31}, X_2;\ X_3, g_{12}) = g_{31}X_2 - X_3g_{12} = \lambda(X_3X_2X_1 - X_1X_3X_2).$$

If $\lambda = 0$, then $S_{312} = 0$. If $\lambda \neq 0$, then, starting with $\mathbf{LM}(S_{312}) = X_3X_2X_1$, the division of S_{312} by $\mathcal{G}$ yields

$$\begin{aligned} S_{312} &= \lambda g_{32}X_1 - \lambda X_1g_{32}, && \omega = 0. \\ S_{312} &= \lambda g_{32}X_1 - \lambda X_1g_{32} - \lambda\omega g_{12}X_3 + \omega\lambda X_2g_{31}, && \omega \neq 0. \end{aligned}$$

It follows that $\overline{S_{312}}^{\mathcal{G}} = 0$. This shows that $\mathcal{G}$ is a Gröbner basis for the ideal $\langle\mathcal{G}\rangle$ by (Ch.3, Theorem 5.1). Hence, by Theorem 3.1 the algebra $A = K\langle X\rangle/\langle\mathcal{G}\rangle$ has the PBW K-basis $\mathscr{B} = \{\overline{X_2}^{\alpha_2}\overline{X_1}^{\alpha_1}\overline{X_3}^{\alpha_3} \mid \alpha_j \in \mathbb{N}\}$.

Next, we observe that with different weight $\mathbb{N}$-gradation for $K\langle X\rangle$, Theorem 2.4 yields different associated graded algebra $G^{\mathbb{N}}(A)$ for the algebra $A = K\langle X\rangle/\langle\mathcal{G}\rangle$, where $G^{\mathbb{N}}(A)$ is determined by the $\mathbb{N}$-filtration FA induced by a fixed weight $\mathbb{N}$-grading filtration of $K\langle X\rangle$. For instance, in the case that $f(X_1)$ has degree $n \geq 3$ if the weight $(1,n,n)$ is used, we see that

$$\mathbf{LH}(\mathcal{G}) = \{X_3X_1 - \lambda X_1X_3,\ X_1X_2 - \lambda X_2X_1,\ X_3X_2 - \omega X_2X_3\}$$

and $G^{\mathbb{N}}(A) \cong K\langle X\rangle/\langle \mathbf{LH}(\mathcal{G})\rangle$. If $\lambda \neq 0$ and $\omega \neq 0$, then we know that $K\langle X\rangle/\langle \mathbf{LH}(\mathcal{G})\rangle$ is just the skew polynomial algebra $\mathcal{O}_3(\lambda_{ji})$ with $\lambda_{31} = \lambda_{21} = \lambda$ and $\lambda_{32} = \omega$ (Ch.1, Section 1). While in the case that $f(X_1)$ has degree ≤ 2, i.e.,

$$f(X_1) = aX_1^2 + bX_1 + c \text{ with } a, b, c \in K,$$

if the weight $(1, 1, 1)$ is used, we see that

$$\mathbf{LH}(\mathcal{G}) = \{X_3X_1 - \lambda X_1X_3,\ X_1X_2 - \lambda X_2X_1,\ X_3X_2 - \omega X_2X_3 + aX_1^2\}$$

and $G^{\mathbb{N}}(A) \cong K\langle X\rangle/\langle \mathbf{LH}(\mathcal{G})\rangle$.

Finally, in the case that $f(X_1)$ has degree ≤ 2, we list three types of popularly studied algebras covered by the above situation.

(a) Let $\zeta \in K^*$. If $\lambda = \zeta^4$, $\omega = \zeta^2$, $\gamma = -(1+\zeta^2)$, $a = 0 = c$, and $b = -\zeta$, then A coincides with S. L. Woronowicz's deformation of $U(\mathsf{sl}(2, K))$ which was introduced in the noncommutative differential calculus [Wor].

(b) If $\lambda\gamma w b \neq 0$ and $c = 0$, then A coincides with L. Le Bruyn's conformal sl_2 enveloping algebra ([Lev], Lemma 2) which provides a special family of E. Witten's deformation of $U(\mathsf{sl}(2, K))$ in quantum group theory, that is, the associated graded algebra $G^{\mathbb{N}}(A)$ of A with respect to the natural $\mathbb{N}$-filtration FA is a three dimensional Auslander regular quadratic algebra with defining relations given by

$$\begin{aligned}\mathbf{LH}(g_{31}) &= X_3X_1 - \lambda X_1X_3,\\ \mathbf{LH}(g_{12}) &= X_1X_2 - \lambda X_2X_1,\\ \mathbf{LH}(g_{32}) &= X_3X_2 - \omega X_2X_3 + aX_1^2.\end{aligned}$$

(c) Suppose that the field K is algebraically closed and set $a = 0 = c$, $b = 1$ in $f(X_1) = aX_1^2 + bX_1 + c$. From [KMP], [CM] and [CS1], if in this case λ, ω are roots of the equation $x^2 - \alpha x - \beta = 0$, then since $\lambda + \omega = \alpha$, $-\lambda\omega = \beta$, it follows that all down-up algebras $A(\alpha, \beta, \gamma)$ in the sense of ([Ben], [BR]) are recaptured by the quadric defining relations

$$\begin{aligned}g_{31} &= X_3X_1 - \lambda X_1X_3 + \gamma X_3,\\ g_{12} &= X_1X_2 - \lambda X_2X_1 + \gamma X_2,\\ g_{32} &= X_3X_2 - \omega X_2X_3 + X_1.\end{aligned}$$

At this stage the reader is invited to review the original defining relations g_1, g_2 of $A(\alpha, \beta, \gamma)$ given in Example 8 of (Ch.3, Section 5).

Example 7. Let $\mathcal{G}$ be the subset of the free K-algebra $K\langle X\rangle = K\langle X_1, X_2, X_3\rangle$ consisting of

$$\begin{aligned}g_{21} &= X_2X_1 - X_1X_2,\\ g_{31} &= X_3X_1 - \lambda X_1X_3 - \mu X_2X_3 - \gamma X_2, \lambda, \mu, \gamma \in K,\\ g_{32} &= X_3X_2 - X_2X_3.\end{aligned}$$

Then, under the $\mathbb{N}$-graded lexicographic ordering $X_1 \prec_{gr} X_2 \prec_{gr} X_3$ with respect to the natural $\mathbb{N}$-gradation of $K\langle X\rangle$, $\mathbf{LM}(g_{ji}) = X_jX_i$, $1 \le i < j \le 3$, and the only nontrivial overlap element is $S_{321} = o(g_{32}, X_1;\ X_3, g_{21}) = -X_2X_3X_1 + X_3X_1X_2$. One checks easily that $\overline{S_{321}}^{\mathcal{G}} = 0$. By (Ch.3, Theorem 5.1), $\mathcal{G}$ is a Gröbner basis for the ideal $\langle \mathcal{G}\rangle$. Hence, by Theorem 3.1 the algebra $A = K\langle X\rangle/\langle \mathcal{G}\rangle$ has the PBW K-basis $\mathscr{B} = \{\overline{X_1}^{\alpha_1}\overline{X_2}^{\alpha_2}\overline{X_3}^{\alpha_3} \mid \alpha_j \in \mathbb{N}\}$. And by Theorem 2.4, with respect to the natural filtration, A has the associated graded algebra $G^{\mathbb{N}}(A) \cong K\langle X\rangle/\langle \mathbf{LH}(\mathcal{G})\rangle$ with

$$\mathbf{LH}(\mathcal{G}) = \{X_2X_1 - X_1X_2,\ X_3X_1 - \lambda X_1X_3 - \mu X_2X_3,\ X_3X_2 - X_2X_3\}.$$

The Jacobi sum and the condition (J)

Looking at Theorem 3.1 further, generally for a subset $\mathcal{G} = \{g_{ji} \mid 1 \le i < j \le n\}$ consisting of the $\frac{n(n-1)}{2}$ elements in the free K-algebra $K\langle X\rangle = K\langle X_1, \ldots, X_n\rangle$, such that $\mathbf{LM}(g_{ji}) = X_jX_i$ with respect to some monomial ordering $\prec$ on $K\langle X\rangle$, though, as the above Example 5 – Example 7 showed, in principle we may use **Algorithm-GB** (or the termination theorem) to check whether $\mathcal{G}$ is a Gröbner basis. It is a hard job (even if a computer is employed) to implement the reduction procedure on overlap elements. However, note that practically if an $\mathbb{N}$-graded monomial ordering with respect to the natural $\mathbb{N}$-gradation of $K\langle X\rangle$ is used (note that the (reverse) lexicographic ordering is not a monomial ordering on $K\langle X\rangle$), then necessarily the set $\mathcal{G}$ consists of elements of the form

$$g_{ji} = X_jX_i - \lambda_{ji}X_iX_j - \sum \lambda_{ji}^{t\ell}X_tX_\ell - \sum \lambda_hX_h - c_{ji},$$

where λ_{ji}, $\lambda_{ji}^{t\ell}$, λ_h, $c_{ji} \in K$, $1 \le i < j \le n$. So, after examining carefully how the overlap elements of the defining relations of an enveloping algebra $U(\mathbf{g})$ (where $\mathbf{g}$ is a finite dimensional Lie algebra) are reduced to zero, we will see that the heavy reduction procedure of **Algorithm-GB** may be reasonably lightened by examining the so-called Jacobi sums.

Example 8. The aim of this example is just for demonstrating how the Jacobi identity in the definition of a K-Lie algebra $\mathbf{g}$ may be used to lighten the reduction procedure in **Algorithm-GB** in order to verify directly that the defining relations of the enveloping algebra $U(\mathbf{g})$ form a Gröbner basis.

Let $\mathbf{g}$ be an n-dimensional K-Lie algebra with the K-basis $\{x_1, \ldots, x_n\}$ and the bracket product $[x_j, x_i] = \sum_{\ell=1}^{n} \lambda_{ji}^{\ell}x_\ell$, $1 \le i < j \le n$. Then $\mathbf{g}$ is, as a K-vector space, isomorphic to the K-subspace of the free K-algebra $K\langle X\rangle = K\langle X_1,\ldots,X_n\rangle$ spanned by $X = \{X_1,\ldots,X_n\}$, which is known as the degree-1 homogeneous part in the natural $\mathbb{N}$-gradation of $K\langle X\rangle$, i.e., $\mathsf{g} \cong K\langle X\rangle_1$. Moreover, we know that the enveloping algebra $U(\mathsf{g})$ of g is realized by the associative K-algebra $K\langle X\rangle/I$, where the ideal I is generated by the subset $\mathcal{G}$ consisting of

$$g_{ji} = X_jX_i - X_iX_j - [X_j, X_i]$$

where $[X_j, X_i] = \sum_{\ell=1}^{n} \lambda_{ji}^{\ell} X_\ell$, $1 \le i < j \le n$. Using the graded lexicographic ordering $X_1 \prec_{gr} X_2 \prec_{gr} \cdots \prec_{gr} X_n$ with respect to the natural $\mathbb{N}$-gradation of $K\langle X\rangle$, we have $\mathbf{LM}(g_{ji}) = X_jX_i$ for $1 \le i < j \le n$, and for ordered pairs in $\mathcal{G}\times\mathcal{G}$, all possible nonzero overlap elements are given by

$$\begin{aligned} S_{kji} &= o(g_{kj}, X_i;\; X_k, g_{ji}) \\ &= g_{kj}X_i - X_kg_{ji} \\ &= X_kX_iX_j - X_jX_kX_i + X_k[X_j, X_i] - [X_k, X_j]X_i, \end{aligned}$$

where $i < j < k$. Obviously, $\mathbf{LM}(S_{kji}) = X_kX_iX_j$. So, always starting with the leading monomial of the remainder at each step, the division of S_{kji} in turn by g_{ki}, g_{ki}, g_{ji} and g_{kj} yields

$$S_{kji} = g_{ki}X_j - X_jg_{ki} - g_{ji}X_k + X_ig_{kj} + \mathbf{J}(X_k, X_j, X_i),$$

where

$$\begin{aligned} \mathbf{J}(X_k, X_j, X_i) = {} & [X_k, X_i]X_j - X_j[X_k, X_i] + X_i[X_k, X_j] \\ & - [X_k, X_j]X_i + X_k[X_j, X_i] - [X_j, X_i]X_k. \end{aligned}$$

Suppose $[X_k, X_i] = \sum_\ell \lambda_{ki}^{\ell} X_\ell$. Then for the fixed X_j,

$$\begin{aligned} [X_k, X_i]X_j - X_j[X_k, X_i] &= \sum_{j<\ell}\lambda_{ki}^{\ell}(X_\ell X_j - X_jX_\ell) - \sum_{\ell<j}\lambda_{ki}^{\ell}(X_jX_\ell - X_\ell X_j) \\ &= \sum_{j<\ell}\lambda_{ki}^{\ell}(g_{\ell j} + [X_\ell, X_j]) - \sum_{\ell<j}\lambda_{ki}^{\ell}(g_{j\ell} + [X_j, X_\ell]) \\ &= \sum_{j<\ell}\lambda_{ki}^{\ell}g_{\ell j} - \sum_{\ell<j}\lambda_{ki}^{\ell}g_{j\ell} + \sum_{\ell}\lambda_{ki}^{\ell}[X_\ell, X_j] \\ &= \sum_{j<\ell}\lambda_{ki}^{\ell}g_{\ell j} - \sum_{\ell<j}\lambda_{ki}^{\ell}g_{j\ell} + [[X_k, X_i], X_j]. \end{aligned}$$

Note that under the K-vector space isomorphism $\mathsf{g} \cong K\langle X\rangle_1$ we have used the properties of Lie-bracket in g:

$$[x_j, x_j] = 0, \quad -[x_j, x_\ell] = [x_\ell, x_j]$$

and

$$\sum_\ell \lambda_{ki}^\ell [x_\ell, x_j] = \left[\sum_\ell \lambda_{ki}^\ell x_\ell,\ x_j\right] = [[x_k, x_i], x_j].$$

Similarly we obtain

$$X_i[X_k, X_j] - [X_k, X_j]X_i = \sum_{i<\ell} \lambda_{kj}^\ell g_{\ell i} - \sum_{\ell<i} \lambda_{kj}^\ell g_{i\ell} + [[X_j, X_k], X_i],$$

$$X_k[X_j, X_i] - [X_j, X_i]X_k = \sum_{k<\ell} \lambda_{ji}^\ell g_{\ell k} - \sum_{\ell<k} \lambda_{ji}^\ell g_{k\ell} + [[X_i, X_j], X_k].$$

Since the Jacobi idetity $[[x_k, x_i], x_j] + [[x_j, x_k], x_i] + [[x_i, x_j], x_k] = 0$ in g entails $[[X_k, X_i], X_j] + [[X_j, X_k], X_i] + [[X_i, X_j], X_k] = 0$ in $K\langle X\rangle_1$, it follows that $\overline{S_{kji}}^{\mathcal{G}} = 0$. This shows that $\mathcal{G}$ is a Gröbner basis for I.

Now, let us fix an admissible system $(\mathcal{B}, \prec_{gr})$ for $K\langle X\rangle = K\langle X_1, \ldots, X_n\rangle$, where $\mathcal{B}$ is the standard K-basis of $K\langle X\rangle$ and $\prec_{gr}$ is an $\mathbb{N}$-*graded monomial ordering on* $\mathcal{B}$ *with respect to the natural* $\mathbb{N}$-*gradation* of $K\langle X\rangle$, and consider a subset $\mathcal{G}$ consisting of the $\frac{n(n-1)}{2}$ elements of the form

$$\text{(1)} \quad \begin{aligned} g_{ji} &= X_jX_i - \lambda_{ji}X_iX_j - \{X_j,\ X_i\}, \text{ where} \\ &\quad \{X_j,\ X_i\} = \textstyle\sum \lambda_{ji}^{t\ell} X_tX_\ell - \sum \lambda_h X_h - c_{ji}, \\ &\quad \lambda_{ji},\ \lambda_{ji}^{t\ell},\ \lambda_h,\ c_{ji} \in K, \end{aligned}$$

such that

$$\text{(2)} \quad \mathbf{LM}(g_{ji}) = X_jX_i, \quad 1 \le i < j \le n.$$

In this case, one verifies that all possible overlap elements of ordered pairs in $\mathcal{G} \times \mathcal{G}$ are given by

$$\text{(3)} \quad \begin{aligned} o(g_{kj}, X_i;\ X_k, g_{ji}) &= g_{kj}X_i - X_kg_{ji} \\ &= \lambda_{ji}X_kX_iX_j - \lambda_{kj}X_jX_kX_i \\ &\quad +X_k\{X_j,\ X_i\} - \{X_k,\ X_j\}X_i, \end{aligned}$$

where $i < j < k$. For convenience, let us fix the $\mathbb{N}$-graded lexicographic ordering

$$X_i \prec_{gr} X_j \prec_{gr} X_k, \quad i < j < k.$$

Also we write

$$\text{(4)} \quad S_{X_kX_jX_i} = o(g_{kj}, X_i;\ X_k, g_{ji})$$

to emphasize the dependence of the overlap elements on the fixed order $\prec_{gr}$ with respect to $i < j < k$. Then, *ignoring* whether the coefficients λ_{ji} and λ_{kj} are zero or not, $S_{X_kX_jX_i}$ can be rewritten, through the division of $\lambda_{ji}X_kX_iX_j - \lambda_{kj}X_jX_kX_i$ in turn by g_{ki}, g_{ki}, g_{ji} and g_{kj}, as follows:

$$\begin{aligned} S_{X_kX_jX_i} &= \lambda_{ji}g_{ki}X_j - \lambda_{kj}X_jg_{ki} - \lambda_{kj}\lambda_{ki}g_{ji}X_k \\ &\quad +\lambda_{ki}\lambda_{ji}X_ig_{kj} + \mathbf{J}(X_k, X_j, X_i), \end{aligned} \tag{5}$$

where

$$\begin{aligned} J(X_k, X_j, X_i) &= \{X_k, X_j\}X_i - \lambda_{ki}\lambda_{ji}X_i\{X_k, X_j\} \\ &\quad -\lambda_{ji}\{X_k, X_i\}X_j + \lambda_{kj}X_j\{X_k, X_i\} \\ &\quad +\lambda_{kj}\lambda_{ki}\{X_j, X_i\}X_k - X_k\{X_j, X_i\}. \end{aligned} \tag{6}$$

By referring to Example 8 above and adopting the definition used in [Ber1], we call $J(X_k, X_j, X_i)$ a *Jacobi sum* of $\mathcal{G}$.

Suppose that the following condition is satisfied:

$$\textbf{(J)}\quad \begin{array}{l} \text{If } S_{X_kX_jX_i} \neq 0, \text{ then} \\ X_kX_iX_j \preceq_{gr} \mathbf{LM}(S_{X_kX_jX_i}) \text{ and } X_jX_kX_i \preceq_{gr} \mathbf{LM}(S_{X_kX_jX_i}). \end{array}$$

By the construction of a Jacobi sum it is clear that if the condition (**J**) is satisfied, then

$$\mathbf{LM}(J(X_k, X_j, X_i)) \preceq_{gr} \mathbf{LM}(S_{X_kX_jX_i}).$$

Thus, to know whether $\mathcal{G}$ is a Gröbner basis, it is sufficient to check whether $\overline{J(X_k, X_j, X_i)}^{\mathcal{G}} = 0$ provided $J(X_k, X_j, X_i) \neq 0$. As one may see from the next example, one of the advantages of using the Jacobi sum is that in practical examples (except for enveloping algebras of Lie algebras) some Jacobi sums may already be zero before doing further reduction.

Example 9. Consider the subset $\mathcal{G}$ of the free K-algebra $K\langle X\rangle = K\langle X_1, X_2, X_3, X_4\rangle$ consisting of

$$\begin{array}{ll} g_{12} = X_1X_2 - q^{-1}X_2X_1, & g_{41} = X_4X_1 - X_1X_4 - (q - q^{-1})X_2X_3, \\ g_{31} = X_3X_1 - qX_1X_3, & g_{42} = X_4X_2 - qX_2X_4, \\ g_{32} = X_3X_2 - X_2X_3, & g_{43} = X_4X_3 - qX_3X_4, \end{array}$$

where $q \in K^*$. If we use the $\mathbb{N}$-graded lexicographic ordering $\prec_{gr}$ with respect to the natural $\mathbb{N}$-gradation of $K\langle X\rangle$ such that

$$X_2 \prec_{gr} X_1 \prec_{gr} X_3 \prec_{gr} X_4,$$

then the condition (2) above on leading monomials of $\mathcal{G}$ is satisfied. By a direct calculation, all possible nonzero Jacobi sums of $\mathcal{G}$ are

$$\begin{aligned} J(X_4, X_1, X_2) &= (q - q^{-1})X_2X_3X_2 + qq^{-1}(q - q^{-1})X_2^2X_3 \\ &= (q - q^{-1})X_2g_{32}, \\ J(X_4, X_3, X_1) &= (1 - q^2)X_2X_3^2 + (q^2 - 1)X_3X_2X_3 \\ &= (q^2 - 1)g_{32}X_3. \end{aligned}$$

We see that if $q^2 \neq 1$ then $J(X_4, X_1, X_2) \neq 0$ and $J(X_4, X_3, X_1) \neq 0$. It is straightforward to check that all three overlap elements $S_{X_4X_3X_1}$, $S_{X_4X_1X_2}$, and $S_{X_3X_1X_2}$ satisfy the condition (**J**) and $\overline{J(X_4, X_1, X_2)}^{\mathcal{G}} = 0 = \overline{J(X_4, X_3, X_1)}^{\mathcal{G}}$. Hence, $\mathcal{G}$ is a Gröbner basis for the ideal $I = \langle \mathcal{G} \rangle$. Consequently, the quadratic algebra $A = K\langle X \rangle / I$ has the PBW K-basis

$$\left\{ \overline{X}_2^{\alpha_2}\overline{X}_1^{\alpha_1}\overline{X}_3^{\alpha_3}\overline{X}_4^{\alpha_4} \;\middle|\; \alpha_i \in \mathbb{N} \right\}.$$

Note that the algebra A obtained obave is known as the coordinate ring $M_q(2) = K[a, b, c, d]$ of the manifold of quantum 2×2 matrices $\begin{pmatrix} a & b \\ c & d \end{pmatrix}$ in the sense of [Man].

Furthermore, note that in the last example we have

$$\begin{aligned} \mathbf{LM}(X_2g_{32}) &= X_2X_3X_2 \prec_{gr} \mathbf{LM}(S_{X_4X_1X_2}) = X_4X_2X_1, \\ \mathbf{LM}(g_{32}X_3) &= X_3X_2X_3 \prec_{gr} \mathbf{LM}(S_{X_4X_3X_1}) = X_4X_1X_3. \end{aligned}$$

So, without doing division by $\mathcal{G}$, it follows immediately from the remark given after (Ch.3, Theorem 5.1) that $\mathcal{G}$ is a Gröbner basis. This is another advantage of using the Jacobi sum.

The foregoing discussion leads to the following proposition.

Proposition 3.2. *Let* $\mathcal{G} = \{g_{ji} \mid 1 \leq i < j \leq n\} \subset K\langle X \rangle = K\langle X_1, \ldots, X_n \rangle$ *be as given by the foregoing* (1) *such that* $\mathbf{LM}(g_{ji}) = X_jX_i$ *with respect to the* $\mathbb{N}$*-graded lexicographic ordering*

$$X_1 \prec_{gr} X_2 \prec_{gr} \cdots \prec_{gr} X_n.$$

With the notations as fixed in (3) – (6) *above, suppose that*
(a) *the condition* (**J**) *holds for each* $S_{X_kX_jX_i} \neq 0$, *where* $X_i \prec_{gr} X_j \prec_{gr} X_k$; and
(b) *each nonzero* $\mathbf{J}(X_k, X_j, X_i)$ *can be expressed as*

$$\mathbf{J}(X_k, X_j, X_i) = \sum \lambda_{pq} g_{pq} + \sum \lambda_{hn\ell} X_h g_{n\ell} + \sum \lambda_{mts} g_{mt} X_s,$$

where $\lambda_{pq}, \lambda_{hn\ell}, \lambda_{mts} \in K^*$, *in which all terms satisfy*

$$\begin{aligned} \mathbf{LM}(g_{pq}) &= X_p X_q \preceq_{gr} \mathbf{LM}(S_{X_k X_j X_i}), \\ \mathbf{LM}(X_h g_{n\ell}) &= X_n X_n X_\ell \preceq_{gr} \mathbf{LM}(S_{X_k X_j X_i}), \\ \mathbf{LM}(g_{mt} X_s) &= X_m X_t X_s \preceq_{gr} \mathbf{LM}(S_{X_k X_j X_i}). \end{aligned}$$

Then $\mathcal{G}$ *is a Gröbner basis for the ideal* $I = \langle \mathcal{G} \rangle$. *Consequently, the algebra* $A = K\langle X \rangle / I$ *has the associated graded algebra* $G^{\mathbb{N}}(A) \cong K\langle X \rangle / \langle \mathbf{LH}(\mathcal{G}) \rangle$ *with*

$$\mathbf{LH}(\mathcal{G}) = \left\{ \mathbf{LH}(g_{ji}) = X_j X_i - \lambda_{ji} X_i X_j - \sum_{t \le \ell} \lambda_{ji}^{t\ell} X_t X_\ell \;\middle|\; 1 \le i < j \le n \right\},$$

and the set $\{X_1^{\alpha_1} X_2^{\alpha_2} \cdots X_n^{\alpha_n} \mid \alpha_j \in \mathbb{N}\}$ *projects to a PBW* K*-basis for* A *and* $G^{\mathbb{N}}(A)$ *respectively.* □

Remark (i) The above proposition is a generalization of ([Ber1], Theorem 2.8.1, Theorem 2.9.2) in the case that K is a field. An earlier version of this proposition was given in ([Li1], Ch.3, Proposition 1.6) but the condition used there is too restrictive.
(ii) As illustrated by Example 9, the delicate point in using Proposition 3.2 is that the overlap element $S_{X_k X_j X_i}$ (and hence $J(X_k, X_j, X_i)$) depends on the chosen $\mathbb{N}$-graded monomial ordering

$$X_{i_1} \prec_{gr} X_{i_2} \prec_{gr} \cdots \prec_{gr} X_{i_n}.$$

As an exercise, the reader is invited to examine Examples 6, 7 by using Proposition 3.2.

In connection with our starting examples in this book (Ch.1, Section 1), we illustrate Proposition 3.2 by quoting two more examples from [Li1].

Example 10. This example provides a family of algebras similar to the n-th Weyl algebra $A_n(K)$. Consider the natural N-gradation of the free algebra $K\langle Y, X \rangle = K\langle Y_n, \ldots, Y_1, X_n, \ldots, X_1 \rangle$ and set on $K\langle Y, X \rangle$ the $\mathbb{N}$-graded lexicographic ordering $\prec_{gr}$ such that

$$X_1 \prec_{gr} Y_1 \prec_{gr} \cdots \prec_{gr} X_{n-1} \prec_{gr} Y_{n-1} \prec_{gr} X_n \prec_{gr} Y_n.$$

Let $\mathcal{G}$ be the subset of $K\langle Y, X\rangle$ consisting of

$$\begin{aligned}
&H_{ji} = X_jX_i - X_iX_j, && 1 \le i < j \le n,\\
&\widetilde{H}_{ji} = X_jY_i - Y_iX_j, && 1 \le i < j \le n,\\
&G_{ji} = Y_jY_i - Y_iY_j, && 1 \le i < j \le n,\\
&\widetilde{G}_{ji} = Y_jX_i - X_iY_j, && 1 \le i < j \le n\\
&R_{jj} = Y_jX_j - q_jX_jY_j - F_{jj}, && 1 \le j \le n,
\end{aligned}$$

where $q_j \in K^*$, $F_{jj} \in K\langle Y, X\rangle$. Then a direct verification shows that all possible nonzero Jacobi sums of $\mathcal{G}$ with respect to the ordering given on generators are

$$\begin{aligned}
&\mathbf{J}(Y_j, X_j, X_i) = F_{jj}X_i - X_iF_{jj}, && 1 \le i < j \le n,\\
&\mathbf{J}(Y_j, X_j, Y_i) = F_{jj}Y_i - Y_iF_{jj}, && 1 \le i < j \le n,\\
&\mathbf{J}(Y_k, Y_j, X_j) = F_{jj}Y_k - Y_kF_{jj}, && 1 \le j < k \le n,\\
&\mathbf{J}(X_k, Y_j, X_j) = F_{jj}X_k - X_kF_{jj}, && 1 \le j < k \le n.
\end{aligned}$$

It is easy to see that in many cases, for instance,

$$(*) \qquad F_{jj} \in K\text{-span}\{X_i^2,\ X_i,\ Y_i,\ 1\},\ 1 \le i < j,$$

all conditions of Proposition 3.2 can be satisfied by $\mathcal{G}$, and hence, $\mathcal{G}$ forms a Gröbner basis for the ideal $I = \langle \mathcal{G}\rangle$. Thus by Theorem 3.1 and Theorem 2.4, such a Gröbner basis $\mathcal{G}$ determines the PBW K-basis

$$\left\{ \overline{X}_1^{\alpha_1} \cdots \overline{X}_n^{\alpha_n} \overline{Y}_1^{\beta_1} \cdots \overline{Y}_n^{\beta_n} \;\middle|\; \alpha_j, \beta_j \in \mathbb{N} \right\}$$

for the algebra $A = K\langle Y, X\rangle / I$ and its associated graded algebra $G^{\mathbb{N}}(A) = K\langle Y, X\rangle / \langle \mathbf{LH}(\mathcal{G})\rangle$.

If in $\mathcal{G}$ we have $F_{jj} = 1$ for $1 \le j \le n$, then A coincides with the additive analogue of the n-th Weyl algebra introduced in quantum physics ([Kur] 1980 and [JBS] 1981). If $F_{jj} = 1$ and $q_1 = \cdots = q_n = q \ne 0$, then A is usually called the algebra of q-differential operators; in particular, if $q = 1$, then $A = A_n(K)$, the n-th Weyl algebra over K.

Example 11. This example provides a family of algebras similar to the Heisenberg enveloping algebra. Let $K\langle X \cup Z \cup Y\rangle$ be the free K algebra generated by $X \cup Z \cup Y = \{X_n, \ldots, X_1, Z_n, \ldots, Z_1, Y_n, \ldots, Y_1\}$, and consider the quotient algebra $A = K\langle X \cup Z \cup Y\rangle / I$, where I is the ideal generated by the subset $\mathcal{G}$ consisting of

$$\begin{array}{ll}
R^x_{ji} = X_jX_i - X_iX_j, & 1 \le i < j \le n, \\
R^y_{ji} = Y_jY_i - Y_iY_j, & 1 \le i < j \le n, \\
R^z_{ji} = Z_jZ_i - Z_iZ_j, & 1 \le i < j \le n, \\
R^{zy}_{ji} = Z_jY_i - \lambda_i^{\delta_{ji}} Y_iZ_j, & 1 \le i,\ j \le n, \\
R^{xz}_{ji} = X_jZ_i - \mu_i^{\delta_{ji}} Z_iX_j, & 1 \le i,\ j \le n, \\
R^{xy}_{ji} = X_jY_i - Y_iX_j, & i \ne j, \\
R^{xy}_{jj} = X_jY_j - q_jY_jX_j - F_{jj}, & 1 \le j \le n,
\end{array}$$

where $\lambda_i, \mu_i, q_j \in K^*$, $F_{jj} \in K\langle X \cup Z \cup Y\rangle$. Consider the natural $\mathbb{N}$-gradation of $K\langle X \cup Z \cup Y\rangle$ and set on $K\langle X \cup Z \cup Y\rangle$ the $\mathbb{N}$-graded lexicographic ordering

$$Y_1 \prec_{gr} \cdots \prec_{gr} Y_n \prec_{gr} Z_1 \prec_{gr} \cdots \prec_{gr} Z_n \prec_{gr} X_1 \prec_{gr} \cdots \prec_{gr} X_n.$$

Then a straightforward verification shows that all possible nonzero Jacobi sums of $\mathcal{G}$ with respect to the ordering given on generators are

$$\begin{array}{ll}
\mathbf{J}(X_k, X_j, Y_j) = F_{jj}X_k - X_kF_{jj}, & 1 \le j < k \le n, \\
\mathbf{J}(X_k, X_j, Y_k) = -F_{kk}X_j + X_jF_{kk}, & 1 \le j < k \le n, \\
\mathbf{J}(X_k, Z_j, Y_k) = -F_{kk}Z_j + Z_jF_{kk}, & 1 \le k,\ j \le n, \\
\mathbf{J}(X_k, Y_k, Y_j) = F_{kk}Y_j - Y_jF_{kk}, & 1 \le j < k \le n, \\
\mathbf{J}(X_j, Y_k, Y_j) = -F_{jj}Y_k + Y_kF_{jj}, & 1 \le j < k \le n.
\end{array}$$

It is easy to see that in many cases, for instance,

$$(**) \qquad F_{jj} \in K\text{-span}\{Y_i^2,\ Y_i,\ Z_p,\ X_p,\ 1\},\ 1 \le i < j,\ 1 \le p \le n,$$

all conditions of Proposition 3.2 are satisfied by $\mathcal{G}$, and hence, $\mathcal{G}$ forms a Gröbner basis for the ideal $I = \langle \mathcal{G}\rangle$. Thus by Theorem 3.1 and Theorem 2.4, such a Gröbner basis $\mathcal{G}$ determines the PBW K-basis

$$\left\{ \overline{Y}_1^{\beta_1} \cdots \overline{Y}_n^{\beta_n} \overline{Z}_1^{\gamma_1} \cdots \overline{Z}_n^{\gamma_n} \overline{X}_1^{\alpha_1} \cdots \overline{X}_n^{\alpha_n} \;\middle|\; \alpha_j, \beta_j, \gamma_j \in \mathbb{N} \right\}$$

for A and its associated graded algebra $G^{\mathbb{N}}(A) = K\langle X \cup Z \cup Y\rangle / \langle \mathbf{LH}(\mathcal{G})\rangle$.

If we take $q_j = 1$ $Z_j = Z$ and $F_{jj} = Z$, $1 \le j \le n$, then the enveloping algebra of the $2n+1$-dimensional Heisenberg Lie algebra is recovered. In the case where $\lambda_i = \mu_i = q \ne 0$, $q_j = q^{-1}$, and $F_{jj} = z_j$, the corresponding algebra A is called the q-Heisenberg algebra in the sense of ([Ber] 1992, [Ros] 1995) which has its root in q-calculus (e.g., [Wal] 1985).

Remark In conclusion, combining (Ch.2, Theorem 1.1, Theorem 4.1) and the fundamental decomposition theorem (Ch.3, Theorem 3.3), so far the results obtained in this chapter have shown that the Gröbner basis theory developed in the context of Ch.3 is essentially an algorithmic realization of a constructive PBW theory (PBW isomorphism plus K-basis structure) in the same context.

4.4 Solvable Polynomial Algebras Revisited

We finish this chapter by revisiting a class of algebras which have the classical PBW K-basis and have an effective Gröbner basis theory for both two-sided and one-sided ideals, namely the class of solvable polynomial algebras in the sense of ([K-RW], 1990). Because of the discussion on getting a PBW K-basis in the last section, the main result of this section is to establish the relation between a solvable polynomial algebra and a Gröbner basis (Proposition 4.2). As a consequence, the argument on solvable polynomial algebras given in ([Li1], Ch.3) is strengthened.

Recall that a finitely generated K-algebra $A = K[a_1, \ldots, a_n]$ is called a *solvable polynomial algebra* if A satisfies the following three conditions:

(S1) A has the PBW K-basis $\mathscr{B} = \{a^\alpha = a_1^{\alpha_1} \cdots a_n^{\alpha_n} \mid \alpha = (\alpha_1, \ldots, \alpha_n) \in \mathbb{N}^n\}$;

(S2) There is a (left) monomial ordering $\prec$ on $\mathscr{B}$ in the sense of (Ch.3, Section 1);

(S3) For any a^α, $a^\beta \in \mathscr{B}$,

$$a^\alpha a^\beta = \lambda_{\alpha,\beta} a^{\alpha+\beta} + f_{\alpha,\beta},$$

satisfying $\lambda_{\alpha,\beta} \in K^*$, and $f_{\alpha,\beta} \in A$ with $\mathbf{LM}(f_{\alpha,\beta}) \prec a^{\alpha+\beta}$.

As before we call the pair $(\mathscr{B}, \prec)$ a (left) *admissible system* of A.

By [K-RW], in a solvable polynomial algebra A every (two-sided, respectively one-sided) ideal has a finite Gröbner basis; in particular, the following result holds.

Theorem 4.1. *Every solvable polynomial algebra A is a Noetherian domain.*

□

To better understand the topic concerning solvable polynomial algebras, the reader is reminded to review the Gröbner basis theory developed in

Ch.3, to examine several typical examples given in (Ch.1, Section 1), and to refer to ([K-RW], [Lev]) for more a detailed algorithmic study on solvable polynomial algebras.

If $A = K[a_1, \ldots a_n]$ is a solvable polynomial algebra with respect to an admissible system $(\mathscr{B}, \prec)$, then it is clear that

$$a_ja_i = \lambda_{ji}a_ia_j + f_{ji},\ 1 \le i < j \le n,\ f_{ji} = \sum_{\alpha} \lambda_\alpha a^\alpha \text{ with } a^\alpha \in \mathscr{B},$$

satisfying $\lambda_{ji}, \lambda_\alpha \in K$, $\lambda_{ji} \ne 0$, and $\mathbf{LM}(f_{ji}) \prec a_ia_j$. Now let $K\langle X\rangle = K\langle X_1, \ldots, X_n\rangle$ be the free K-algebra generated by $X = \{X_1, \ldots, X_n\}$, and consider the subset $\mathcal{G}$ of $K\langle X\rangle$ consisting of

$$g_{ji} = X_jX_i - \lambda_{ji}X_iX_j - F_{ji},\ 1 \le i < j \le n,$$

where $F_{ji} = \sum_\alpha \lambda_\alpha w_\alpha$ is determined by f_{ji} subject to the correspondence

$$a^\alpha = a_1^{\alpha_1}a_2^{\alpha_2}\cdots a_n^{\alpha_n} \longleftrightarrow X_1^{\alpha_1}X_2^{\alpha_2}\cdots X_n^{\alpha_n} = w_\alpha.$$

Proposition 4.2. *Let A, $\mathcal{G}$ and the related notations be as fixed above. If there is some monomial ordering $\prec_X$ on the standard K-basis $\mathcal{B}$ of $K\langle X\rangle$ such that $\mathbf{LM}(g_{ji}) = X_jX_i$, $1 \le i < j \le n$, then the following two statements hold.*
(i) *$A \cong K\langle X\rangle/I$, where $I = \langle \mathcal{G}\rangle$.*
(ii) *$\mathcal{G}$ is a Gröbner basis for the ideal $I = \langle \mathcal{G}\rangle$ with respect to $(\mathcal{B}, \prec_X)$.*

Proof. (i). Since $\mathbf{LM}(g_{ji}) = X_jX_i$ by the assumption, $1 \le i < j \le n$, for each element $f \in K\langle X\rangle$, the division of f by $\mathcal{G}$ yields $f = h + \overline{f}^{\mathcal{G}}$, where $h \in I$ and $\overline{f}^{\mathcal{G}}$ is a linear combination of monomials of the form $X_1^{\alpha_1}X_2^{\alpha_2}\cdots X_n^{\alpha_n}$. So the desired isomorphism is induced by the canonical algebra epimorphism $K\langle X\rangle \to A$ with $X_i \mapsto a_i$, $1 \le i \le n$.

(ii). It follows from the isomorphism in (i) that $K\langle X\rangle/I$ has the PBW K-basis $\{\overline{X}_1^{\alpha_1}\overline{X}_2^{\alpha_2}\cdots\overline{X}_n^{\alpha_n} \mid \alpha_j \in \mathbb{N}\}$, where each $\overline{X}_i$ is the canonical image of X_i in $K\langle X\rangle/I$. Hence, by Theorem 3.1, $\mathcal{G}$ is a Gröbner basis with respect to $(\mathcal{B}, \prec_X)$. □

Let $K\langle X\rangle = K\langle X_1, \ldots, X_n\rangle$ be the free K-algebra of n generators with an admissible system $(\mathcal{B}, \prec)$. In view of Theorem 3.1 and the above Proposition 4.2, we describe below a computational way in getting (verifying) a solvable polynomial algebra of the form $A = K\langle X\rangle/\langle\mathcal{G}\rangle$.

Step 1. Look for a Gröbner basis $\mathcal{G} \subset K\langle X\rangle$ with respect to $(\mathcal{B}, \prec)$, which has necessarily $\frac{n(n-1)}{2}$ elements of the form

$$g_{ji} = X_jX_i - \lambda_{ji}X_iX_j - F_{ji},\ 1 \le i < j \le n,$$

satisfying $\mathbf{LM}(g_{ji}) = X_jX_i$, $\lambda_{ji} \in K^*$, and $F_{ji} = \sum \lambda_\alpha w_\alpha$ with $\lambda_\alpha \in K$, $w_\alpha = X_1^{\alpha_1}X_2^{\alpha_2}\cdots X_n^{\alpha_n}$, $\alpha_j \in \mathbb{N}$, so that the algebra $A = K\langle X\rangle/I$ with $I = \langle \mathcal{G}\rangle$ has the PBW K-basis

$$\mathscr{B} = \overline{N(I)} = \{\overline{X}_1^{\alpha_1}\overline{X}_2^{\alpha_2}\cdots \overline{X}_n^{\alpha_n} \mid \alpha_j \in \mathbb{N}\}.$$

Step 2. Look for a monomial ordering $\prec_{_I}$ on $\mathscr{B}$ such that

$$\mathbf{LM}(\overline{F}_{ji}) \prec_{_I} \overline{X}_i\overline{X}_j,\ 1 \le i < j \le n,\ \text{where } \overline{F}_{ji} = \sum \lambda_\alpha \overline{w}_\alpha.$$

Remark Here we should point out that although the monomial ordering $\prec$ on the standard K-basis $\mathcal{B}$ of $K\langle X\rangle$ induces a well-ordering on the PBW K-basis $\mathscr{B}$ of $A = K\langle X\rangle/I$, in general $\prec$ does not necessarily always induce a monomial ordering on $\mathscr{B}$. For instance, let $\mathcal{G}$ be the subset of the free K-algebra $K\langle X\rangle = K\langle X_1, X_2, X_3\rangle$ consisting of

$$\begin{aligned} g_{21} &= X_2X_1 - X_1X_2, \\ g_{31} &= X_3X_1 - X_1X_3 - X_2X_3 - X_2, \\ g_{32} &= X_3X_2 - X_2X_3. \end{aligned}$$

Then, under the $\mathbb{N}$-graded lexicographic ordering $X_1 \prec_{gr} X_2 \prec_{gr} X_3$ with respect to the natural $\mathbb{N}$-gradation of $K\langle X\rangle$, $\mathbf{LM}(g_{ji}) = X_jX_i$, $1 \le i < j \le 3$, and the only nontrivial overlap element of ordered pairs in $\mathcal{G} \times \mathcal{G}$ is $S_{321} = o(g_{32}, X_1;\ X_3, g_{21}) = -X_2X_3X_1 + X_3X_1X_2$. Also we have

$$J(X_3, X_2, X_1) = -X_2X_3X_2 + X_2^2X_3 = -X_2g_{32}.$$

Hence, by Proposition 3.2, $\mathcal{G}$ is a Gröbner basis for the ideal $I = \langle \mathcal{G}\rangle$. Writing a_i for the canonical image of X_i in the quotient algebra $A = K\langle X\rangle/I$, $1 \le i \le 3$, then $A = K[a_1, a_2, a_3]$ which has the PBW K-basis $\mathscr{B} = \{a_1^{\alpha_1}a_2^{\alpha_2}a_3^{\alpha_3} \mid \alpha_j \in \mathbb{N}\}$. If we use the ordering $a_1 \prec_{gr} a_2 \prec_{gr} a_3$ on $\mathscr{B}$ induced by $X_1 \prec_{gr} X_2 \prec_{gr} X_3$ on $K\langle X\rangle$, then since

$$a_3a_1 = a_1a_3 + a_2a_3 + a_2 \text{ with } a_1a_3 \prec_{gr} a_2a_3,$$

A is not a solvable polynomial algebra. However, if we use the lexicographic ordering $a_2 \prec_{lex} a_1 \prec_{lex} a_3$ on $\mathcal{B}$, then

$$\begin{aligned} a_1a_2 &= a_2a_1, \\ a_3a_1 &= a_1a_3 + a_2a_3 + a_2 \text{ with } a_2a_3 \prec a_1a_3, \\ a_3a_2 &= a_2a_3. \end{aligned}$$

Hence A is a solvable polynomial algebra.

A similar illustration is provided by Example 9 in Section 3. The reader is invited to check what will happen if the graded lexicographic monomial ordering $X_1 \prec_{gr} X_2 \prec_{gr} X_3 \prec_{gr} X_4$ is used there.

Remark The above example also shows that the condition of using a graded monomial ordering for getting a quadric solvable polynomial algebra posed in ([Li1], Ch.3) is too restrictive.

Bearing in mind the remark above, we now use the foregoing Step 1 – Step 2 to verify that the Gröbner bases constructed in Examples 5, 6 of Section 3 define solvable polynomial algebras. To this end, we need a little more preparation.

Let $A = K[a_1, \ldots, a_n]$ be a K-algebra with the PBW K-basis $\mathscr{B} = \{a^\alpha = a_1^{\alpha_1} \cdots a_n^{\alpha_n} \mid \alpha = (\alpha_1, \ldots, \alpha_n) \in \mathbb{N}^n\}$. Note that if each generator a_i of A is assigned a positive degree n_i, $1 \leq i \leq n$, then we also have the notion of a weight $\mathbb{N}$-graded monomial ordering on $\mathscr{B}$ (though A does not necessarily have a weight $\mathbb{N}$-gradation determined by the weight $\{n_i\}_{i=1}^n$). This may be done as follows. For $a^\alpha = a_1^{\alpha_1} a_2^{\alpha_2} \cdots a_n^{\alpha_n} \in \mathscr{B}$, put $|\alpha| = n_1\alpha_1 + n_2\alpha_2 + \cdots + n_i\alpha_n$. If $\prec$ is a well-ordering on $\mathscr{B}$ and if the order $\prec_{gr}$ defined for $a^\alpha, a^\beta \in \mathscr{B}$ subject to the rule

$$a^\alpha \prec_{gr} a^\beta \Leftrightarrow |\alpha| = \textstyle\sum_{i=1}^n n_i\alpha_i < \sum_{i=1}^n n_i\beta_i = |\beta| \quad \text{or} \quad |\alpha| = |\beta| \text{ but } a^\alpha \prec a^\beta$$

is a monomial ordering, then $\prec_{gr}$ is called a *weight* $\mathbb{N}$*-graded monomial ordering* on $\mathcal{B}$ determined by the given weight $\{n_i\}_{i=1}^n$.

Example 1. Let I be the ideal of the free K-algebra $K\langle X\rangle = K\langle X_1, X_2\rangle$ generated by the single element

$$g_{21} = X_2X_1 - qX_1X_2 - \alpha X_2 - f(X_1),$$

where $q, \alpha \in K$, and $f(X_1)$ is a polynomial in the variable X_1. Assigning X_1 the degree 1, it is proved (Section 3, Example 5) that in either of the following two cases:

(a) $\deg f(X_1) \leq 2$, and X_2 is assigned the degree 1;

(b) $\deg f(X_1) = n \geq 3$, and X_2 is assigned the degree n,

$\mathcal{G} = \{g_{21}\}$ forms a Gröbner basis for I, where in both cases the monomial ordering used is the $\mathbb{N}$-graded lexicographic ordering $X_1 \prec_{gr} X_2$ with respect to the natural $\mathbb{N}$-gradation of $K\langle X\rangle$, respectively the weight $\mathbb{N}$-gradation of $K\langle X\rangle$ with weight $\{1,n\}$. Consider the algebra $A=K\langle X\rangle/I=K[a_1,a_2]$, where $a_1=\overline{X}_1$, $a_2=\overline{X}_2$. If $q\neq 0$, then one checks easily that the order $X_1\prec_{gr}X_2$ on $K\langle X\rangle$ induces on the PBW K-basis $\mathscr{B}=\{a_1^{\alpha_1}a_2^{\alpha_2}\mid \alpha_1,\alpha_2\in\mathbb{N}\}$ of A the weight $\mathbb{N}$-graded monomial ordering $a_1\prec_{gr}a_2$, such that A forms a solvable polynomial algebra.

Example 2. Let $\mathcal{G}$ be the subset of the free K-algebra $K\langle X\rangle = K\langle X_1,X_2,X_3\rangle$ consisting of

$$\begin{aligned} g_{31} &= X_3X_1-\lambda X_1X_3+\gamma X_3,\\ g_{12} &= X_1X_2-\lambda X_2X_1+\gamma X_2,\\ g_{32} &= X_3X_2-\omega X_2X_3+f(X_1), \end{aligned}$$

where $\lambda,\gamma,\omega\in K$, and $f(X_1)$ is a polynomial in the variable X_1. Assigning X_1 the degree 1, it is proved (Section 3, Example 6) that in either of the following two cases:
(a) $\deg f(X_1)\le 2$, and X_2, X_3 are assigned the degree 1;
(b) $\deg f(X_1)=n\ge 3$, and X_2, X_3 are assigned the degree n,
$\mathcal{G}=\{g_{31},g_{12},g_{32}\}$ forms a Gröbner basis for the ideal $I=\langle\mathcal{G}\rangle$, where in both cases the monomial ordering used is the $\mathbb{N}$-graded lexicographic ordering $X_2\prec_{gr}X_1\prec_{gr}X_3$ with respect to the natural $\mathbb{N}$-gradation of $K\langle X\rangle$, respectively the weight $\mathbb{N}$-gradation of $K\langle X\rangle$ with weight $\{1,n,n\}$. Consider the algebra $A=K\langle X\rangle/I=K[a_1,a_2,a_3]$, where $a_i=\overline{X}_i$, $1\le i\le 3$. If $\lambda\neq 0$, $\omega\neq 0$, then one checks easily that the order $X_2\prec_{gr}X_1\prec_{gr}X_3$ on $K\langle X\rangle$ induces on the PBW K-basis $\mathscr{B}=\{a_2^{\alpha_2}a_1^{\alpha_1}a_3^{\alpha_3}\mid \alpha_j\in\mathbb{N}\}$ of A the weight $\mathbb{N}$-graded monomial ordering $a_2\prec_{gr}a_1\prec_{gr}a_3$, such that A forms a solvable polynomial algebra.

By referring to Example 6(c) of Section 3, we see particularly that all down-up algebras are solvable polynomial algebras provided that $\lambda\neq 0$, $\omega\neq 0$.

Chapter 5

Using $A_{\rm LH}^{\mathcal{B}}$ in Terms of Gröbner Bases

Let the free K-algebra $K\langle X\rangle = K\langle X_1, \ldots, X_n\rangle$ of n generators be equipped with a positive weight $\mathbb{N}$-gradation, that is, each X_i is assigned a positive degree n_i, $1 \le i \le n$, and let I be an ideal of $K\langle X\rangle$, $A = K\langle X\rangle/I$. Bearing in mind the task posed at the beginning of Ch.3, if $\mathcal{G}$ is a Gröbner basis of I and if the $\mathbb{N}$-filtration FA of A induced by the weight $\mathbb{N}$-grading filtration $FK\langle X\rangle$ of $K\langle X\rangle$ is considered, then, the significant combinatorial-algorithmic results concerning monomial algebras ([An1, 2, 3, 4], [Bor], [G-IL], [G-I1, 2], [Nor2], [Ok], [Uf1, 2]) and the results obtained in Ch.3 – Ch.4 enable us now to study the algebra A, the associated $\mathbb{N}$-graded algebra $G^{\mathbb{N}}(A)$ of A, and the Rees algebra $\widetilde{A} = \oplus_{p\in\mathbb{N}}F_pA$ of A (see Ch.7), via the monomial algebra $A_{\rm LH}^{\mathcal{B}} = K\langle X\rangle/\langle \mathbf{LM}(\mathcal{G})\rangle$ in a computational way. This forms the theme of the present chapter and the next two chapters.

Note that in view of (Ch.3, Section 4), any nonempty subset Ω of monomials in $K\langle X\rangle$ itself is a Gröbner basis, and that Ω stands for a family of Gröbner bases that have the same set of leading monomials. So, the results presented in this and the next two chapters also provide interesting families of algebras with better global structural properties.

All conventions and notations used in previous chapters are maintained.

5.1 The Working Strategy

We start this chapter by picturing clearly our working strategy of using Gröbner bases in the structure theory of associative algebras.

The case that $\mathbf{LH}(\mathcal{G})$ *is immediately good*

Let $R = \oplus_{\gamma\in\Gamma} R_\gamma$ be a Γ-graded K-algebra, where Γ is a totally ordered semigroup with the total ordering $<$. Consider an arbitrary ideal I of R and the quotient algebra $A = R/I$. Then by (Ch.2, Theorem 3.2),

$$(1) \qquad A^{\Gamma}_{\rm LH} = R/\langle \mathbf{LH}(I)\rangle \cong G^{\Gamma}(A),$$

where $A^{\Gamma}_{\rm LH}$ is the Γ-leading homogeneous algebra of A defined by the set of Γ-leading homogeneous elements of I taking with respect to the Γ-gradation of R, and $G^{\Gamma}(A)$ is the associated Γ-graded algebra of A defined by the Γ-filtration $F^{\Gamma}A$ induced by the Γ-grading filtration $F^{\Gamma}R$ of R. If Γ is a well-ordered monoid with the well-ordering $<$, then in view of (Ch.2, Section 4), the above formula (1) tells us that A may be studied in terms of the algebra $A^{\Gamma}_{\rm LH} = R/\langle \mathbf{LH}(I)\rangle$ via the commutative diagram

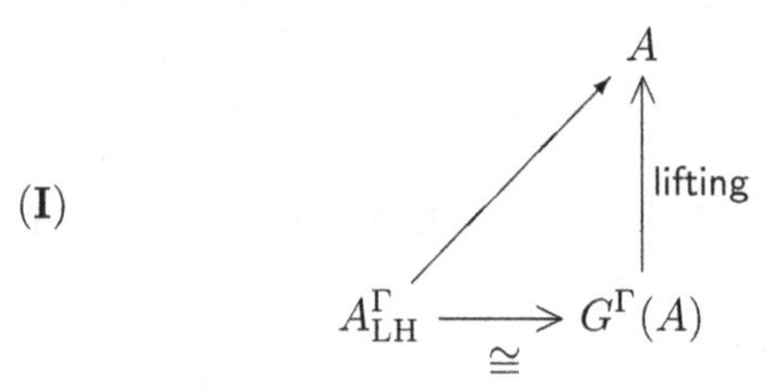

To see why Gröbner bases can make the above diagram (**I**) work effectively, let us assume that R has a skew multiplicative K-basis $\mathcal{B}$ consisting of Γ-*homogeneous elements*, and that there is a Γ-*graded monomial ordering* $\prec_{gr}$ on $\mathcal{B}$. Then, R has a Gröbner basis theory in the sense of (Ch.3, Section 4). Considering a Gröbner basis $\mathcal{G}$ of the ideal I in R with respect to the data $(\mathcal{B}, \prec_{gr})$, it follows from (Ch.4, Proposition 2.2, Theorem 2.3) that

$$(2) \qquad A^{\Gamma}_{\rm LH} = R/\langle \mathbf{LH}(I)\rangle = R/\langle \mathbf{LH}(\mathcal{G})\rangle \cong G^{\Gamma}(A).$$

The actual benefit of the formula (2) may be better understood by taking R to be the free K-algebra $K\langle X\rangle = K\langle X_1, \ldots, X_n\rangle$ with a fixed weight $\mathbb{N}$-gradation. That is, if the $\mathbb{N}$-leading homogeneous algebra $A^{\mathbb{N}}_{\rm LH}$ of A defined by the set $\mathbf{LH}(\mathcal{G})$ of $\mathbb{N}$-leading homogeneous elements of $\mathcal{G}$ happens to be some important algebra we are familiar with, or an algebra with some nice structural properties we may derive in terms of $\mathbf{LH}(\mathcal{G})$, as illustrated by examples given in (Ch.4, Sections 3, 4), then the lifting principle of "from $G^{\mathbb{N}}(A)$ to A" shown by the diagram (**I**) will work very well.

In practice, however, the $\mathbf{LH}(\mathcal{G})$ is, of course, not always immediately good as we expected. Moreover, if I is a Γ-graded ideal of R, then we have $A = R/I = R/\langle \mathbf{LH}(I)\rangle = A^{\Gamma}_{\rm LH}$, and starting with a homogenous generating set of I we obtain a homogeneous Gröbner basis $\mathcal{G}$ that gives rise to $\mathbf{LH}(\mathcal{G}) = \mathcal{G}$. Nevertheless, it follows from a Γ-analogue of (Ch.3, Lemma 6.2(i)) that by using $\prec_{gr}$, the equality $\mathbf{LM}(\mathcal{G}) = \mathbf{LM}(\mathbf{LH}(\mathcal{G}))$ holds for any Gröbner basis $\mathcal{G}$ of I. This motivates us to clarify, on the basis of (Ch.2, Section 4) and (Ch.4, Theorem 2.1), how to make effective use of the algebra $R/\langle\mathbf{LM}(\mathcal{G})\rangle$ in the next part.

***The case of using* $\mathbf{LM}(\mathcal{G})$**

Let $R = \oplus_{\gamma\in\Gamma}R_\gamma$ be a Γ-graded K-algebra, where Γ is a totally ordered semigroup with the total ordering $<$. Suppose that R has an admissible system $(\mathcal{B}, \prec)$, in which $\mathcal{B}$ is a skew multiplicative K-basis of R, and $\prec$ is a monomial ordering on $\mathcal{B}$, then R has a Gröbner basis theory in the sense of (Ch.3, Section 4), and R is also graded by the $\prec$-ordered $\mathcal{B}$ (as being used in Section 2 of Ch.4), i.e., $R = \oplus_{u\in\mathcal{B}}R_u$ with $R_u = Ku$. Consider an arbitrary ideal I of R and the quotient algebra $A = R/I$. It follows that the $\mathcal{B}$-leading homogeneous algebra $A^{\mathcal{B}}_{\rm LH}$ of A is well defined with respect to the $\mathcal{B}$-gradation of R, and in this case we have

$$A^{\mathcal{B}}_{\rm LH} = R/\langle\mathbf{LM}(I)\rangle. \tag{3}$$

Furthermore, assume that $\mathcal{B}$ consists of Γ-*homogeneous elements* of R and that there is a Γ-*graded monomial ordering* $\prec_{gr}$ on $\mathcal{B}$. Then by a Γ-analogue of (Ch.3, Lemma 6.2(i)) and (Ch.4, Proposition 2.2(i)) the following corollary is clear.

Corollary 1.1. *With notation as fixed above, the following statements hold.*
(i) *The algebra A and its Γ-leading homogeneous algebra $A^{\Gamma}_{\rm LH} = R/\langle\mathbf{LH}(I)\rangle$ have the same $\mathcal{B}$-leading homogeneous algebra, that is,*

$$R/\langle\mathbf{LM}(I)\rangle = A^{\mathcal{B}}_{\rm LH} = \left(A^{\Gamma}_{\rm LH}\right)^{\mathcal{B}}_{\rm LH} = R/\langle\mathbf{LM}(\langle\mathbf{LH}(I)\rangle)\rangle. \tag{4}$$

(ii) *If $\mathcal{G}$ is a Gröbner basis of I, then $\langle\mathbf{LM}(I)\rangle = \langle\mathbf{LM}(\mathcal{G})\rangle$ and hence*

$$A^{\mathcal{B}}_{\rm LH} = R/\langle\mathbf{LM}(\mathcal{G})\rangle = (A^{\Gamma}_{\rm LH})^{\mathcal{B}}_{\rm LH}. \tag{5}$$

□

Since $A^{\mathcal{B}}_{\rm LH} = R/\langle\mathbf{LM}(I)\rangle$ is a kind of "monomial algebra" (or a kind of semigroup algebra) with the set of defining relations $\mathbf{LM}(\mathcal{G})$ provided $\mathcal{G}$ is a Gröbner basis of I, in many practical (commutative or noncommutative) cases, $A^{\mathcal{B}}_{\rm LH}$ can be studied effectively in a combinatorial-computational way. Thus, combining the foregoing (1), (2), (3), (4) and (5), we have every reason to expect, at a modest level, that certain nice properties of $A^{\mathcal{B}}_{\rm LH}$ can be transferred to $A = R/I$ and $A^{\Gamma}_{\rm LH} = R/\langle \mathbf{LH}(I)\rangle \cong G^{\Gamma}(A)$, as indicated by the diagram

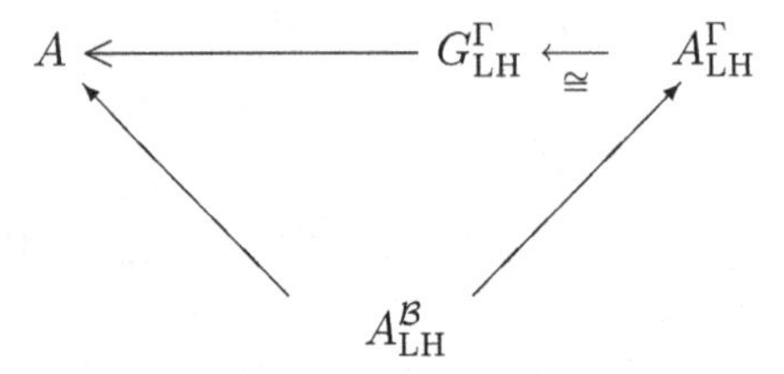

For instance, with respect to any admissible system $(\mathcal{B}, \prec)$ of R, we know from (Ch.3, Theorem 3.3, Proposition 4.8) that

$$\dim_K A = \dim_K A^{\mathcal{B}}_{\rm LH} = |N(I)|$$

with $N(I) = \{u \in \mathcal{B} \mid \mathbf{LM}(f) \not| \; u,\ f \in I\}$. Hence, using a Γ-graded monomial ordering $\prec_{gr}$ on the skew multiplicative K-basis $\mathcal{B}$ consisting of Γ-homogeneous elements, we immediately have the following result.

Theorem 1.2. [Compare with (Ch.2, Theorem 4.1(i))] *With notation and the assumption as above, $N(I)$ projects to a K-basis for A, $A^{\Gamma}_{\rm LH}$ and $A^{\mathcal{B}}_{\rm LH}$ respectively, and hence*

$$dim_K A = \ dim_K G^{\Gamma}(A) = \ dim_K A^{\mathcal{B}}_{\rm LH} = |N(I)|.$$

If $\mathcal{G}$ is a Gröbner basis of the ideal I, then the set $N(I)$ can be obtained by the division of monomials by $\mathbf{LM}(\mathcal{G})$, i.e., $N(I) = \{u \in \mathcal{B} \mid \mathbf{LM}(g) \not| \; u,\ g \in \mathcal{G}\}$.

□

Applying Theorem 1.2 to $\mathbb{N}$-filtered algebras, the next two results are derived.

Proposition 1.3. *Let R be one of the following algebras:* (a) *the free K-algebra $K\langle X\rangle = K\langle X_1, \ldots, X_n\rangle$;* (b) *the commutative polynomial K-algebra $K[x_1, \ldots, x_n]$; and* (c) *the path algebra KQ defined by a finite directed graph Q over K, and let I be an ideal of R, $A = R/I$. Consider the $\mathbb{N}$-filtration FA of A induced by the $\mathbb{N}$-grading filtration FR of R with respect to a fixed weight $\mathbb{N}$-gradation of R, and let $G^{\mathbb{N}}(A)$ be the associated $\mathbb{N}$-graded algebra*

of A determined by FA. With notation as before, the following statements hold.
(i) *$dim_K A = c = dim_K G^{\mathbb{N}}(A)$, where c denotes the cardinal number of a K-basis for A and $G^{\mathbb{N}}(A)$ respectively.*
(ii) ([MR], Proposition 8.6.5) *Writing GK.dim for the Gelfand-Kirillov dimension of K-algebras (see the definition in Section 3 later), then*

$$GK.dimA = GK.dimG^{\mathbb{N}}(A).$$

In the case that R is the commutative polynomial algebra $K[x_1, \dots, x_n]$, the Gelfand-Kirillov dimension in the above equality is replaced by the Krull dimension.

Proof. Let $\mathcal{B}$ be the standard K-basis of R.Then an $\mathbb{N}$-graded monomial ordering $\prec_{gr}$ on $\mathcal{B}$ does exist by (Ch.3, Section 1). It follows from Theorem 1.2 that

$$\dim_K A = \dim_K G^{\mathbb{N}}(A) = \dim_K A^{\mathcal{B}}_{\rm LH} = |N(I)|.$$

Moreover, since $N(I)$ projects to a K-basis for A, $A^{\mathbb{N}}_{\rm LH}$ and $A^{\mathcal{B}}_{\rm LH}$ respectively, it is easy to see that all three algebras have the same dimension function (usually referred to as Hilbert function) measuring their growth with respect to the natural $\mathbb{N}$-filtration (see Section 3 later), and hence, they have the same growth, or equivalently, they have the same Gelfand-Kirillov dimension. □

Remark For a later usage, it is convenient to note that the proof of the last proposition indeed tells us that

$$\text{GK.dim}A^{\mathcal{B}}_{\rm LH} = \text{GK.dim}A = \text{GK.dim}A^{\mathbb{N}}_{\rm LH}.$$

Proposition 1.4. *Let R be as in Proposition 1.3, and let $(\mathcal{B}, \prec_{gr})$ be an admissible system of R, in which $\prec_{gr}$ is an $\mathbb{N}$-graded monomial ordering on the standard K-basis $\mathcal{B}$ of R (note that an $\mathbb{N}$-graded monomial ordering always exists for the R considered). If J is an $\mathbb{N}$-graded ideal of R, then the $\mathbb{N}$-graded K-algebra $A = R/J$ and the $\mathbb{N}$-graded K-algebra $A^{\mathcal{B}}_{\rm LH} = R/\langle \mathbf{LM}(J)\rangle$ have the same Hilbert series. Hence, for an arbitrary ideal I of R, the $\mathbb{N}$-graded algebras $A^{\mathcal{B}}_{\rm LH}$ and $G^{\mathbb{N}}(A)$ have the same Hilbert series.*

Proof. Note that the set $N(I)$ of normal monomials in $\mathcal{B}$ (mod I) consists of $\mathbb{N}$-homogeneous elements of R. The assertion follows from the definition of Hilbert series for an $\mathbb{N}$-graded algebra (see Section 8 later) and the proof of Proposition 1.3 above. □

For quotient algebras of the free algebra $K\langle X\rangle = K\langle X_1, \ldots, X_n\rangle$, a combinatorial-computational approach to the determination of Gelfand-Kirillov dimension and Hilbert series is discussed in the subsequent Section 3 and Section 8 respectively.

The case of using $G^{\mathcal{B}}(A)$

Let $R = \oplus_{\gamma\in\Gamma} R_\gamma$ be a Γ-graded K-algebra, where Γ is a totally ordered semigroup with the total ordering $<$. Suppose that R has an admissible system $(\mathcal{B}, \prec)$, in which $\mathcal{B}$ is a skew multiplicative K-basis of R, and $\prec$ is a monomial ordering on $\mathcal{B}$, then R has a Gröbner basis theory in the sense of (Ch.3, Section 4), and R is also graded by the $\prec$-ordered $\mathcal{B}$, i.e., $R = \oplus_{u\in\mathcal{B}} R_u$ with $R_u = Ku$. To transfer more properties of $A^{\mathcal{B}}_{\mathbf{LH}}$ to A and $G^{\Gamma}(A)$, we assume further that *the $\mathcal{B}$-grading filtration $F^{\mathcal{B}}R$ of R exists* (for instance, if R does not have divisors of zero). Then, considering a proper ideal I of R and the quotient algebra $A = R/I$, by (Ch.2, Theorem 3.2) we have

$$A^{\mathcal{B}}_{\mathrm{LH}} = R/\langle \mathbf{LM}(I)\rangle \cong G^{\mathcal{B}}(A), \tag{6}$$

where $G^{\mathcal{B}}(A)$ is the associated $\mathcal{B}$-graded algebra of A defined by the $\mathcal{B}$-filtration $F^{\mathcal{B}}A$ induced by the $\mathcal{B}$-grading filtration $F^{\mathcal{B}}R$ of R, and by (Ch.2, Section 4), the above formula (6) tells us that A may be studied in terms of the algebra $A^{\mathcal{B}}_{\mathbf{LH}}$ via the commutative diagram

(II)

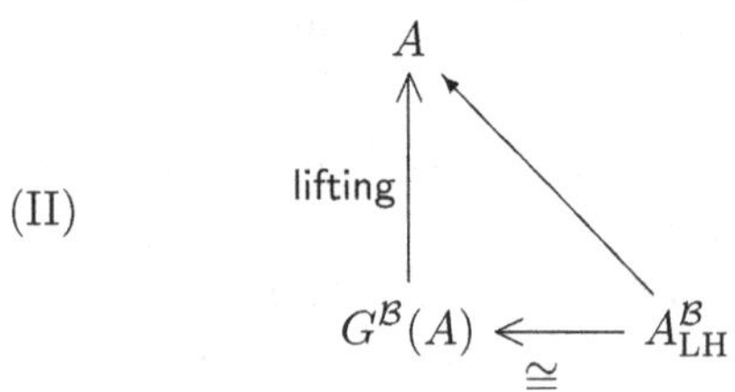

Since R has a Gröbner basis theory, the effectiveness of the diagram **(II)** above lies in the effective study of $A^{\mathcal{B}}_{\mathrm{LH}} = R/\langle \mathbf{LM}(\mathcal{G})\rangle$ provided $\mathcal{G}$ is a Gröbner basis of I.

Conclusion

Let $R = \oplus_{\gamma\in\Gamma} R_\gamma$ be a Γ-graded K-algebra, where Γ is a totally ordered semigroup with the total ordering $<$, and suppose that R has a Gröbner basis theory with respect to an admissible system $(\mathcal{B}, \prec)$, where $\mathcal{B}$ is a skew multiplicative K-basis and $\prec$ is a monomial ordering on $\mathcal{B}$. Consider a proper ideal I of R and the quotient algebra $A = R/I$. If $\mathcal{G}$ is a Gröbner basis for I, then, with notation as before, the working strategy we have described in this section may be summarized by the following diagram:

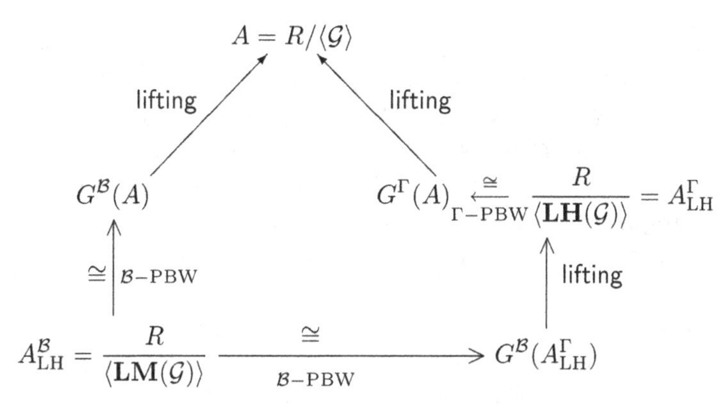

in which, the route

$$A^{\mathcal{B}}_{\rm LH} = R/\langle\langle \mathbf{LM}(\mathcal{G})\rangle \overset{\cong}{\longrightarrow} G^{\mathcal{B}}(A^{\Gamma}_{\rm LH}) \longrightarrow A^{\Gamma}_{\rm LH} = R/\langle \mathbf{LH}(\mathcal{G})\rangle \longrightarrow A = R/\langle \mathcal{G}\rangle$$

works provided $\mathcal{B}$ consists of Γ-homogeneous elements, and $\prec$ is a Γ-graded monomial ordering on $\mathcal{B}$.

5.2 Ufnarovski Graph

Let $K\langle X\rangle = K\langle X_1, \ldots, X_n\rangle$ be the free K-algebra of n generators. Starting from this section we focus on the computational study of quotient algebras of $K\langle X\rangle$ by using Gröbner bases as indicated in the last section.

The aim of this section is to introduce the Ufnarovski graph associated to a finite Gröbner basis in $K\langle X\rangle$.

Let $\mathcal{B}$ be the standard K-basis of $K\langle X\rangle$. First recall from (Ch.3, Section 2) that a subset $\Omega \subset \mathcal{B}$ is said to be *reduced* if $u, v \in \Omega$ and $u \neq v$ implies $u \nmid v$; and that a Gröbner basis $\mathcal{G}$ is called *LM-reduced* if $\mathbf{LM}(\mathcal{G})$ is reduced.

Now, let $\Omega = \{u_1, \ldots, u_s\}$ be a reduced finite subset of $\mathcal{B}$. For each $u_i \in \Omega$, say $u_i = X_{i_1}^{\alpha_1} \cdots X_{i_s}^{\alpha_s}$ with $X_{i_j} \in X$ and $\alpha_j \in \mathbb{N}$, as before we write $l(u_i) = \alpha_1 + \cdots + \alpha_s$ for the length of u_i . Put

$$\ell = \max\left\{ l(u_i) \;\middle|\; u_i \in \Omega \right\}.$$

Then the *Ufnarovski graph* of Ω (in the sense of [Uf1]), denoted by $\Gamma(\Omega)$, is defined as a directed graph, in which the set of vertices V is given by

$$V = \left\{ v_i \;\middle|\; v_i \in \mathcal{B} - \langle \Omega \rangle,\ l(v_i) = \ell - 1 \right\},$$

and the set of edges E contains the edge $v_i \to v_j$ if and only if there exist $X_i, X_j \in X$ such that $v_i X_i = X_j v_j \in \mathcal{B} - \langle \Omega \rangle$.

Since any Gröbner basis in $K\langle X \rangle$ may be reduced to an LM-reduced Gröbner basis (see Ch.3, Section 5), for a finite Gröbner basis $\mathcal{G} = \{g_1, \ldots, g_s\} \subset K\langle X \rangle$, the *Ufnarovski graph of* $\mathcal{G}$ is defined to be the Ufnarovski graph of the unique reduced monomial generating set of the monomial ideal $\langle \mathbf{LM}(\mathcal{G}) \rangle$, and is denoted by $\Gamma(\mathbf{LM}(\mathcal{G}))$ without confusion.

Remark To better understand the practical application of $\Gamma(\Omega)$ in the subsequent sections, it is essential to notice that a Ufnarovski graph is defined by using the *length* $l(u)$ *of the monomial (word)* $u \in \mathcal{B}$ instead of using the *degree of* u *as an* $\mathbb{N}$*-homogeneous element in* $K\langle X \rangle$ whenever a weight $\mathbb{N}$-gradation of $K\langle X \rangle$ is used, though both notions coincide when each X_i is assigned the degree 1.

Example 1. Consider in $K\langle X_1, X_2 \rangle$ the subset of monomials $\Omega = \{X_1^2 X_2,\ X_1 X_2^2\}$. Then

$$\begin{aligned} &\ell = \max\{l(u_i) \mid u_i \in \Omega\} = 3, \\ &V = \{v_1 = X_1X_2,\ v_2 = X_2X_1,\ v_3 = X_1^2,\ v_4 = X_2^2\}. \end{aligned}$$

So $\Gamma(\Omega)$ is presented by

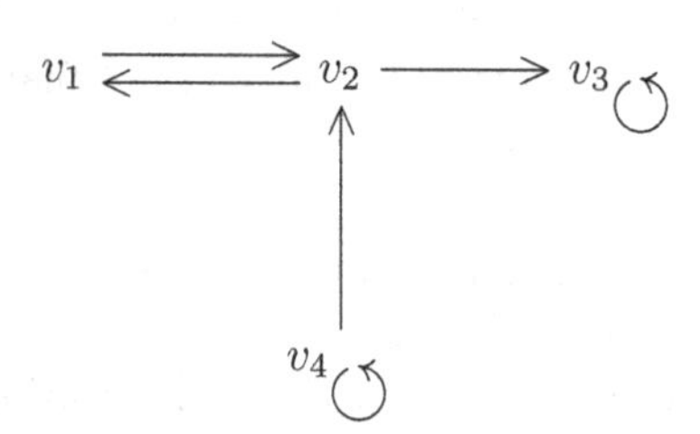

Example 2. Consider in $K\langle X_1, X_2, X_3, X_4 \rangle$ the subset of monomials $\Omega = \{X_j X_i \mid 1 \le i < j \le 4\}$. Then

$$\begin{aligned} &\ell = \max\{l(u_i) \mid u_i \in \Omega\} = 2, \\ &V = \{v_1 = X_1, v_2 = X_2, v_3 = X_3, v_4 = X_4\}. \end{aligned}$$

It turns out that $\Gamma(\Omega)$ is presented by

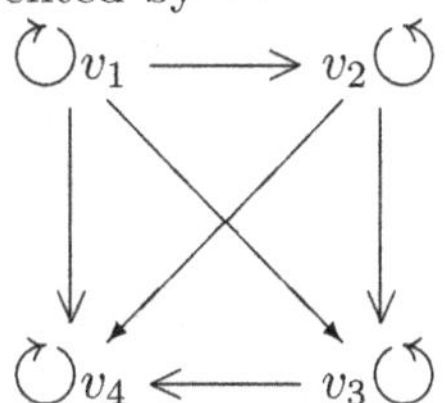

Similarly, if in place of V we use successively the subset of vertices $V_{ijkm} = \{v_1 = X_i,\ v_2 = X_j,\ v_3 = X_k,\ v_4 = X_m\}$, $1 \le i < j < k < m \le n$, then we get the graph $\Gamma(\Omega)$ for the subset of monomials $\Omega = \{X_jX_i \mid 1 \le i < j \le n\}$ in $K\langle X_1, \ldots, X_n\rangle$.

Example 3. Let n be a fixed positive integer and consider in $K\langle X_1, X_2\rangle$ the subset Ω consisting of the single monomial $X_2^nX_1$. Then

$$\begin{aligned} &\ell = l(X_2^nX_1) = n+1,\\ &V = \{X_1^{n-i}X_2^i,\ X_2^{n-i}X_1^i \mid 0 \le i \le n\}, \end{aligned}$$

and it is straightforward to verify that for $n = 1$ the Ufnarovski graph $\Gamma(\Omega)$ is presented by

$$\circlearrowleft X_1 \longrightarrow X_2 \circlearrowright$$

for $n = 2$, the Ufnarovski graph $\Gamma(\Omega)$ is presented by

$$\circlearrowleft X_1^2 \longleftarrow X_2X_1 \rightleftarrows X_1X_2 \longrightarrow X_2^2 \circlearrowright$$

and for $n \ge 3$, the Ufnarovski graph $\Gamma(\Omega)$ is presented by

$$\begin{array}{ccccccc} & & & & & & X_2^n \circlearrowright \\ & & & & & & \uparrow \\ X_1^{n-1}X_2 & \rightarrow & X_1^{n-2}X_2^2 & \rightarrow & \cdots & \rightarrow & X_1X_2^{n-1} \\ \uparrow & & & & & & \downarrow \\ X_2X_1^{n-1} & \leftarrow & \cdots & \leftarrow & X_2^{n-2}X_1^2 & \leftarrow & X_2^{n-1}X_1 \\ \downarrow & & & & & & \\ \circlearrowleft X_1^n & & & & & & \end{array}$$

Basic notions from classical graph theory fully suit an Ufnarofski graph $\Gamma(\Omega)$. For instance, a *route of length* m in $\Gamma(\Omega)$ with $m \geq 1$ is a sequence of edges

$$v_0 \to v_1 \to v_2 \to \cdots \to v_{m-1} \to v_m.$$

If in a route no edge appears repeatedly then it is called a *simple route.* A simple route with $m \geq 1$ and $v_0 = v_m$ is called a *cycle.* If, as an undirected graph, there is a route between any two distinct vertices of $\Gamma(\Omega)$, then $\Gamma(\Omega)$ is called *connected.* A *connected component* of $\Gamma(\Omega)$ (as an undirected graph) is a connected subgraph which is connected to no additional vertices (cf. `http://en.wikipedia.org/wiki/Graph_theory`).

For a given reduced finite subset $\Omega \subset \mathcal{B}$, the following property builds the essential link between the Ufnarovski graph $\Gamma(\Omega)$ and the monomial K-algebra $R/\langle\Omega\rangle$.

Proposition 2.1 ([Uf1]). *There is a one-to-one correspondence between the monomials (words) of length* $\geq \ell - 1$ *in* $\mathcal{B} - \mathbf{LM}(\langle\Omega\rangle)$ *and the routes in* $\Gamma(\Omega)$, *that is, if* $u \in \mathcal{B} - \mathbf{LM}(\langle\Omega\rangle)$ *and* $u = X_{i_1} \cdots X_{i_s}$ *with* $s \geq \ell - 1$, *then* u *is mapped to the route*

$$R(u): \quad v_0 \to v_1 \to \cdots \to v_m,$$

where

$$\begin{aligned} &m = s - \ell + 1, \text{ and} \\ &v_j = X_{i_{j+1}} X_{i_{j+2}} \cdots X_{i_{j+\ell-1}}, \; 0 \leq j \leq m. \end{aligned}$$

Moreover, under this correspondence, a cycle beginning at v *is represented by a monomial of the form* $vw = tv$ *for some* $w, t \in \mathcal{B}$.

□

Note that the Ufnarovski graph $\Gamma(\Omega)$ of a reduced finite subset $\Omega \subset \mathcal{B}$ is a finite directed graph. As we will see soon after this section, once a certain structural property P of the monomial algebra $K\langle X\rangle/\langle\Omega\rangle$ can be translated via Proposition 2.1 into a property on the graph $\Gamma(\Omega)$, then, with the aid of normal elements and the division algorithm for monomials, the property P may become algorithmically realizable.

5.3 Determination of Gelfand-Kirillov Dimension

Let the free K-algebra $K\langle X\rangle = K\langle X_1,\ldots,X_n\rangle$ of n generators be equipped with a weight $\mathbb{N}$-gradation by assigning each X_i to a positive degree n_i, and let I be an ideal of $K\langle X\rangle$, $A = K\langle X\rangle/I$. Considering the $\mathbb{N}$-filtration FA of A induced by the weight $\mathbb{N}$-grading filtration $FK\langle X\rangle$ of $K\langle X\rangle$, it follows from Proposition 1.3 of Section 1 that thc algebra A and the associated $\mathbb{N}$-graded algebra $G^{\mathbb{N}}(A)$ of A have the same Gelfand-Kirillov dimension. In this section we show that if I is generated by a finite Gröbner basis $\mathcal{G}$ with respect to some admissible system $(\mathcal{B},\prec)$ of $K\langle X\rangle$, where $\prec$ is *any fixed* monomial ordering on the standard K-basis $\mathcal{B}$ of $K\langle X\rangle$, then the Gelfand-Kirillov dimension of A, and hence the Gelfand-Kirillov dimension of $G^{\mathbb{N}}(A)$, is determined by the Ufnarovski graph $\Gamma(\mathbf{LM}(\mathcal{G}))$ of $\mathcal{G}$ introduced in the last section.

For a general theory on the Gelfand-Kirillov dimension of algebras we refer to ([KL]) or ([M-R], Chapter 8). Here we recall only the basic notion concerned in our text. Let A be an associative K-algebra generated by a finite dimensional K-vector space V. Consider for each $i\in\mathbb{N}$ the subspace V^i of A spanned by all the products of i elements of V, and note that $V^0 = K$. Then the function

$$f(n) = \dim_K\left(\sum_{i=0}^{n} V^i\right),\quad n\in\mathbb{N},$$

measures the rate of growth of A. The algebra A is said to have *polynomial growth of degree* d if there is a polynomial $p(x)$ of degrre d with real coefficients such that

$$f(n)\le p(n) \text{ for } n\gg 0;$$

and A is said to have *exponential growth* if there exists a real number $\varepsilon > 1$ such that

$$f(n)\ge \varepsilon^n \text{ for } n\gg 0.$$

The asymptotic behavior of the seuqnece $\{f(n)\}_{n\in\mathbb{N}}$ is indicated by the *Gelfand-Kirillov dimension*, or *GK dimension* for short, which is defined as

$$\text{GK.dim}A = \lim_{n\to\infty}\sup\frac{\log f(n)}{\log n}.$$

The GK dimension of A defined above does not depend on the choice of the generating subspace V. Moreover, if A has polynomial growth of degree d, then $\text{GK.dim}A = d$, and if A has exponential growth, then $\text{GK.dim}A = \infty$.

Now, let $\Omega = \{u_1, \ldots, u_s\}$ be a reduced finite subset of monomials in the free K-algebra $K\langle X\rangle = K\langle X_1, \ldots, X_n\rangle$, and let $\Gamma(\Omega)$ be the Ufnarovski graph of Ω as defined in the last section. The first effective application of $\Gamma(\Omega)$ was made to determine the growth of the monomial algebra $K\langle X\rangle/\langle\Omega\rangle$.

Theorem 3.1 ([Uf1]). (i) *The growth of $K\langle X\rangle/\langle\Omega\rangle$ is alternative, that is, it is exponential if and only if there are two different cycles with a common vertex in the graph $\Gamma(\Omega)$; otherwise, $K\langle X\rangle/\langle\Omega\rangle$ has polynomial growth of degree d, or equivalently, the GK dimension of $K\langle X\rangle/\langle\Omega\rangle$ is equal to d, where d is, among all routes of $\Gamma(\Omega)$, the largest number of distinct cycles occurring in a single route.*
(ii) *$K\langle X\rangle/\langle\Omega\rangle$ is a finite dimensional K-vector space if and only if $\Gamma(\Omega)$ does not contain any cycle.*

□

Combining Theorem 3.1 with the proof of Proposition 1.3 given in Section 1, the following result is immediately clear now.

Theorem 3.2. *With the convention made at the beginning of this section, let $\mathcal{G} = \{g_1, \ldots, g_s\}$ be an LM-reduced finite Gröbner basis in $K\langle X\rangle$ with respect to some fixed admissible system $(\mathcal{B}, \prec)$, $A = K\langle X\rangle/I$ where $I = \langle\mathcal{G}\rangle$. With $A^{\mathcal{B}}_{\rm LH} = K\langle X\rangle/\langle\mathbf{LM}(I)\rangle$ and $A^{\mathbb{N}}_{\rm LH} = K\langle X\rangle/\langle\mathbf{LH}(I)\rangle \cong G^{\mathbb{N}}(A)$ as before, the following statements hold.*
(i) *All three algebras A, $A^{\mathcal{B}}_{\rm LH}$ and $G^{\mathbb{N}}(A)$ have exponential growth if and only if the graph $\Gamma(\mathbf{LM}(\mathcal{G}))$ has two different cycles with a common vertex. Otherwise, they all have polynomial growth of degree d, or equivalently, their GK dimension is equal to d, where d is, among all routes of $\Gamma(\mathbf{LM}(\mathcal{G}))$, the largest number of distinct cycles occurring in a single route.*
(ii) *All three algebras A, $A^{\mathcal{B}}_{\rm LH}$ and $G^{\mathbb{N}}(A)$ are finite dimensional K-vector spaces of the same dimension if and only if $\Gamma(\mathbf{LM}(\mathcal{G}))$ does not contain any cycle.*

□

Remark For an LM-reduced finite Gröbner basis $\mathcal{G}$ in $K\langle X\rangle$, the application of $\Gamma(\mathbf{LM}(\mathcal{G}))$ to the determination of the growth of $A = K\langle X\rangle/\langle\mathcal{G}\rangle$ was first observed in ([G-IL], Proposition 2).

In all examples given below, the convention made at the beginning of this section is maintained. We also remind the reader to notice that if I is generated by a Gröbner basis $\mathcal{G}$ with respect to some admissible system $(\mathcal{B}, \prec)$ but in which $\prec$ is not an $\mathbb{N}$-graded monomial ordering, then in general $\langle \mathbf{LH}(I)\rangle \neq \langle \mathbf{LH}(\mathcal{G})\rangle$, i.e., $G^{\mathbb{N}}(A) \cong A^{\mathbb{N}}_{\rm LH} = K\langle X\rangle/\langle \mathbf{LH}(I)\rangle \neq K\langle X\rangle/\langle \mathbf{LH}(\mathcal{G})\rangle$ (see Ch.4, Section 2).

Example 1. Consider in the free K-algebra $K\langle X\rangle = K\langle X_1, X_2, X_3\rangle$ the subset of monomials $\Omega = \{X_1X_2, X_3X_1, X_2^2\}$. Then

$$\ell = \max\{l(u_i) \mid u_i \in \Omega\} = 2,$$
$$V = \{v_1 = X_1, v_2 = X_2, v_3 = X_3\},$$

and the Ufnarovski graph $\Gamma(\Omega)$ is presented by

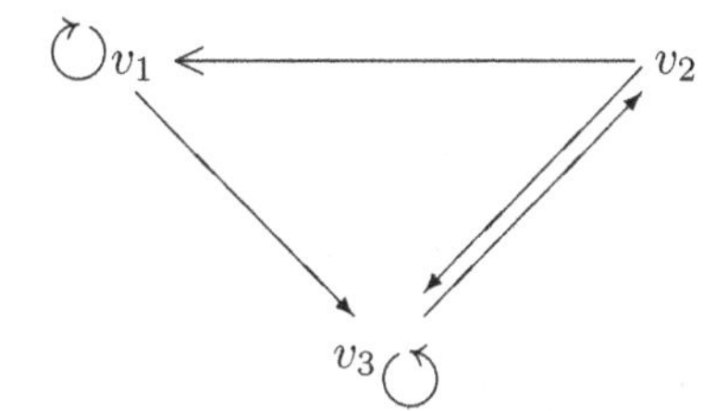

Hence, by Theorem 3.2, if $\mathcal{G} = \{g_1, g_2, g_3\} \subset K\langle X\rangle$ is any Gröbner basis with respect to some fixed monomial ordering $\prec$ such that $\mathbf{LM}(g_1) = X_1X_2$, $\mathbf{LM}(g_2) = X_3X_1$ and $\mathbf{LM}(g_3) = X_2^2$, then for $I = \langle \mathcal{G}\rangle$, all three algebras $A = K\langle X\rangle/I$, $A^{\mathcal{B}}_{\rm LH} = K\langle X\rangle/\langle \mathbf{LM}(I)\rangle$ and $G^{\mathbb{N}}(A) \cong A^{\mathbb{N}}_{\rm LH} = K\langle X\rangle/\langle \mathbf{LH}(I)\rangle$ have exponential growth, i.e., they all have infinite GK dimension.

Example 2. Let $(\mathcal{B}, \prec)$ be an admissible system of the free K-algebra $K\langle X\rangle = K\langle X_1, X_2\rangle$. If $\mathcal{G} = \{g_1, g_2\}$ is any Gröbner basis in $K\langle X\rangle$ such that $\mathbf{LM}(g_1) = X_1^2X_2$ and $\mathbf{LM}(g_2) = X_1X_2^2$, then for $I = \langle \mathcal{G}\rangle$, it follows from Example 1 of the last section and Theorem 3.2 that all three algebras $A = K\langle X\rangle/I$, $A^{\mathcal{B}}_{\rm LH} = K\langle X\rangle/\langle \mathbf{LM}(I)\rangle$ and $G^{\mathbb{N}}(A) \cong A^{\mathbb{N}}_{\rm LH} = K\langle X\rangle/\langle \mathbf{LH}(I)\rangle$ have polynomial growth of degree 3 and therefore have GK dimension 3.

In particular, all the algebras (including all down-up algebras $A = A(\alpha, \beta, \gamma)$) considered in Example 8 of (Ch.3, Section 5) have GK dimension 3 (see also Example 2 of (Ch.4, Section 2)).

Example 3. Consider in the free K-algebra $K\langle X\rangle = K\langle X_1, \ldots, X_n\rangle$ a subset of monomials $\Omega \subseteq \{X_jX_i \mid 1 \le i < j \le n\}$ or $\Omega \subseteq \{X_iX_j \mid 1 \le$

$i < j \le n\}$. If $\mathcal{G}$ is any Gröbner basis with respect to some fixed monomial ordering $\prec$ such that $\mathbf{LM}(\mathcal{G}) = \Omega$, then for $I = \langle \mathcal{G} \rangle$, it follows from Example 2 of the last section (or the construction of $\Gamma(\Omega)$) and Theorem 3.2 that the algebras $A = K\langle X \rangle / I$, $A^{\mathcal{B}}_{\rm LH} = K\langle X \rangle / \langle \mathbf{LM}(I) \rangle$ and $G^{\mathbb{N}}(A) \cong A^{\mathbb{N}}_{\rm LH} = K\langle X \rangle / \langle \mathbf{LH}(I) \rangle$, all have polynomial growth of degree $\le n$ and therefore have GK dimension $\le n$. In the case that $\Omega = \{X_j X_i \mid 1 \le i < j \le n\}$ or $\Omega = \{X_i X_j \mid 1 \le i < j \le n\}$, all three algebras considered above have GK dimension n.

At this stage, the reader is invited to review all algebras constructed in (Ch.4, Sections 3, 4) which have the set of defining relations given by a quadric Gröbner basis.

Example 4. Consider in the free K-algebra $K\langle X \rangle = K\langle X_1, \ldots, X_n \rangle$ the subset of monomials

$$\Omega = \left\{ X_{k_1}^p, \ldots, X_{k_s}^p,\ X_j X_i \;\middle|\; 1 \le i < j \le n \right\}$$

where $p \ge 2$ and $1 \le s \le n$ are fixed integers. Then

$$\begin{aligned} &\ell = \max\{l(u_i) \mid u_i \in \Omega\} = p, \\ &V = \{v = X_1^{\alpha_1} X_2^{\alpha_2} \cdots X_n^{\alpha_n} \mid \alpha_j \in \mathbb{N},\ \alpha_1 + \alpha_2 + \cdots + \alpha_n = p - 1\}. \end{aligned}$$

It is straightforward to verify that all edges in the Ufnarovski graph $\Gamma(\Omega)$ are of the form

$$X_i w \longrightarrow w X_j \text{ with } w \in \mathcal{B} - \langle \Omega \rangle \text{ and } j \ge i,$$

$\Gamma(\Omega)$ has only $n - s$ cycles which are of the form

$$v_j \circlearrowleft \text{ with } v_j = X_j^{p-1} \text{ and } X_j \notin \{X_{k_1}, \ldots, X_{k_s}\},$$

and all $n - s$ cycles occur in the single route

$$\begin{array}{ccccccc} X_1^{p-1} & \rightarrow X_1^{p-2} X_2 \rightarrow & \cdots & & \rightarrow X_1 X_2^{p-2} \\ & & & & \downarrow \\ X_{n-1}^{p-1} & \leftarrow & \cdots & \leftarrow & X_2^{p-2} X_3 & \leftarrow & X_2^{p-1} \\ \downarrow & & & & & & \\ X_{n-1}^{p-2} X_n \rightarrow & & \cdots & & \rightarrow X_{n-1} X_n^{p-2} & \rightarrow & X_n^{p-1} \end{array}$$

For instance, if $n = 4$, $\Omega = \{X_1^3, X_2^3,\ X_j X_i \mid 1 \le i < j \le 4\}$, then $\Gamma(\Omega)$

has the presentation

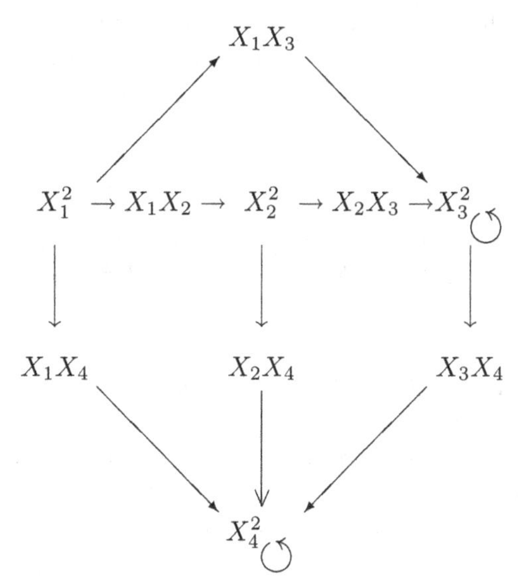

So, if $\mathcal{G}$ is any Gröbner basis with respect to some fixed monomial ordering $\prec$ such that $\mathbf{LM}(\mathcal{G}) = \Omega$, then for $I = \langle\mathcal{G}\rangle$, it follows from the construction of $\Gamma(\Omega)$ and Theorem 3.2 that all three algebras $A = K\langle X\rangle/I$, $A_{\mathrm{LH}}^{\mathcal{B}} = K\langle X\rangle/\langle\mathbf{LM}(I)\rangle$ and $G^{\mathbb{N}}(A) \cong A_{\mathrm{LH}}^{\mathbb{N}} = K\langle X\rangle/\langle\mathbf{LH}(I)\rangle$ have polynomial growth of degree $n - s$ and therefore have GK dimension $n - s$.

We see that this example applies to all algebras given in Examples 6, 7 of (Ch.3, Section 5), and that it especially recaptures the classical result on the finite dimensionality of exterior algebras and Clifford algebras.

Example 5. Let $(\mathcal{B}, \prec)$ be an admissible system of the free K-algebra $K\langle X\rangle = K\langle X_1, X_2\rangle$. If $\mathcal{G}$ is any Gröbner basis in $K\langle X\rangle$ consisting of a single element g such that $\mathbf{LM}(g) = X_2^n X_1$ with $n \geq 1$ fixed, then for $I = \langle\mathcal{G}\rangle$, it follows from Example 3 of the last section and Theorem 3.2 that the algebras $A = K\langle X\rangle/I$, $A_{\mathrm{LH}}^{\mathcal{B}} = K\langle X\rangle/\langle\mathbf{LM}(I)\rangle$ and $G^{\mathbb{N}}(A) \cong A_{\mathrm{LH}}^{\mathbb{N}} = K\langle X\rangle/\langle\mathbf{LH}(I)\rangle$, all have polynomial growth of degree 2 and therefore have GK dimension 2.

5.4 Recognizing Noetherianity

Let $K\langle X\rangle = K\langle X_1, \ldots, X_n\rangle$ be the free K-algebra of n generators, and $\mathcal{B}$ the standard K-basis of $K\langle X\rangle$. From (Ch.4, Section 4) we know that if, with respect to some admissible system $(\mathcal{B}, \prec)$ of $K\langle X\rangle$, $\mathcal{G}$ is a Gröbner basis in $K\langle X\rangle$ such that $A = K\langle X\rangle/\langle\mathcal{G}\rangle$ forms a solvable polynomial algebra, then A is a Noetherian domain. In this section we show that the Ufnarovski graph introduced in Section 2 may be used to determine Noetherian algebras of special type. Notations are kept as in previous sections.

Theorem 4.1 ([Uf2], [Nor2]). *Let $\Omega = \{u_1, \ldots, u_s\}$ be a reduced finite subset of $\mathcal{B}$ and $A = K\langle X\rangle/\langle\Omega\rangle$. If there is no edge entering (leaving) any cycle of the Ufnarovski graph $\Gamma(\Omega)$, then A is left (right) Noetherian.*

□

Theorem 4.2. *Let $(\mathcal{B}, \prec)$ be an admissible system for $K\langle X\rangle$, and I an ideal of $K\langle X\rangle$. Suppose that $\langle\mathbf{LM}(I)\rangle$ contains a reduced finite subset of monomials Ω, such that there is no edge entering (leaving) any cycle of the Ufnarovski graph $\Gamma(\Omega)$. Then the monomial algebra $A_{\mathrm{LH}}^{\mathcal{B}} = K\langle X\rangle/\langle\mathbf{LM}(I)\rangle$ is left (right) Noetherian of GK dimension not exceeding 1, and hence $A = K\langle X\rangle/I$ is left (right) Noetherian of GK dimension not exceeding 1. Considering the $\mathbb{N}$-filtration FA of A induced by the weight $\mathbb{N}$-grading filtration $FK\langle X\rangle$ of $K\langle X\rangle$ with respect to a fixed weight $\mathbb{N}$-gradation of $K\langle X\rangle$, if an $\mathbb{N}$-graded monomial ordering $\prec_{gr}$ is used , then, under the same assumption the associated $\mathbb{N}$-graded algebra $G^{\mathbb{N}}(A)$ of A is left (right) Noetherian of GK dimension not exceeding 1 as well.*

Proof. Since $\langle\Omega\rangle \subset \langle\mathbf{LM}(I)\rangle$ by the assumption, it follows from Theorem 4.1 that $A_{\mathrm{LH}}^{\mathcal{B}}$ is left (right) Noetherian. Thus A is left (right) Noetherian by (Ch.2, Theorem 4.2). Note that $G^{\mathbb{N}}(A) \cong A_{\mathrm{LH}}^{\mathbb{N}} = K\langle X\rangle/\langle\mathbf{LH}(I)\rangle$ by (Ch.2, Theorem 3.2). If an $\mathbb{N}$-graded monomial ordering $\prec_{gr}$ is used, then, under the same assumption the left (right) Noetherianity of $A_{\mathrm{LH}}^{\mathcal{B}}$ can be lifted not only to A but also to $G^{\mathbb{N}}(A)$ by Corollary 1.1 and (Ch.2, Theorem 4.2). Finally, the statement on GK dimension follows from the assumption on Ω and Theorem 3.2 (see also [Ok], Proposition 7).

□

Example 1. As a small example let us look at the monomial algebra $S = K\langle X, Y\rangle/\langle X^2, YX\rangle$. Put $\Omega = \{X^2, YX\}$. It is easy to see that the Ufnarovski graph $\Gamma(\Omega)$ of Ω is of the form

$$X \longrightarrow Y \circlearrowleft$$

and there is no edge leaving the only cycle of $\Gamma(\Omega)$. Hence S is right Noetherian of GK dimension 1. So Theorem 4.2 can be applied to any algebra $A = K\langle X,Y\rangle/I$ with X^2, $YX \in \langle \mathbf{LM}(I)\rangle$, for instance, $I = \langle X^2 + aY^2 + b,\ YX + cX + d, h(X,Y)\rangle$, where $a,b,c,d \in K$, $h(X,Y) \in K\langle X,Y\rangle$, and the monomial ordering used may be the $\mathbb{N}$-graded lexicographic ordering $Y \prec_{gr} X$.

Indeed, by referring to Example 4 of Section 3 it is a good exercise to show that if $p \geq 2$ is a fixed integer and I is an ideal of $K\langle X\rangle = K\langle X_1,\ldots,X_n\rangle$ such that $\mathbf{LM}(I)$ contains $\{X_1^p,\ldots,X_{n-1}^p,\ X_jX_i \mid 1 \leq i < j \leq n\}$, then all three algebras $A_{\mathrm{LH}}^{\mathcal{B}} = K\langle X\rangle/\langle \mathbf{LM}(I)\rangle$, $A = K\langle X\rangle/I$ and $A_{\mathrm{LH}}^{\mathbb{N}} = K\langle X\rangle/\langle \mathbf{LH}(I)\rangle \cong G^{\mathbb{N}}(A)$, are right Noetherian of GK dimension not exceeding 1.

Remark Recall that an algebra is called *weak Noetherian* if it satisfies the ascending chain condition for ideals. If $\Omega \subset \mathcal{B}$ is a reduced finite subset, then it was proved in [Nor2] that the monomial algebra $K\langle X\rangle/\langle\Omega\rangle$ is weak Noetherian if and only if the Ufnarofski graph $\Gamma(\Omega)$ does not contain any cycle with edges both entering and leaving it. In a similar way one may get an analogue of Theorem 4.2 on the weak Noetherianity of $A = K\langle X\rangle/I$ and $A_{\mathrm{LH}}^{\mathbb{N}} = K\langle X\rangle/\langle \mathbf{LH}(I)\rangle \cong G^{\mathbb{N}}(A)$.

For an algorithmic recognition of the Noetherian (semi-)prime PI algebra of GK dimension 1 and Krull dimension 1, we refer to the next section.

5.5 Recognizing (Semi-)Primeness and PI-Property

Let $K\langle X\rangle = K\langle X_1,\ldots,X_n\rangle$ be the free K-algebra of n generators, and $\mathcal{B}$ the standard K-basis of $K\langle X\rangle$. Fixing an admissible system $(\mathcal{B},\prec)$ for $K\langle X\rangle$, let $I = \langle\mathcal{G}\rangle$ be an ideal of $K\langle X\rangle$ generated by a finite Gröbner basis $\mathcal{G}$, $A = K\langle X\rangle/I$. In this section we show how to recognize the semi-primeness (primeness) of $A_{\mathrm{LH}}^{\mathcal{B}} = K\langle X\rangle/\langle \mathbf{LM}(I)\rangle$, A and $G^{\mathbb{N}}(A)$ by using the Ufnarovski graph $\Gamma(\mathbf{LM}(\mathcal{G}))$, where $G^{\mathbb{N}}(A)$ is the associated $\mathbb{N}$-graded algebra of A defined by the $\mathbb{N}$-filtration FA of A induced by the weight $\mathbb{N}$-grading filtration $FK\langle X\rangle$ of $K\langle X\rangle$. Moreover, we show how to recognize the PI-property (see the definition given before Theorem 5.5 later) of $A_{\mathrm{LH}}^{\mathcal{B}}$, A and $G^{\mathbb{N}}(A)$ by means of $\Gamma(\mathbf{LM}(\mathcal{G}))$ in certain special cases.

Let $\Omega = \{u_1,\ldots,u_s\}$ be a reduced finite subset of $\mathcal{B}$, and let $\Gamma(\Omega)$ be the Ufnarovski graph of Ω with $\ell = \max\{l(u_i) \mid u_i \in \Omega\}$. Assume $l(u_i) \geq 2$, $i = 1, \ldots, s$. Recall from [Uf1] and [G-I1] that a vertex v of $\Gamma(\Omega)$ is called a *cyclic vertex* if it belongs to a cyclic route of $\Gamma(\Omega)$; and a monomial $u \in \mathcal{B} - \langle\Omega\rangle$, $u \neq 1$, is called a *cyclic monomial* if $l(u) \leq \ell - 1$ and u is a suffix of some cyclic vertex of $\Gamma(\Omega)$, or, if $l(u) > \ell - 1$ and the associated route $R(u)$ of u (see Proposition 2.1 of Section 2) is a subroute of some cyclic route.

Concerning the algorithmic realization of the semi-primeness (primeness) and the Jacobson semi-simplicity of a finitely presented monomial algebra, the following two results are fundamental.

Theorem 5.1. *With notation as above, let J be the Jacobson radical of the monomial algebra $K\langle X\rangle/\langle\Omega\rangle$. Then the following statements hold.*
(i) ([G-IL]) *J coincides with the upper nil-radical of $K\langle X\rangle/\langle\Omega\rangle$.*
(ii) ([G-I1]) *J is the K-linear span of the set of all noncyclic monomials in $\mathcal{B} - \langle\Omega\rangle$, and moreover, J is finitely generated as an ideal by the set of all noncyclic monomials of degree $\leq \ell$.*

□

Recall that a ring with zero Jacobson radical is called Jacobson semi-simple (see e.g., [Lam]). By the above theorem it is clear that the monomial algebra $K\langle X\rangle/\langle\Omega\rangle$ is Jacobson semi-simple if and only if it is semi-prime. There is also an algorithmic characterization of this property.

Theorem 5.2 ([G-I1]). *With notation as before, the following statements hold.*
(i) *The monomial algebra $K\langle X\rangle/\langle\Omega\rangle$ is Jacobson semi-simple, or equivalently semi-prime, if and only if each monomial $u \in \mathcal{B} - \langle\Omega\rangle$ with $1 \leq l(u) \leq \ell$ is cyclic.*
(ii) *The monomial algebra $K\langle X\rangle/\langle\Omega\rangle$ is prime if and only if the following two conditions are satisfied:*
(a) *Any $u \in \mathcal{B} - \langle\Omega\rangle$ with $l(u) < \ell - 1$ is a suffix of some vertex of $\Gamma(\Omega)$;*
(b) *For any two vertices v_i and v_j of $\Gamma(\Omega)$, there exists a route from v_i to v_j and vice versa.*

□

Concerning the algorithmic criterion on semi-primeness of algebras defined by finite Gröbner bases, we derive the following result. The discussion on primeness is postponed till Theorem 5.7.

Theorem 5.3. *With the convention made at the beginning of this section, let $\mathcal{G} = \{g_1, \ldots, g_s\}$ be an LM-reduced finite Gröbner basis in $K\langle X\rangle$ with respect to some fixed admissible system $(\mathcal{B}, \prec)$, and let $\Gamma(\mathbf{LM}(\mathcal{G}))$ be the Ufnarovski graph of $\mathcal{G}$. Consider the quotient algebra $A = K\langle X\rangle/\langle I\rangle$ with $I = \langle \mathcal{G}\rangle$, and bear in mind $A^{\mathcal{B}}_{\rm LH} = K\langle X\rangle/\langle \mathbf{LM}(I)\rangle$, $A^{\mathbb{N}}_{\rm LH} = K\langle X\rangle/\langle \mathbf{LH}(I)\rangle \cong G^{\mathbb{N}}(A)$. With $\ell = \max\{l(u_i) \mid u_i \in \mathbf{LM}(\mathcal{G})\}$, if each monomial $u \in \mathcal{B} - \langle \mathbf{LM}(\mathcal{G})\rangle$ with $1 \le l(u) \le \ell$ is cyclic, then the monomial algebra $A^{\mathcal{B}}_{\rm LH}$ is semi-prime, and hence A is semi-prime. If an $\mathbb{N}$-graded monomial ordering $\prec_{gr}$ is used, then, under the same condition the associated $\mathbb{N}$-graded algebra $G^{\mathbb{N}}(A)$ of A is semi-prime as well.*

Proof. Under the given condition $A^{\mathcal{B}}_{\rm LH}$ is semi-prime by Theorem 5.2(i). It follows from (Ch.2, Theorem 4.1(iii)) that A is semi-prime. If an $\mathbb{N}$-graded monomial ordering $\prec_{gr}$ is used, then, under the same condition the semi-primeness of $A^{\mathcal{B}}_{\rm LH}$ can be lifted not only to A but also to $G^{\mathbb{N}}(A)$ by Corollary 1.1 and (Ch.2, Theorem 4.1(iii)). □

By means of the above theorem, we obtain a family of semi-prime algebras including all down-up algebras, as follows.

Corollary 5.4. *Let the free K-algebra $K\langle X\rangle = K\langle X_1, X_2\rangle$ be equipped with a positive weight $\mathbb{N}$-gradation, and let $\mathcal{G} = \{g_1, g_2\}$ be any Gröbner basis in $K\langle X\rangle$ with respect to some graded monomial ordering $\prec_{gr}$, such that $\mathbf{LM}(g_1) = X_1^2X_2$ and $\mathbf{LM}(g_2) = X_1X_2^2$. Consider the K-algebra $A = K\langle X\rangle/\langle I\rangle$ with $I = \langle\mathcal{G}\rangle$. With the notation in Theorem 5.3, the monomial algebra $A^{\mathcal{B}}_{\rm LH} = K\langle X\rangle/\langle \mathbf{LM}(I)\rangle$ is Jacobson semi-simple, and hence, the algebras A and $A^{\mathbb{N}}_{\rm LH} = K\langle X\rangle/\langle \mathbf{LH}(I)\rangle \cong G^{\mathbb{N}}(A)$ are semi-prime.*

The above result holds for the algebras given in Example 8 of (Ch.3, Section 5), in particular, any down-up algebra $A = A(\alpha, \beta, \gamma) = K\langle X\rangle/\langle\mathcal{G}\rangle$ and its two associated graded algebras $A^{\mathcal{B}}_{\rm LH} = K\langle X\rangle/\langle \mathbf{LM}(\mathcal{G})\rangle$ and $A^{\mathbb{N}}_{\rm LH} = K\langle X\rangle/\langle \mathbf{LH}(\mathcal{G})\rangle \cong G^{\mathbb{N}}(A)$, both cubic algebras (see also Example 2 of (Ch.4, Section 2)), have the property mentioned above.

Proof. Put $\Omega = \{X_1^2X_2,\ X_1X_2^2\}$. Then $\ell = \max\{l(u_i) \mid u_i \in \Omega\} = 3$ and

$$\{u \mid u \in \mathcal{B} - \langle\Omega\rangle,\ 1 \le l(u) \le \ell\}$$
$$= \{X_1,\ X_2,\ X_1X_2,\ X_2X_1,\ X_1^3,\ X_2^3\}.$$

If we verify the graph $\Gamma(\mathbf{LM}(\mathcal{G}))$ given in Example 1 of Section 2, then the condition of Theorem 5.2(i) is satisfied. Hence, the monomial algebra $A_{\mathrm{LH}}^{\mathcal{B}}$ is semi-prime (Jacobson semi-simple). It follows from Theorem 5.3 that both algebras A and $G^{\mathbb{N}}(A)$ are semi-prime. □

Recall that a ring A is said to be a *PI-ring* if it satisfy a polynomial identity $f(X_1,\ldots,X_n)$ in the free $\mathbb{Z}$-algebra $\mathbb{Z}\langle X_1,\ldots,X_n\rangle$, i.e., $f(a_1,\ldots,a_n) = 0$ for all $a_i \in A$ (see e.g., [M-R], Chapter 13). In connection with the matrix representation of associative algebras, the PI-property of a finitely presented monomial algebra was realized via the Ufnarovski graph by V. V. Borisenko.

Theorem 5.5 ([Bo]). *Let $\Omega = \{u_1,\ldots,u_s\}$ be a reduced finite subset of monomials in $K\langle X\rangle$. Then the algebra $A = K\langle X\rangle/\langle\Omega\rangle$ satisfies a polynomial identity, i.e., A is a PI-algebra, if and only if it has the polynomial growth and if and only if in the Ufnarovski graph $\Gamma(\Omega)$ any two cycles do not have a common vertex.*

□

Although any algebra $A = K\langle X\rangle/I$ has the same growth as its associated $\mathcal{B}$-leading homogeneous algebra $A_{\mathrm{LH}}^{\mathcal{B}} = K\langle X\rangle/\langle\mathbf{LM}(I)\rangle$ (see the remark given before Proposition 1.4 of Section 1), in general we do not know if it is always possible to lift the PI-property of $A_{\mathrm{LH}}^{\mathcal{B}}$ (if it has such a property) to A. Nevertheless, a celebrated result of [SSW] tells us that

- any affine algebra A of GK dimension 1 is a PI algebra; and if furthermore A is semi-prime, then A is Noetherian of Krull dimension 1.

Moreover, J.P. Bell and P. Peckcagliyan proved in [BP] that

- a prime finitely presented monomial algebra is either primitive or it has GK dimension 1 (hence satisfies a polynomial identity).

Thus, it turns out that if $A = K\langle X\rangle/\langle\mathcal{G}\rangle$ is defined by a finite Gröbner basis $\mathcal{G}$, then there is an algorithmic way to check whether $A_{\mathrm{LH}}^{\mathcal{B}} = K\langle X\rangle/\langle\mathbf{LM}(\mathcal{G})\rangle$ is prime or semi-prime PI of GK dimension 1, and then we may lift the obtained property to both A and $G^{\mathbb{N}}(A)$. We summarize the corresponding results in the next two theorems.

Theorem 5.6. *With the convention made at the beginning of this section, let $\mathcal{G} = \{g_1, \ldots, g_s\}$ be an LM-reduced finite Gröbner basis in $K\langle X\rangle$ with respect to some fixed admissible system $(\mathcal{B}, \prec)$, and let $\Gamma(\mathbf{LM}(\mathcal{G}))$ be the Ufnarovski graph of $\mathcal{G}$. Consider the quotient algebra $A = K\langle X\rangle/\langle I\rangle$ with $I = \langle \mathcal{G}\rangle$, and bear in mind $A^{\mathcal{B}}_{\rm LH} = K\langle X\rangle/\langle \mathbf{LM}(I)\rangle$, $A^{\mathbb{N}}_{\rm LH} = K\langle X\rangle/\langle \mathbf{LH}(I)\rangle \cong G^{\mathbb{N}}(A)$. With $\ell = \max\{l(u_i) \mid u_i \in \mathbf{LM}(\mathcal{G})\}$, the following statements hold.*
(i) *If in the graph $\Gamma(\mathbf{LM}(\mathcal{G}))$ any connected component contains at most one cycle, and there is at least one connected component containing exactly one cycle, then all three algebras $A^{\mathcal{B}}_{\rm LH}$, A and $G^{\mathbb{N}}(A)$ are PI-algebras of GK dimension 1.*
(ii) *Under the assumption of (i) above if furthermore each monomial $u \in \mathcal{B} - \langle \mathbf{LM}(\mathcal{G})\rangle$ with $1 \le l(u) \le \ell$ is cyclic, then $A^{\mathcal{B}}_{\rm LH}$ is a Noetherian semi-prime PI-algebra of GK dimension 1 and of Krull dimension 1, and hence A has the same properties. If an $\mathbb{N}$-graded monomial ordering $\prec_{gr}$ is used, then, under the same condition the associated $\mathbb{N}$-graded algebra $G^{\mathbb{N}}(A)$ of A is also a Noetherian semi-prime PI-algebra of GK dimension 1 and of Krull dimension 1.*

Proof. (i). By the assumption, this assertion follows from Theorem 3.2 and [SSW].
(ii). This follows from (i), Theorem 5.3 and [SSW]. □

Theorem 5.7. *With the convention made at the beginning of this section, let $\mathcal{G} = \{g_1, \ldots, g_s\}$ be an LM-reduced finite Gröbner basis in $K\langle X\rangle$ with respect to some fixed admissible system $(\mathcal{B}, \prec)$, and let $\Gamma(\mathbf{LM}(\mathcal{G}))$ be the Ufnarovski graph of $\mathcal{G}$. Consider the quotient algebra $A = K\langle X\rangle/\langle I\rangle$ with $I = \langle \mathcal{G}\rangle$, and bear in mind $A^{\mathcal{B}}_{\rm LH} = K\langle X\rangle/\langle \mathbf{LM}(I)\rangle$, $A^{\mathbb{N}}_{\rm LH} = K\langle X\rangle/\langle \mathbf{LH}(I)\rangle \cong G^{\mathbb{N}}(A)$. With $\ell = \max\{l(u_i) \mid u_i \in \mathbf{LM}(\mathcal{G})\}$, if the following two conditions are satisfied:*

(a) *any $u \in \mathcal{B} - \langle \mathbf{LM}(\mathcal{G})\rangle$ with $l(u) < \ell - 1$ is a suffix of some vertex of $\Gamma(\mathbf{LM}(\mathcal{G}))$;*
(b) *for any two vertices v_i and v_j of $\Gamma(\mathbf{LM}(\mathcal{G}))$, there exists a route from v_i to v_j and vice versa,*

then the monomial algebra $A^{\mathcal{B}}_{\rm LH}$ is prime, and hence A is prime or a Noetherian prime PI-algebra of GK dimension 1 and of Krull dimension 1. Also if an $\mathbb{N}$-graded monomial ordering $\prec_{gr}$ is used, then, under the same condition the associated $\mathbb{N}$-graded algebra $G^{\mathbb{N}}(A)$ of A is prime or a Noetherian *prime PI-algebra of GK dimension 1 and of Krull dimension 1.*

Proof. Under the assumption $A_{\rm LH}^{\mathcal{B}}$ is prime by Theorem 5.2(ii). So the theorem follows from (Ch.2, Theorem 4.1(iii), Theorem 4.2(i)), (Ch.4, Theorem 2.4), the foregoing Corollary 1.1, Theorem 3.2, Theorem 5.5, [BP], and [SSW]. $\square$

Example 1. Consider in $K\langle X\rangle = \langle X_1, X_2, X_3\rangle$ the subset of monomials $\Omega = \{X_1X_2, X_3X_1, X_2^2\}$. Then $\ell = \max\{l(u_i) \mid u_i \in \Omega\} = 2$ and

$$\begin{aligned}&\{u \mid u \in \mathcal{B} - \langle\Omega\rangle,\ 1 \le l(u) \le \ell\}\\ &= \left\{X_1, X_2, X_3, X_1^2, X_3^2, X_2X_1, X_1X_3, X_2X_3, X_3X_2\right\}.\end{aligned}$$

By referring to the Ufnarovski graph $\Gamma(\Omega)$ of Ω given in Example 1 of Section 3, it follows from Proposition 5.2(i) that the monomial algebra $K\langle X\rangle/\langle\Omega\rangle$ is semi-prime and hence Jacobson semi-simple as well. Indeed, by Proposition 5.2(ii), the graph $\Gamma(\Omega)$ shows that $K\langle X\rangle/\langle\Omega\rangle$ is a prime algebra. So, the foregoing theorems are applied to any algebra $A = K\langle X\rangle/\langle\mathcal{G}\rangle$ with $\mathcal{G}$ a Gröbner basis such that $\mathbf{LM}(\mathcal{G}) = \Omega$, but note that all three algebras $A_{\rm LH}^{\mathcal{B}}$, A and $G^{\mathbb{N}}(A)$ have infinite GK dimension.

Example 2. Consider in the free K-algebra $K\langle X\rangle = K\langle X_1, X_2\rangle$ the subsets of monomials $\Omega_1 = \{X_1^2, X_2^2\}$, $\Omega_2 = \{X_1X_2, X_2X_1\}$. Then, the Ufnarovski graph $\Gamma(\Omega_1)$ of Ω_1 is presented by

$$X_1 \rightleftarrows X_2$$

and the Ufnarovski graph $\Gamma(\Omega_2)$ of Ω_2 is presented by

$$\circlearrowleft X_1 \qquad X_2 \circlearrowright$$

By Theorem 5.2, the monomial algebra $K\langle X\rangle/\langle\Omega_1\rangle$ is prime, the monomial algebra $K\langle X\rangle/\langle\Omega_2\rangle$ is semi-prime, and both algebras are Noetherian semi-prime PI-algebras of GK dimension 1 and of Krull dimension 1. Consequently, for any Gröbner basis $\mathcal{G} \subset K\langle X\rangle$ with respect to some graded monomial ordering $\prec_{gr}$ such that $\mathbf{LM}(\mathcal{G}) = \Omega_1$, respectively $\mathbf{LM}(\mathcal{G}) = \Omega_2$, all the three corresponding algebras $A = K\langle X\rangle/\langle\mathcal{G}\rangle$, $A_{\rm LH}^{\mathcal{B}} = K\langle X\rangle/\langle\mathbf{LM}(\mathcal{G})\rangle$ and $A_{\rm LH}^{\mathbb{N}} = K\langle X\rangle/\langle\mathbf{LH}(\mathcal{G})\rangle \cong G^{\mathbb{N}}(A)$ are Noetherian prime PI algebras of GK dimension 1 and of Krull dimension 1, respectively Noetherian semi-prime PI algebras of GK dimension 1 and of Krull dimension 1. For instance, under the $\mathbb{N}$-graded lexicographic ordering $X_1 \prec_{gr} X_2$ we may take, for $a, b, c, d, e, f, g \in K$, a Gröbner basis of the form

$$\mathcal{G} = \{g_1 = X_1^2 + aX_1 + bX_2 + c,\ g_2 = X_2^2 + dX_1X_2 + eX_1 + fX_2 + g\}.$$

5.6 Anick's Resolution over Monomial Algebras

Let the free K-algebra $K\langle X\rangle = K\langle X_1, \ldots, X_n\rangle$ of n generators be equipped with the augmentation map ε sending each X_i to zero. For any ideal J contained in the augmentation ideal $\langle X_1, \ldots, X_n\rangle$ (i.e., the kernel of ε), D. J. Anick constructed in [An3] a free resolution of the trivial module K over the quotient algebra $K\langle X\rangle/J$, which gave rise to several efficient applications to the homological aspects of associative algebras ([An3], [An4]). Instead of recalling in detail the construction of the Anick resolution of K over $K\langle X\rangle/J$, for the purpose of subsequent sections and Ch.6 we demonstrate in this section the procedure constructing the Anick resolution of K over a monomial algebra $K\langle X\rangle/\langle\Omega\rangle$ in terms of n-chains of the graph $\Gamma_{\rm C}(\Omega)$ (see the definition below), where Ω is a nonempty subset of the standard K-basis $\mathcal{B}$ of $K\langle X\rangle$.

The graph $\Gamma_{\rm C}(\Omega)$ of n-chains

Let $\Omega \subset \mathcal{B}$ be a *reduced* (finite or infinite) subset of monomials. To better understand and use Anick's resolution from a computational viewpoint, V. Ufnarovski constructed in [Uf1] the *graph of n-chains* of Ω as a directed graph $\Gamma_{\rm C}(\Omega)$, in which the set of vertices V is defined as

$$V = \{1\} \cup X \cup \{\text{all proper suffixes of } u \in \Omega\},$$

and the set of edges E consists of all edges

$$1 \longrightarrow X_i \text{ for every } X_i \in X$$

and edges defined subject to the rule: for $u, v \in V - \{1\}$,

$$u \longrightarrow v \text{ in } E \Leftrightarrow \text{there is a } w = X_{i_1}\cdots X_{i_{m-1}}X_{i_m} \in \Omega \text{ such that}$$
$$uv = \begin{cases} w \in \Omega, \text{ or} \\ sw \text{ with } s \in \mathcal{B} \text{ and } sX_{i_1}\cdots X_{i_{m-1}} \in \mathcal{B} - \langle\Omega\rangle. \end{cases}$$

As Ω is reduced, one easily sees that if a monomial w (as described in the definition above) exists, it must be *unique*. For $n \geq -1$, an *n-chain* in $\Gamma_{\rm C}(\Omega)$ is a monomial (word) $v = v_1 \cdots v_n v_{n+1}$ which can be read out through a route of length $n+1$ starting from 1:

$$1 \rightarrow v_1 \rightarrow \cdots \rightarrow v_n \rightarrow v_{n+1}.$$

Writing C_n for the set of all n-chains of Ω, it is clear that $C_{-1} = \{1\}$, $C_0 = X$, and $C_1 = \Omega$.

Since any Gröbner basis in $K\langle X\rangle$ may be reduced to an LM-reduced Gröbner basis (see Ch.3, Section 5), for a (finite or infinite) Gröbner basis $\mathcal{G}$ of $K\langle X\rangle$, the *graph of n-chains of $\mathcal{G}$* is defined to be the graph of n-chains of the unique reduced monomial generating set of the monomial ideal $\langle \mathbf{LM}(\mathcal{G})\rangle$, and is denoted by $\Gamma_{\rm C}(\mathbf{LM}(\mathcal{G}))$ without confusion.

Remark As with the Ufnarovski graph $\Gamma(\Omega)$ defined in Section 2, to better understand the practical application of the graph $\Gamma_{\rm C}(\Omega)$ of n-chains determined by Ω in the subsequent sections and the next chapter, it is essential to notice that *an n-chain is defined by a route of length $n+1$ starting with 1, as described above, instead of by the degree of the* $\mathbb{N}$*-homogeneous element $v = v_1 \cdots v_n v_{n+1}$ read out of that route.*

Example 1. In the free K-algebra $K\langle X_1, X_2\rangle$, $\mathcal{G} = \{X_1^2 - X_2^2,\ X_1X_2^2 - X_2^2X_1\}$ is a Gröbner basis with respect to the $\mathbb{N}$-graded lexicographic ordering $X_2 \prec_{gr} X_1$ (check it!), and hence $\mathbf{LM}(\mathcal{G}) = \{X_1^2,\ X_1X_2^2\}$. Thus, with

$$V = \{1\} \cup \{X_1, X_2\} \cup \{X_2^2\},$$

the graph $\Gamma_{\rm C}(\mathbf{LM}(\mathcal{G}))$ is given by

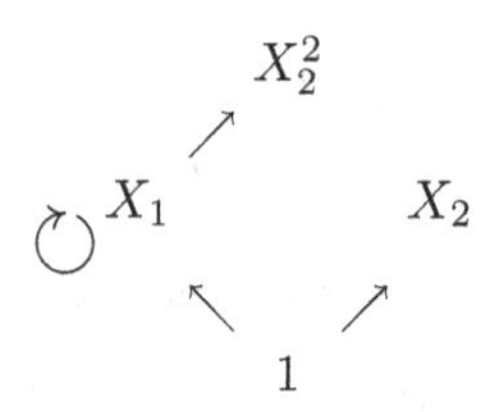

Accordingly, $C_{-1} = \{1\}$, $C_0 = \{X_1,\ X_2\}$, $C_1 = \{X_1^2,\ X_1X_2^2\}$, and for $n \geq 2$, $C_n = \{X_1^{n+1},\ X_1^nX_2^2\}$.

Example 2. By Example 3 of (Ch.3, Section 5), in the free K-algebra $K\langle X_1, X_2\rangle$, $\mathcal{G} = \{X_1X_2^nX_1 - X_1X_2^{n+1} \mid n \in \mathbb{N}\}$ is an infinite Gröbner basis and $\mathbf{LM}(\mathcal{G}) = \{X_1X_2^nX_1 \mid n \in \mathbb{N}\}$ with respect to the $\mathbb{N}$-graded lexicographic ordering $X_2 \prec_{gr} X_1$. Hence, with

$$V = \{1\} \cup \{X_1, X_2\} \cup \{X_2^nX_1 \mid n \geq 1\},$$

the graph $\Gamma_{\rm C}(\mathbf{LM}(G))$ may be sketched as

$$\begin{array}{llll} & X_2 & & \\ & \nearrow & & \\ 1 & & & \\ & \searrow & & \\ & X_1 \to X_2^i X_1 \rightleftarrows X_2^i X_1 \ (i \geq 1) \\ & \uparrow\downarrow \qquad \uparrow\downarrow \\ & X_1 \to X_2^j X_1 \rightleftarrows X_2^j X_1 \ (j \geq 1,\ j \neq i) \end{array}$$

Accordingly, $C_{-1} = \{1\}$, $C_0 = \{X_1,\ X_2\}$, $C_1 = \{X_1^2,\ X_1X_2^iX_1 \mid i \geq 1\}$, and for $n \geq 2$, $C_n = \{X_1X_2^{k_1}X_1X_2^{k_2}\cdots X_1X_2^{k_n}X_1 \mid k_i \geq 0\}$.

Anick's resolution of K over $K\langle X\rangle/\langle\Omega\rangle$

Let Ω be a reduced subset of monomials in the standard K-basis $\mathcal{B}$ of the free K-algebra $K\langle X\rangle = K\langle X_1, \ldots, X_n\rangle$, and let $\overline{A} = K\langle X\rangle/\langle\Omega\rangle$. Since every $u \in \mathcal{B}$ is an $\mathbb{N}$-homogeneous element with respect to any fixed weight $\mathbb{N}$-gradation of $K\langle X\rangle$ with positive weight $\{n_i\}_{i=1}^n$, $\overline{A}$ is doubly $\mathbb{N}$-$\mathcal{B}$-graded, and consequently, the trivial $\overline{A}$-module $K \cong \overline{A}/(\oplus_{u\neq 1}\overline{A}_u)$ is doubly $\mathbb{N}$-$\mathcal{B}$-graded as well. By means of the n-chains read out of the graph $\Gamma_{\rm C}(\Omega)$, we construct below Anick's resolution for the module K over the monomial algebra $\overline{A}$.

Let C_n be the set of all n-chains of $\Gamma_{\rm C}(\Omega)$ as defined in the last part, and consider the K-vector space spanned by C_n, again denoted C_n. Noticing that $\mathcal{B} - \langle\Omega\rangle$ projects to a K-basis of $\overline{A}$, let us denote this basis again by $\mathcal{B} - \langle\Omega\rangle$. We see that $C_n \otimes_K \overline{A}$ is a free right $\overline{A}$-module that has the $\overline{A}$-basis $\{u \otimes 1 \mid u \in C_n\}$ and the K-basis $\{u \otimes v \mid u \in C_n,\ v \in \mathcal{B} - \langle\Omega\rangle\}$. Moreover, $C_n \otimes_K \overline{A}$ is $\mathcal{B}$-graded, that is, $C_n \otimes_K \overline{A} = \oplus_{w\in\mathcal{B}}(C_n \otimes_K \overline{A})_w$ with

$$(C_n \otimes_K \overline{A})_w = K\text{-Span}\left\{u \otimes v \;\middle|\; u \in C_n,\ v \in \mathcal{B} - \langle\Omega\rangle,\ uv = w\right\},\quad w \in \mathcal{B},$$

which, in turn, gives rise the $\mathbb{N}$-gradation of $C_n \otimes_K A$, i.e., $C_n \otimes_K A = \oplus_{p\in\mathbb{N}}(C_n \otimes_K A)_p$ with

$$(C_n \otimes_K A)_p = \bigoplus_{l(w)=p,\ w\in\mathcal{B}} (C_n \otimes_K A)_w.$$

The Anick resolution of K over $\overline{A}$ is the free $\overline{A}$-resolution:

$$(\mathcal{A},\ \delta) \qquad \cdots \to C_n \otimes_K \overline{A} \xrightarrow{\delta_n} C_{n-1} \otimes_K \overline{A} \xrightarrow{\delta_{n-1}} \cdots$$

$$\longrightarrow C_1 \otimes_K \overline{A} \xrightarrow{\delta_1} C_0 \otimes_K \overline{A} \xrightarrow{\delta_0} \overline{A} \xrightarrow{\varepsilon} K \to 0$$

where ε is the projection of $\overline{A}$ onto K, which is clearly $\mathbb{N}$-graded as well as $\mathcal{B}$-graded, and the δ_n are constructed inductively, as follows.

First, let us observe a useful fact about n-chains, that is,

- for each $u \in C_n$, there is a *unique* $(n-1)$-chain $u(n-1)$ and some $t \in B - \langle\Omega\rangle - \{1\}$ such that $u = u(n-1) \cdot t$.

To better understand the resolution, we begin the inductive construction procedure by a clear step-by-step description of the initial exact sequence

$$C_2 \otimes_K \overline{A} \xrightarrow{\delta_2} C_1 \otimes_K \overline{A} \xrightarrow{\delta_1} C_0 \otimes_K \overline{A} \xrightarrow{\delta_0} \overline{A} \xrightarrow{\varepsilon} K \to 0.$$

(i) The construction of the sequence

$$C_0K \otimes_K \overline{A} \xrightarrow{\delta_0} \overline{A} \xrightarrow{\varepsilon} K \to 0$$

and its exactness.

Define

$$\begin{aligned} \delta_0 : C_0K \otimes_K \overline{A} &\longrightarrow \overline{A} \\ x \otimes 1 \quad &\mapsto \quad x \ \ x \in X. \end{aligned}$$

Then δ_0 is a $\mathcal{B}$-graded $\overline{A}$-homomorphism and hence an $\mathbb{N}$-graded $\overline{A}$-homomorphism as well, and $\varepsilon\delta_0 = 0$. Furthermore, noticing that $\mathrm{Ker}\varepsilon = \oplus_{v\in\mathcal{B}-\langle\Omega\rangle-\{1\}}\overline{A}_v$, if we define

$$\begin{aligned} i_0 : \mathrm{Ker}\varepsilon &\longrightarrow C_0K \otimes_K \overline{A} \\ v \quad &\mapsto \quad x \otimes v' \quad\quad v = xv' \in \mathcal{B} - \langle\Omega\rangle - \{1\},\ x \in X \end{aligned}$$

then $\delta_0 i_0 = 1_{\mathrm{Ker}\varepsilon}$. Hence $\mathrm{Im}\delta_0 = \mathrm{Ker}\varepsilon$.

(ii) The construction of the sequence

$$C_1K \otimes_K \overline{A} \xrightarrow{\delta_1} C_0K \otimes_K \overline{A} \xrightarrow{\delta_0} \overline{A}$$

and its exactness.

Note that by the above observation ($\bullet$), each $u \in C_1$ has a unique representation $u = xt$ for some $t \in \mathcal{B} - \langle\Omega\rangle$. Define

$$\begin{aligned} \delta_1 : C_1K \otimes_K \overline{A} &\longrightarrow C_0K \otimes_K \overline{A} \\ u \otimes 1 \quad &\mapsto \quad x \otimes t. \end{aligned}$$

Then δ_1 is a $\mathcal{B}$-graded $\overline{A}$-homomorphism and hence an $\mathbb{N}$-graded $\overline{A}$-homomorphism as well, and by the definition of a 1-chain, $(\delta_0\delta_1)(u\otimes 1) = 0$. Furthermore, note that every element of $C_0K \otimes_K \overline{A}$ has a unique representation $\sum \lambda x \otimes v$. If $\delta_0(\sum \lambda x \otimes v) = \sum \lambda xv = 0$, then $\delta_0(x \otimes v) = xv = 0$.

Since $v \in \mathcal{B} - \langle\Omega\rangle$, this means that v has a unique decomposition $v = v_1v_2$ such that xv_1 is a 1-chain. We may define

$$\begin{array}{rcl} i_1 : \mathrm{Ker}\delta_0 & \longrightarrow & C_1K \otimes_K \overline{A} \\ x \otimes v & \mapsto & xv_1 \otimes v_2. \end{array}$$

Clearly $\delta_1 i_1 = 1_{\mathrm{Ker}\delta_0}$. Hence $\mathrm{Im}\delta_1 = \mathrm{Ker}\delta_0$.

(iii) The construction of the sequence

$$C_2K \otimes_K \overline{A} \xrightarrow{\delta_2} C_1K \otimes_K \overline{A} \xrightarrow{\delta_1} C_0K \otimes_K \overline{A}$$

and its exactness.

By the previous observation (•), each $u \in C_2$ has a unique representation $u = u(1)t$ for some $u(1) \in C_1$ and $t \in \mathcal{B} - \langle\Omega\rangle$. Define

$$\begin{array}{rcl} \delta_2 : C_2K \otimes_K \overline{A} & \longrightarrow & C_1K \otimes_K \overline{A} \\ u \otimes 1 & \mapsto & u(1) \otimes t. \end{array}$$

Then δ_1 is a $\mathcal{B}$-graded $\overline{A}$-homomorphism and hence an $\mathbb{N}$-graded $\overline{A}$-homomorphism as well, and by the definition of a 2-chain, $(\delta_1\delta_2)(u\otimes 1) = 0$. Furthermore, note that every element of $C_1K \otimes_K \overline{A}$ has a unique representation $\sum \lambda w \otimes v$ with $w = xt$. If $\delta_1(\sum \lambda w \otimes v) = \sum \lambda x \otimes tv = 0$, then $\delta_1(w \otimes v) = \delta_1(xt \otimes v) = x \otimes tv = 0$. As $v \in \mathcal{B} - \langle\Omega\rangle$, this means that v has a unique decomposition $v = v_1v_2$ such that $tv_1 \in \Omega$. Hence $wv_1 = xtv_1$ is a 2-chain. Thus, we may define

$$\begin{array}{rcl} i_2 : \mathrm{Ker}\delta_1 & \longrightarrow & C_2K \otimes_K \overline{A} \\ w \otimes v & \mapsto & wv_1 \otimes v_2. \end{array}$$

Clearly $\delta_2 i_2 = 1_{\mathrm{Ker}\delta_1}$. Hence $\mathrm{Im}\delta_2 = \mathrm{Ker}\delta_1$.

In general for $n \geq 1$, since each $u \in C_n$ has a unique representation $u = u(n-1)t$ with $u(n-1) \in C_{n-1}$ and $t \in \mathcal{B} - \langle\Omega\rangle$, we may define

$$\begin{array}{rcl} \delta_n : C_nK \otimes_K \overline{A} & \longrightarrow & C_{n-1}K \otimes_K \overline{A} \\ u \otimes 1 & \mapsto & u(n-1) \otimes t \end{array}$$

and construct the sequence of $\mathcal{B}$-graded $\overline{A}$-modules and $\mathcal{B}$-graded homomorphisms

$$C_nK \otimes_K \overline{A} \xrightarrow{\delta_n} C_{n-1}K \otimes_K \overline{A} \xrightarrow{\delta_{n-1}} C_{n-2}K \otimes_K \overline{A}$$

which is also an $\mathbb{N}$-graded sequence. Furthermore, note that every element of $C_{n-1}K \otimes_K \overline{A}$ has a unique representation $\sum \lambda w \otimes v$ with $w = w(n-2)t$. If $\delta_{n-1}(\sum \lambda w \otimes v) = \sum \lambda w(n-2) \otimes tv = 0$, then $\delta_{n-1}(w \otimes v) = \delta_{n-1}(w(n-2)t \otimes v) = w(n-2) \otimes tv = 0$. As $v \in \mathcal{B} - \langle\Omega\rangle$, this means that v has a unique decomposition $v = v_1v_2$ such that $tv_1 \in \Omega$. Hence $wv_1 = w(n-2)tv_1$ is an n-chain. It follows that the exactness is verified by using the K-linear mapping

$$\begin{array}{rcl} i_n : \mathrm{Ker}\delta_{n-1} & \longrightarrow & C_nK \otimes_K \overline{A} \\ w \otimes v & \mapsto & wv_1 \otimes v_2. \end{array}$$

5.7 Recognizing Finiteness of Global Dimension

Let I be an arbitrary ideal of the free K-algebra $K\langle X\rangle = K\langle X_1,\ldots,X_n\rangle$, that is, *$I$ is not necessarily contained in the augmentation ideal $\langle X_1,\ldots,X_n\rangle$* of $K\langle X\rangle$, and consider the K-algebra $A = K\langle X\rangle/I$, $A^{\mathbb{N}}_{\rm LH} = K\langle X\rangle/\langle \mathbf{LH}(I)\rangle \cong G^{\mathbb{N}}(A)$, and $A^{\mathcal{B}}_{\rm LH} = K\langle X\rangle/\langle \mathbf{LM}(I)\rangle$ as before, where $\mathbf{LH}(I)$ is taken with respect to a fixed positive weight $\mathbb{N}$-gradation of $K\langle X\rangle$, $G^{\mathbb{N}}(A)$ is determined by the $\mathbb{N}$-filtration FA induced by the weight $\mathbb{N}$-grading filtration $FK\langle X\rangle$ of $K\langle X\rangle$, and $\mathbf{LM}(I)$ is taken with respect to a fixed monomial ordering $\prec$ on the standard K-basis $\mathcal{B}$ of $K\langle X\rangle$. In this section we demonstrate how to recognize the finiteness of global homological dimension of A and $G^{\mathbb{N}}(A)$ via n-chains of the graph $\Gamma_C(\Omega)$ that define the Anick resolution of the trivial module K over the monomial algebra $A^{\mathcal{B}}_{\rm LH}$, where Ω is the reduced monomial generating set of $\langle \mathbf{LM}(I)\rangle$. Moreover, under the condition that gl.dim$A^{\mathcal{B}}_{\rm LH} < \infty$ and that $A^{\mathcal{B}}_{\rm LH}$ has polynomial growth (both properties are recognizable by means of $\Gamma_C(\Omega)$ and $\Gamma(\Omega)$ respectively), we also derive the equality

$$\text{gl.dim}G^{\mathbb{N}}(A) = \text{gl.dim}A^{\mathcal{B}}_{\rm LH}.$$

With notation in the last section, if we start by tensoring Anick's resolution $(\mathcal{A},\delta)$ of K over $\overline{A} = K\langle X\rangle/\langle\Omega\rangle$ with K, we obtain the derived sequence

$$\begin{array}{rcl} \cdots \longrightarrow C_n \otimes_K \overline{A} \otimes_{\overline{A}} K & \overset{\delta_n\otimes 1}{\longrightarrow} & C_{n-1} \otimes_K \overline{A} \otimes_{\overline{A}} K \longrightarrow \cdots \\ u \otimes 1 \otimes 1 & \mapsto & u(n-1) \otimes t \otimes 1. \end{array}$$

Note that $K = \overline{A}/(\oplus_{u\neq 1, u\in\mathcal{B}}\overline{A}_u)$ and for $u \in C_n$ we have $u = u(n-1)\cdot t$ with $u(n-1) \in C_{n-1}$, $t \in \mathcal{B} - \langle\Omega\rangle - \{1\}$. It follows that $\delta_n \otimes 1 = 0$. Hence, concerning the corresponding homology groups of the $\overline{A}$-module K, we have the $\mathcal{B}$-graded and hence $\mathbb{N}$-graded isomorphism

$$\mathrm{Tor}^{\overline{A}}_n(K,K) = \frac{\mathrm{Ker}(\delta_{n-1}\otimes 1)}{\mathrm{Im}(\delta_n \otimes 1)} = \frac{C_{n-1}\otimes_K K}{\{0\}} \cong C_{n-1},\ n = 1,2,\ldots$$

Theorem 7.1 ([An3]). *Let $\Omega\subset\mathcal{B}$ be a reduced subset with $\Omega\cap X=\emptyset$. Then gl.dim$K\langle X\rangle/\langle\Omega\rangle\le d$ if and only if the graph $\Gamma_{\rm C}(\Omega)$ of n-chains of Ω does not have any d-chains.*

□

Theorem 7.2 ([An4]). *Let $\Omega\subset\mathcal{B}$ be a reduced subset with $\Omega\cap X=\emptyset$. Suppose $K\langle X\rangle/\langle\Omega\rangle$ has finite global dimension m. If $K\langle X\rangle/\langle\Omega\rangle$ does not contain a free subalgebra of two generators, then the following statements hold.*
(i) *$K\langle X\rangle/\langle\Omega\rangle$ is finitely presented, that is, $\langle\Omega\rangle$ is finitely generated.*
(ii) *$K\langle X\rangle/\langle\Omega\rangle$ has the polynomial growth of degree m.*
(iii) *The Hilbert series of $K\langle X\rangle/\langle\Omega\rangle$ is of the form $H_{K\langle X\rangle/\langle\Omega\rangle}(t)=\prod_{i=1}^m(1-t^{e_i})^{-1}$, where each e_i is a positive integer, $1\le i\le m$.*

□

By using the above theorem, the following result was derived by T. Gateva-Ivanova in [G-I2].

Theorem 7.3 ([G-I2]). *Let J be an $\mathbb{N}$-graded ideal of $K\langle X\rangle$ and $R=K\langle X\rangle/J$ the corresponding $\mathbb{N}$-graded algebra defined by J. Suppose that $\mathbf{LM}(J)\cap X=\emptyset$ and that the associated monomial algebra $\overline{R}=K\langle X\rangle/\langle\mathbf{LM}(J)\rangle$ of R has finite global dimension and the polynomial growth of degree m, where $\mathbf{LM}(J)$ is taken with respect to a fixed $\mathbb{N}$-graded monomial ordering $\prec_{gr}$ on $\mathcal{B}$. Then the following statements hold.*
(i) *gl.dim$R=$ gl.dim$\overline{R}=m$.*
(ii) *The ideal J has a finite Gröbner basis.*
(iii) *The Hilbert series of R is of the form $H_R(t)=\prod_{i=1}^m(1-t^{e_i})^{-1}$, where each e_i is a positive integer, $1\le i\le n$.*

□

Remark The determination of Hilbert series concerned in Theorem 7.2 and Theorem 7.3 above will be discussed further in the next section and (Ch.7, Section 3 for Rees algebras).

Let $(\mathcal{B},\prec)$ be an arbitrarily fixed admissible system of $K\langle X\rangle$. Consider an ideal I of $K\langle X\rangle$, the quotient algebra $A=K\langle X\rangle/\langle I\rangle$, and bear in mind $A^{\mathcal{B}}_{\rm LH}=K\langle X\rangle/\langle\mathbf{LM}(I)\rangle$, $A^{\mathbb{N}}_{\rm LH}=K\langle X\rangle/\langle\mathbf{LH}(I)\rangle\cong G^{\mathbb{N}}(A)$, where $G^{\mathbb{N}}(A)$ is defined by the $\mathbb{N}$-filtration FA of A induced by the weight $\mathbb{N}$-grading filtration $FK\langle X\rangle$ of $K\langle X\rangle$. Then by (Ch.2, Theorem 4.3) we always have

$$(*)\quad \text{gl.dim}A \le \text{gl.dim}A^{\mathcal{B}}_{\rm LH},\quad \text{gl.dim}A \le \text{gl.dim}A^{\mathbb{N}}_{\rm LH} = \text{gl.dim}G^{\mathbb{N}}(A).$$

Now, combining (Ch.3, Theorem 4.4) we are able to derive the main result of this section.

Theorem 7.4. *Let $(\mathcal{B},\prec)$, I and $A = K\langle X\rangle/I$ be as above, and let Ω be the reduced monomial generating set of $\langle\mathbf{LM}(I)\rangle$. Suppose that $\Omega\cap X=\emptyset$. The following satements hold.*
(i) *If the graph $\Gamma_{\rm C}(\Omega)$ does not have any d-chain, then*

$$gl.dimA \le gl.dimA^{\mathcal{B}}_{\rm LH} \le d.$$

(ii) *If an $\mathbb{N}$-graded monomial ordering $\prec_{gr}$ on $\mathcal{B}$ is used, then, under the condition of* (i) *above we also have*

$$gl.dimG^{\mathbb{N}}(A) \le gl.dimA^{\mathcal{B}}_{\rm LH} \le d;$$

and if furthermore $A^{\mathcal{B}}_{\rm LH}$ has the polynomial growth of degree m, then

$$gl.dimG^{\mathbb{N}}(A) = gl.dimA^{\mathcal{B}}_{\rm LH} = m.$$

(iii) *In the case of* (ii) *above, the ideal I has a finite Gröbner basis $\mathcal{G}$ such that $\mathbf{LH}(\mathcal{G})$ is a finite Gröbner basis of $\langle\mathbf{LH}(I)\rangle$, and thereby $K\langle X\rangle/\langle\mathbf{LH}(\mathcal{G})\rangle \cong G^{\mathbb{N}}(A)$.*

Proof. The assertion of (i) follows from Theorem 7.1 and the formula $(*)$ above. Since $A^{\mathbb{N}}_{\rm LH}\cong G^{\mathbb{N}}(A)$, if an $\mathbb{N}$-graded monomial ordering on $\mathcal{B}$ is used, then the assertions claimed by (ii) and (iii) follow from Theorem 7.1, Corollary 1.1 of Section 1, the formula $(*)$ above and Theorem 7.3. □

By (Ch.4, Theorem 2.1, Proposition 2.2, Theorem 2.4), it is clear that if the monomial ideal $\langle\mathbf{LM}(I)\rangle$ is finitely generated, or equivalently, if $\mathcal{G} = \{g_1,\dots,g_s\}$ is a finite Gröbner basis of I, then Theorem 7.4 will work more effectively via the finite directed graph $\Gamma_{\rm C}(\mathbf{LM}(\mathcal{G}))$, as illustrated by the next two examples which are of independent interest (see also Examples 2, 4 in Section 3, and the assertion of Corollary 5.4).

Corollary 7.5. *Let the free K-algebra $K\langle X\rangle = K\langle X_1, X_2\rangle$ be equipped with a positive weight $\mathbb{N}$-gradation, and let $\mathcal{G} = \{g_1, g_2\}$ be any Gröbner basis with respect to some admissible system $(\mathcal{B},\prec)$, such that $\mathbf{LM}(g_1) =$*

$X_1^2X_2$ and $\mathbf{LM}(g_2) = X_1X_2^2$. Consider the K-algebra $A = K\langle X\rangle/I$ with $I = \langle I\rangle$ and its associated algebras, $A_{\rm LH}^{\mathcal{B}} = K\langle X\rangle/\langle\mathbf{LM}(I)\rangle$ and $A_{\rm LH}^{\mathbb{N}} = K\langle X\rangle/\langle\mathbf{LH}(I)\rangle \cong G^{\mathbb{N}}(A)$. Then

$$gl.dimA \le\ gl.dimA_{\rm LH}^{\mathcal{B}} = 3.$$

If an $\mathbb{N}$-graded monomial ordering $\prec_{gr}$ on $\mathcal{B}$ is used such that $\mathbf{LM}(g_1) = X_1^2X_2$ and $\mathbf{LM}(g_2) = X_1X_2^2$, then

$$gl.dimG^{\mathbb{N}}(A) =\ gl.dimA_{\rm LH}^{\mathcal{B}} = 3.$$

The above result holds for the algebras given in Example 8 of (Ch.3, Section 5), in particular, for any down-up algebra $A = A(\alpha,\beta,\gamma) = K\langle X\rangle/\langle\mathcal{G}\rangle$ and its two associated graded algebras, $A_{\rm LH}^{\mathcal{B}} = K\langle X\rangle/\langle\mathbf{LM}(\mathcal{G})\rangle$ and $A_{\rm LH}^{\mathbb{N}} = K\langle X\rangle/\langle\mathbf{HT}(\mathcal{G})\rangle \cong G^{\mathbb{N}}(A)$, both cubic algebras (see also Example 2 of (Ch.4, Section 2)), we have $gl.dimA \le 3$, $gl.dimG^{\mathbb{N}}(A) = gl.dimA_{\rm LH}^{\mathcal{B}} = 3$.

Proof. By Example 2 of Section 3, in both cases all algebras concerned have polynomial growth of degree 3. Put $\Omega = \{X_1^2X_2,\ X_1X_2^2\}$. Then

$$V = \{1\} \cup \{X_1, X_2\} \cup \{X_1X_2, X_2^2\},$$

and the graph $\Gamma_{\rm C}(\Omega)$ is presented by

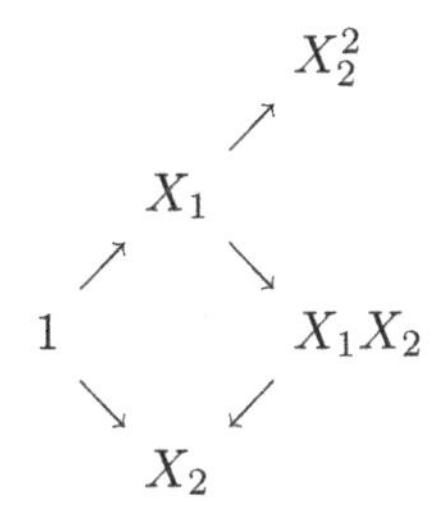

from which we see that $C_{-1} = \{1\}$, $C_0 = \{X_1, X_2\}$, $C_1 = \{X_1^2X_2, X_1X_2^2\}$, $C_2 = \{X_1^2X_2^2\}$, and $\Gamma_C(\Omega)$ does not contain any d-chain for $d \ge 3$. Hence the assertion holds by Theorem 7.4. □

Corollary 7.6. *Let the free K-algebra $K\langle X\rangle = K\langle X_1,\dots,X_n\rangle$ be equipped with a positive weight $\mathbb{N}$-gradation, and let $\mathcal{G}$ be a Gröbner basis in $K\langle X\rangle$ with respect to some admissible system $(\mathcal{B},\prec)$, such that $\mathbf{LM}(\mathcal{G}) = \Omega$ where $\Omega \subseteq \{X_jX_i \mid 1 \le i < j \le n\}$ or $\Omega \subseteq \{X_iX_j \mid 1 \le i < j \le n\}$. Consider the algebra $A = K\langle X\rangle/I$ with $I = \langle\mathcal{G}\rangle$ and its associated graded algebras, $A_{\rm LH}^{\mathcal{B}} = K\langle X\rangle/\langle\mathbf{LM}(I)\rangle$ and $A_{\rm LH}^{\mathbb{N}} = K\langle X\rangle/\langle\mathbf{LH}(I)\rangle \cong G^{\mathbb{N}}(A)$. Then*

$$gl.dimA \le\ gl.dimA_{\rm LH}^{\mathcal{B}} \le n.$$

If an $\mathbb{N}$-graded monomial ordering $\prec_{gr}$ on $\mathcal{B}$ is used such that $\mathbf{LM}(\mathcal{G}) = \Omega$, *then*

$$gl.dimG^{\mathbb{N}}(A) \le \; gl.dimA^{\mathcal{B}}_{\mathrm{LH}} \le n,$$

and the equality $gl.dimG^{\mathbb{N}}(A) = \; gl.dimA^{\mathcal{B}}_{\mathrm{LH}} = n$ *holds provided* $\mathbf{LM}(\mathcal{G}) = \Omega = \{X_jX_i \mid 1 \le i < j \le n\}$, *or* $\mathbf{LM}(\mathcal{G}) = \Omega = \{X_iX_j \mid 1 \le i < j \le n\}$.

Proof. Combining Example 2 of Section 2 with Example 3 of Section 3, we see that in both cases all algebras concerned have polynomial growth of degree $\le n$. It is a good exercise to verify that the graph $\Gamma_{\mathrm{C}}(\Omega)$ does not contain any d-chain for $d \ge n$. For instance, if $n = 3$ and $\Omega = \{X_2X_1, X_3X_1, X_3X_2\}$, then we have $V = \{1\} \cup \{X_1, X_2, X_3\}$ and the graph $\Gamma_{\mathrm{C}}(\Omega)$ is presented by

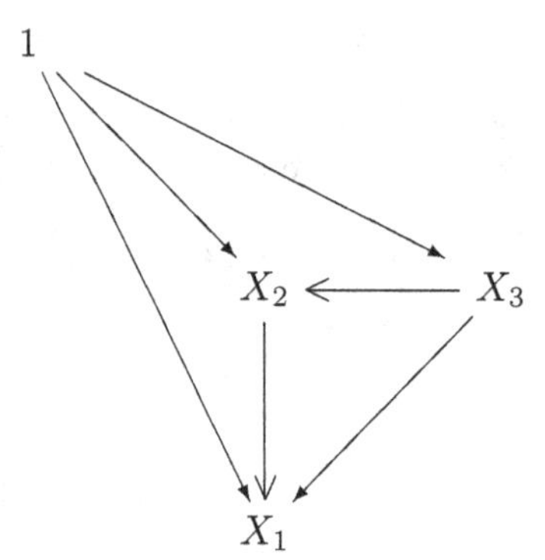

Clearly, $\Gamma_{\mathrm{C}}(\Omega)$ does not contain 3-chains. Hence the assertion holds by Theorem 7.4. □

At this stage, the reader should review all algebras with a quadric Gröbner basis as the set of defining relations given in (Ch.3, Section 5) and (Ch.4, Sections 3, 4).

The algebras considered in the next corollary will be used in (Ch.6, Proposition 3.5).

Corollary 7.7. *Let the free K-algebra $K\langle X\rangle = K\langle X_1, X_2\rangle$ be equipped with a positive weight $\mathbb{N}$-gradation, and for any fixed positive integer n, let $\mathcal{G} = \{g\}$ where $g = X_2^nX_1 - qX_1X_2^n - F$ with $q \in K$, $F \in K\langle X\rangle$. Suppose that $\mathcal{G}$ forms a Gröbner basis with respect to some admissible system $(\mathcal{B}, \prec)$ of $K\langle X\rangle$ such that* $\mathbf{LM}(g) = X_2^nX_1$. *Then, considering the algebra $A =$*

$K\langle X\rangle/I$ and its associated algebras, $A^{\mathcal{B}}_{\rm LH} = K\langle X\rangle/\langle \mathbf{LM}(I)\rangle$ and $A^{\mathbb{N}}_{\rm LH} = K\langle X\rangle/\langle \mathbf{LH}(I)\rangle \cong G^{\mathbb{N}}(A)$, we have

$$gl.dim A \le\ gl.dim A^{\mathcal{B}}_{\rm LH} = 2.$$

If an $\mathbb{N}$-graded monomial ordering $\prec_{gr}$ on $\mathcal{B}$ is used such that $\mathbf{LM}(g) = X_2^nX_1$, then

$$gl.dim G^{\mathbb{N}}(A) =\ gl.dim A^{\mathcal{B}}_{\rm LH} = 2.$$

Proof. By Example 5 of Section 3, in both cases all algebras concerned have polynomial growth of degree 2. Furthermore, putting $\Omega = \{X_2^nX_1\}$, it is straightforward to verify that the graph $\Gamma_{\rm C}(\Omega)$ of n-chains is presented by

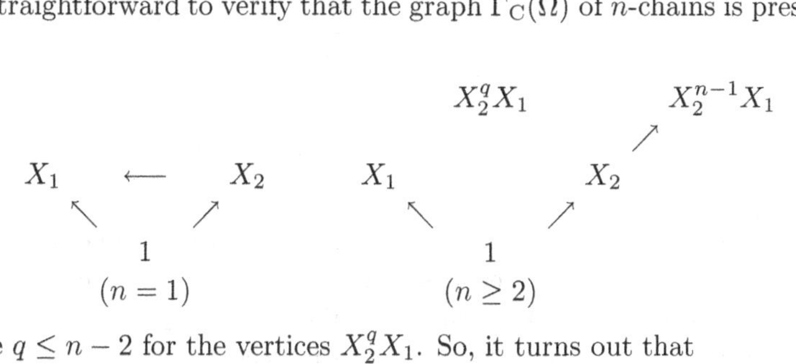

where $q \le n-2$ for the vertices $X_2^qX_1$. So, it turns out that

$$C_{i-1} = \begin{cases} \{X_1,\ X_2\},\ i = 1, \\ \{X_2^nX_1\}, \quad i = 2, \\ \emptyset, \qquad\quad\ \ i \ge 3. \end{cases}$$

Thus the assertion holds by Theorem 7.4. □

Remark We have remarked after Example 8 of (Ch.3, Section 5) that whether a given subset $\mathcal{G}$ in $K\langle X\rangle = K\langle X_1,\ldots,X_n\rangle$ is a Gröbner basis depends heavily on the choice of a monomial ordering. Here let us point out that the effectiveness of using the monomial algebra $A^{\mathcal{B}}_{\rm LH} = K\langle X\rangle/\langle \mathbf{LM}(\mathcal{G})\rangle$ is not only affected by the choice of the monomial ordering $\prec$ but also by the choice of the $\mathbb{N}$-gradation. For instance, consider again the ideal $I = \langle g\rangle$ of $K\langle X_1, X_2\rangle$ generated by the single element $g = X_1^2 - X_1X_2$. We know from Example 3 of (Ch.3, Section 5) that $\mathcal{G} = \{X_1X_2^nX_1 - X_1X_2^{n+1} \mid n \in \mathbb{N}\}$ is an infinite Gröbner basis for I with respect to the $\mathbb{N}$-graded lexicographic ordering $X_2 \prec_{gr} X_1$, where the $\mathbb{N}$-gradation used is the natural $\mathbb{N}$-gradation (i.e., both X_1 and X_2 have degree 1). So we cannot derive the growth of the algebra $A = K\langle X_1, X_2\rangle/I$ by means of the Ufnarovski graph. Moreover, by Example 2 of Section 6 and Theorem 7.1, gl.dim$K\langle X_1, X_2\rangle/\langle \mathbf{LM}(\mathcal{G})\rangle = \infty$, thus we cannot read out any information about the finiteness of gl.dim$A$. However, if we use the $\mathbb{N}$-graded monomial ordering $X_1 \prec_{gr} X_2$ with respect to the natural $\mathbb{N}$-gradation for $K\langle X_1, X_2\rangle$, then $\mathcal{G} = \{g = X_1^2 - X_1X_2\}$ is a Gröbner basis for I with $\mathbf{LM}(\mathcal{G}) = X_1X_2$. It follows that $\mathcal{G}$ has the Ufnarovski graph

$$\Gamma(\mathbf{LM}(\mathcal{G})) : \quad \circlearrowleft X_1 \longleftarrow X_2 \circlearrowright$$

and the graph of n-chains

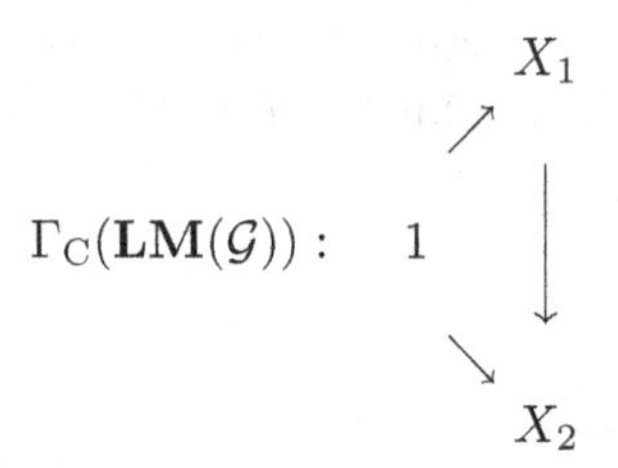

Hence in this case we have GK.dim$A = 2$, and moreover, gl.dim$A = 2 =$ gl.dim$K\langle X_1, X_2\rangle/\langle X_1X_2\rangle$. Yet as the reader may see that if $K\langle X_1, X_2\rangle$ is equipped with the weight $\mathbb{N}$-gradation by assigning $d(X_2) > d(X_1) \geq 1$ and if any $\mathbb{N}$-graded monomial ordering $\prec_{gr}$ with respect to this $\mathbb{N}$-gradation is used, then A is no longer $\mathbb{N}$-graded and we can only derive an inequality gl.dim$A \leq 2$ instead of an equality.

5.8 Determination of Hilbert Series

Let the free K-algebra $K\langle X\rangle = K\langle X_1, \ldots, X_n\rangle$ of n generators be equipped with a positive weight $\mathbb{N}$-gradation by assigning to each X_i a positive degree n_i, and let J be an ideal of $K\langle X\rangle$, $A = K\langle X\rangle/J$. If J is an $\mathbb{N}$-*graded ideal* of $K\langle X\rangle$, then, determining the Hilbert series (see the definition below) of the $\mathbb{N}$-graded algebra A is well-known to be important in the structure theory of A and related topics. Note that if we consider the $\mathbb{N}$-filtration FA of A induced by the weight $\mathbb{N}$-grading filtration $FK\langle X\rangle$ of $K\langle X\rangle$, then since J is graded and thus $\langle \mathbf{LH}(J)\rangle = J$, it follows from (Ch.2, Theorem 3.2) that $G^{\mathbb{N}}(A) \cong A^{\mathbb{N}}_{\rm LH} = K\langle X\rangle/\langle \mathbf{LH}(J)\rangle = A$. Hence, in connection with Proposition 1.4 of Section 1, the problem we just mentioned has the following more general version:

- For an *arbitrary ideal* I of $K\langle X\rangle$, determine the Hilbert series of the $\mathbb{N}$-graded algebra $G^{\mathbb{N}}(A)$, where $A = K\langle X\rangle/I$ and the associated $\mathbb{N}$-graded algebra $G^{\mathbb{N}}(A)$ of A is defined by the $\mathbb{N}$-filtration FA induced by the weight $\mathbb{N}$-grading filtration $FK\langle X\rangle$ of $K\langle X\rangle$.

In this section, by making use of the techniques developed in previous sections and several well-known results ([Gov], [An1,2]), we present a solution to the task posed above in the case that I is generated by a finite Gröbner basis.

We start by recalling the notion of the Hilbert series of $R = K\langle X\rangle/J$, where J is an $\mathbb{N}$-graded ideal of $K\langle X\rangle$. Since $R = \oplus_{p\in\mathbb{N}}R_p$ with $R_p = (K\langle X\rangle_p + J)/J$, each homogeneous component R_p of R is a finite dimensional K-vector space. The *Hilbert series* of R, denoted $H_R(t)$, is defined to be the generating function for the K-dimensions of R_p, that is,

$$H_R(t) = \sum_{p\in\mathbb{N}}(\dim_K R_p)t^p.$$

For instance, if each X_i has degree (weight) 1, then the free algebra $K\langle X\rangle = \oplus_{p\in\mathbb{N}}K\langle X\rangle_p$ has $\dim_K K\langle X\rangle_p = n^p$, so

$$H_{K\langle X\rangle}(t) = \sum_{p\in\mathbb{N}} n^p t^p = \frac{1}{1-nt}.$$

If $R = K\langle X\rangle/J$ with $J = \langle X_jX_i - X_iX_j \mid 1 \le i < j \le n\rangle$, then $R \cong K[x_1,\dots,x_n]$, the commutative polynomial algebra in n-variables, and since $\dim_K R_p = \binom{p+n-1}{n-1}$ for each $p \in \mathbb{N}$,

$$H_R(t) = \sum_{p\in\mathbb{N}}\binom{p+n-1}{n-1}t^p = \frac{1}{(1-t)^n}.$$

More generally, it is well known that if $A = K[x_1,\dots,x_n]/J$, where J is an $\mathbb{N}$-graded ideal of $K[x_1,\dots,x_n]$, then $H_A(t) = p(t)/(1-t)^n$ for some polynomial $p(t) \in \mathbb{Z}[t]$ (see e.g., [Mat]).

Note that all three functions $\frac{1}{1-nt}$, $\frac{1}{(1-t)^n}$, and $\frac{p(t)}{(1-t)^n}$ given above are rational functions of t. However, as shown in [Sh], not every noncommutative algebra $R = K\langle X\rangle/J$ with J finitely generated has the Hilbert series which is a rational function of t.

The first result of this section determines the rationality of $H_{G^{\mathbb{N}}(A)}(t)$ in the case that $A = K\langle X\rangle/I$ with I generated by a finite Gröbner basis $\mathcal{G}$.

Theorem 8.1 ([Gov]). *Let Ω be a finite subset of monomials in the free K-algebra $K\langle X\rangle = K\langle X_1, \ldots, X_n\rangle$. Then the Hilbert series $H_R(t)$ of the monomial algebra $R = K\langle X\rangle/\langle\Omega\rangle$ is a rational function in t.*

□

Corollary 8.2. *Let the free K-algebra $K\langle X\rangle = K\langle X_1, \ldots, X_n\rangle$ be equipped with a fixed positive weight $\mathbb{N}$-gradation and let $\mathcal{G}$ be a finite Gröbner basis in $K\langle X\rangle$ with respect to some fixed admissible system $(\mathcal{B}, \prec_{gr})$, in which $\prec_{gr}$ is an $\mathbb{N}$-graded monomial ordering on the standard K-basis $\mathcal{B}$ of $K\langle X\rangle$ subject to the fixed $\mathbb{N}$-gradation of $K\langle X\rangle$. Consider the algebra $A = K\langle X\rangle/\langle\mathcal{G}\rangle$ and the associated $\mathbb{N}$-graded algebra $G^{\mathbb{N}}(A)$ of A defined by the $\mathbb{N}$-filtration FA induced by the weight $\mathbb{N}$-grading filtration $FK\langle X\rangle$. Then the Hilbert series $H_{G^{\mathbb{N}}(A)}(t)$ is a rational function of t.*

In particular, the Hilbert series of every $\mathbb{N}$-graded algebra of the form $K\langle X\rangle/\langle\mathcal{G}\rangle$ with $\mathcal{G}$ an $\mathbb{N}$-homogeneous finite Gröbner basis is a rational function of t.

Proof. Since we are using the $\mathbb{N}$-graded monomial ordering $\prec_{gr}$ on $\mathcal{B}$, by (Ch.4, Proposition 2.2, Theorem 2.4),

$$G^{\mathbb{N}}(A) \cong K\langle X\rangle/\langle \mathbf{HT}(\mathcal{G})\rangle = A^{\mathbb{N}}_{\mathrm{LH}},$$

where $A^{\mathbb{N}}_{\mathrm{LH}}$ is the $\mathbb{N}$-leading homogeneous algebra of A. Also by Corollary 1.1 and Proposition 1.4 of Section 1, $A^{\mathbb{N}}_{\mathrm{LH}}$ has the $\mathcal{B}$-leading homogeneous algebra $(A^{\mathbb{N}}_{\mathrm{LH}})^{\mathcal{B}}_{\mathrm{LH}} = K\langle X\rangle/\langle \mathbf{LM}(\mathcal{G})\rangle = A^{\mathcal{B}}_{\mathrm{LH}}$ and $H_{A^{\mathbb{N}}_{\mathrm{LH}}}(t) = H_{A^{\mathcal{B}}_{\mathrm{LH}}}(t)$. The corollary follows now from Theorem 8.1. □

The next two results, due to D. J. Anick, show that the Hilbert series may be used to judge if a noncommutative graded algebra has global homological dimension 2, which, in turn, reveal some general *undecidability* concerning global dimension and Hilbert series (see the explanation given after Theorem 8.4 below).

Theorem 8.3 ([An1]). *Let $\mathcal{X}$ be a minimal generating set of an $\mathbb{N}$-graded algebra R and $\mathcal{R}$ a nonempty minimal set of homogeneous defining relations. Then*

$$gl.dim R = 2 \text{ if and only if } H_R(t) = (1 - H_{\mathcal{X}}(t) + H_{\mathcal{R}}(t))^{-1},$$

where $H_{\mathcal{X}}(t) = \sum_{x\in\mathcal{X}} N(x)t^{d(x)}$ is the generating function of the number $N(x)$ of generators with given degree $d(x)$, and $H_{\mathcal{R}}(t) = \sum_{r\in\mathcal{R}} N(r)t^{d(r)}$ is *the generating function of the number $N(r)$ of defining relations with given degree $d(r)$.*

□

Theorem 8.4 ([An2]). *For every system of diophantine equations $S = 0$, there exists a finitely presented algebra R (which can be constructively expressed in terms of the coefficients of S) such that R has global dimension 2 if and only if the system $S = 0$ has no solutions. Moreover, this algebra is the universal enveloping algebra of a Lie superalgebra, defined by quadratic relations only.*

□

Based on Theorem 8.3 and Theorem 8.4, Ufnarofski argued in [Uf3] that the following attempts are impossible:

- To find an algorithm that calculates the global dimension of an arbitrary finitely presented algebra;
- To find an algorithm that takes relations as input and gives the Hilbert series as output. Furthermore, one cannot decide in general whether the Hilbert series of a given algebra is equal to a given series;
- To predict the behavior of a Hilbert series, knowing only a finite number of its coefficients.

Yet on the other hand, we have pretty good news from [An3] and [CU] for algorithmically calculating the Hilbert series of $\mathbb{N}$-graded algebras defined by finite Gröbner bases, that is, this job can be done via the efficient computation of the Poincaré series of the associated monomial algebras ($\mathcal{B}$-leading homogeneous algebras in the language used in this book). To see this clearly, we introduce below the Poincaré series of an $\mathbb{N}$-graded algebra.

For any $\mathbb{N}$-graded K-algebra $R = \oplus_{p\in\mathbb{N}} R_p$ with $R_0 = K$, the standard text on homological algebra tells us that there exists a free R-resolution of the trivial $\mathbb{N}$-graded R-module $K \cong R/(\oplus_{p\geq 1} R_p)$:

$$(\mathcal{R},\delta) \qquad \cdots \xrightarrow{\delta_n} L_n \xrightarrow{\delta_{n-1}} \cdots \xrightarrow{\delta_3} L_2 \xrightarrow{\delta_2} L_1 \xrightarrow{\delta_1} L_0 \xrightarrow{\varepsilon} K$$

where each L_i is an $\mathbb{N}$-graded R-module and each δ_i is an $\mathbb{N}$-graded R-module homomorphism. Consider the homology of the derived sequence

$$\cdots \xrightarrow{\delta_n\otimes 1_K} L_n\otimes_R K \xrightarrow{\delta_{n-1}\otimes 1_K} \cdots \xrightarrow{\delta_2\otimes 1_K} L_1\otimes_R K \xrightarrow{\delta_1\otimes 1_K} L_0\otimes_R K \longrightarrow 0.$$

If the K-module $\mathrm{Tor}_i^R(K,K)$ is finite dimensional for $i = 1, 2, \ldots,$ then the *Poincaré series* $P_R(t)$ of R is defined to be the generating function for the K-dimensions of $\mathrm{Tor}_i^R(K,K)$, that is,

$$P_R(t)=\sum_{i\geq 0}\dim_K\mathrm{Tor}_i^R(K,K)\cdot t^i.$$

If furthermore the $\mathbb{N}$-graded structure $\mathrm{Tor}_i^R(K,K)=\oplus_{p\in\mathbb{N}}\mathrm{Tor}_i^R(K,K)_p$ is taken into account, then the double Poincaré series of R is defined to be

$$P_R(s,t)=\sum_{i,p\geq 0}\dim_K\mathrm{Tor}_i^R(K,K)_p\cdot s^it^p.$$

Now, let us return to the $\mathbb{N}$-graded algebra $R=K\langle X\rangle/J$ defined by an $\mathbb{N}$-graded ideal J of $K\langle X\rangle=K\langle X_1,\ldots,X_n\rangle$. Then, with respect to a fixed admissible system $(\mathcal{B},\prec_{gr})$ of $K\langle X\rangle$, the $\mathcal{B}$-leading homogeneous algebra $R^{\mathcal{B}}_{\rm LH}=K\langle X\rangle/\langle\mathbf{LM}(J)\rangle$ of R is well defined. Since we already know that $H_R(t)=H_{R^{\mathcal{B}}_{\rm LH}}(t)$ when either side is defined, a wonderful result obtained in [An3] is the algorithmic computation of $H_R(t)$ via the Anick resolution of K over $R^{\mathcal{B}}_{\rm LH}$, or more precisely, via the $\mathbb{N}$-$\mathcal{B}$-graded isomorphism

$$\mathrm{Tor}_i^{R^{\mathcal{B}}_{\rm LH}}(K,K)\cong C_{i-1},\ \ i=1,2,...,$$

where C_{i-1} stands for the $\mathbb{N}$-$\mathcal{B}$-graded K-vector space spanned by all $(i-1)$-chains in the graph $\Gamma_{\rm C}(\Omega)$ with Ω a reduced monomial generating set of $\langle\mathbf{LM}(J)\rangle$ (see Section 7).

Theorem 8.5 ([An3], formula (16)). *With notation as fixed above,*

$$H_R(t)=P_{R^{\mathcal{B}}_{\rm LH}}(-1,t)^{-1}=\left(1-\sum_{i=0}(-1)^iH_{C_i}(t)\right)^{-1},$$

where $H_{C_i}(t)$ denotes the Hilbert series of the $\mathbb{N}$-graded K-module C_i.

□

Considering the algebra $A=K\langle X\rangle/\langle\mathcal{G}\rangle$ defined by an arbitrary LM-reduced finite Gröbner basis $\mathcal{G}\subset K\langle X\rangle$ with respect to some fixed admissible system $(\mathcal{B},\prec_{gr})$, we are able to compute the Hilbert series $H_{G^{\mathbb{N}}(A)}(t)$ of the associated $\mathbb{N}$-graded algebra $G^{\mathbb{N}}(A)$ of A defined by the $\mathbb{N}$-filtration FA induced by the previously fixed weight $\mathbb{N}$-grading filtration $FK\langle X\rangle$.

Theorem 8.6. *With notation as fixed above, the Hilbert series of $G^{\mathbb{N}}(A)$ may be computed by means of the formula*

$$H_{G^{\mathbb{N}}(A)}(t)=P_{A^{\mathcal{B}}_{\rm LH}}(-1,t)^{-1}=\left(1-\sum_{i=0}(-1)^iH_{C_i}(t)\right)^{-1},$$

where $H_{C_i}(t)$ denotes the Hilbert series of the $\mathbb{N}$-graded K-module spanned by the set C_i of i-chains in the graph $\Gamma_{\rm C}(\mathbf{LM}(\mathcal{G}))$, $A^{\mathcal{B}}_{\rm LH} = K\langle X\rangle/\langle\mathbf{LM}(\mathcal{G})\rangle$.

Proof. Again since we are using an $\mathbb{N}$-graded monomial ordering $\prec_{gr}$ on the standard K-basis $\mathcal{B}$ of $K\langle X\rangle$, by (Ch.4, Proposition 2.2, Theorem 2.4),

$$G^{\mathbb{N}}(A) \cong K\langle X\rangle/\langle\mathbf{HT}(\mathcal{G})\rangle = A^{\mathbb{N}}_{\rm LH},$$

where $A^{\mathbb{N}}_{\rm LH}$ is the $\mathbb{N}$-leading homogeneous algebra of A. Also by Corollary 1.1 and Proposition 1.4 of Section 1, $A^{\mathbb{N}}_{\rm LH}$ has the $\mathcal{B}$-leading homogeneous algebra $(A^{\mathbb{N}}_{\rm LH})^{\mathcal{B}}_{\rm LH} = K\langle X\rangle/\langle\mathbf{LM}(\mathcal{G})\rangle = A^{\mathcal{B}}_{\rm LH}$ and $H_{A^{\mathbb{N}}_{\rm LH}}(t) = H_{A^{\mathcal{B}}_{\rm LH}}(t)$. It follows from Theorem 8.5 that the assertion holds. □

Corollary 8.7. *Let $A = K\langle X\rangle/\langle\mathcal{G}\rangle$, $A^{\mathcal{B}}_{\rm LH} = K\langle X\rangle/\langle\mathbf{LM}(\mathcal{G})\rangle$, and $G^{\mathbb{N}}(A)$ be as in Corollary 8.2 and Theorem 8.6. Then the following property is algorithmically decidable: gl.dim$G^{\mathbb{N}}(A) = 2$ if and only if gl.dim$A^{\mathcal{B}}_{\rm LH} = 2$ if and only if*

$$H_{G^{\mathbb{N}}(A)}(t) = (1 - H_{\mathcal{X}}(t) + H_{\mathcal{R}}(t))^{-1} = H_{A^{\mathcal{B}}_{\rm LH}}(t),$$

where $H_{\mathcal{X}}(t)$ and $H_{\mathcal{R}}(t)$ are as defined in Theorem 8.3.

□

Example 1. Given degree $d(X_1) = d(X_2) = 1$ for the free algebra $K\langle X\rangle = K\langle X_1, X_2\rangle$, if F is any homogeneous element of degree 2 with respect to the natural $\mathbb{N}$-gradation of $K\langle X\rangle$, such that $\mathbf{LM}(F) = X_2X_1$ with respect to some monomial ordering $\prec$ on $K\langle X\rangle$, then $\{F\}$ is a Gröbner basis by Example 1 of (Ch.4, Section 3). Hence the algebra $R = K\langle X\rangle/\langle F\rangle$ has global dimension 2, for, now R has a PBW K-basis and

$$H_R(t) = (1 - H_{\mathcal{X}}(t) + H_{\mathcal{R}}(t))^{-1} = \frac{1}{(1-t)^2}.$$

Corollary 8.8. *Let the free algebra $K\langle X\rangle = K\langle X_1, X_2\rangle$ be equipped with the natural $\mathbb{N}$-gradation, i.e., both X_1 and X_2 are of degree 1, and $\mathcal{G} = \{g_1, g_2\}$ be a Gröbner basis in $K\langle X\rangle$ with respect to some fixed admissible system $(\mathcal{B}, \prec_{gr})$, such that $\mathbf{LM}(g_1) = X_1^2X_2$ and $\mathbf{LM}(g_2) = X_1X_2^2$. Consider the K-algebra $A = K\langle X\rangle/\langle\mathcal{G}\rangle$ and its associated $\mathbb{N}$-graded algebras $G^{\mathbb{N}}(A)$ defined by the natural $\mathbb{N}$-filtration FA. By referring to the graph of n-chains $\Gamma_{\rm C}(\mathbf{LM}(\mathcal{G}))$ constructed in previous Corollary 7.5, it follows from Theorem 8.6 above that*

$$H_{G^{\mathbb{N}}(A)}(t) = P_{A^{\mathcal{B}}_{\rm LH}}(-1,t)^{-1} = \frac{1}{1-2t+2t^3-t^4},$$

where $A_{\rm LH}^{\mathcal{B}} = K\langle X\rangle/\langle \mathbf{LM}(G)\rangle$.

In particular, if we look at the associated $\mathbb{N}$*-graded algebra* $G^{\mathbb{N}}(A)$ *of any down-up algebra* $A = A(\alpha,\beta,\gamma)$ *given in Example 8 of (Ch.3, Section 5), then*

$$\begin{aligned} G^{\mathbb{N}}(A) = K\langle X_1, X_2\rangle/\langle \mathbf{HT}(g_1), \mathbf{HT}(g_2)\rangle \ &with \\ \mathbf{HT}(g_1) = X_1^2X_2 - \alpha X_1X_2X_1 - \beta X_2X_1^2&, \\ \mathbf{HT}(g_2) = X_1X_2^2 - \alpha X_2X_1X_2 - \beta X_2^2X_1&, \end{aligned}$$

and $G^{\mathbb{N}}(A)$ *has the Hilbert series* $\frac{1}{1-2t+2t^3-t^4}$. □

Note that the Hilbert series obtained in Corollary 8.8 above has the form

$$\frac{1}{1-2t+2t^3-t^4} = \frac{1}{(1-t)^2(1-t^2)},$$

as asserted by Theorem 7.2(iii) (or Theorem 7.3(iii)). But the next example indicates that this is not always true (see also Corollary 3.7 of Ch.7).

Example 2. Let the free K-algebra $K\langle X\rangle = K\langle X_1, X_2\rangle$ be equipped with a positive weight $\mathbb{N}$-gradation, and for any fixed positive integer n, let $\mathcal{G} = \{g\}$ with $g = X_2^nX_1 - qX_1X_2^n - F$ with $q \in K$, $F \in K\langle X\rangle$. Suppose that $\mathcal{G}$ forms a Gröbner basis with respect to some fixed admissible system $(\mathcal{B}, \prec_{gr})$, such that $\mathbf{LM}(g) = X_2^nX_1$ (for instance if both X_1 and X_2 are assigned the degree 1 and if the $\mathbb{N}$-graded monomial ordering $X_1 \prec_{gr} X_2$ is used). Considering the K-algebra $A = K\langle X\rangle/\langle\mathcal{G}\rangle$ and its associated graded algebra $G^{\mathbb{N}}(A)$ defined by the $\mathbb{N}$-filtration FA which is induced by the weight $\mathbb{N}$-grading filtration of $K\langle X\rangle$, then by referring to the graph $\Gamma_{\rm C}(\Omega)$ of n-chains constructed in the proof of previous Corollary 7.7, we obtain

$$H_{G^{\mathbb{N}}(A)}(t) = \frac{1}{1-2t+t^{n+1}}$$

and we see that in the case of $n \geq 2$ it cannot be the form as asserted by Theorem 7.2(iii) (or Theorem 7.3(iii)).

Chapter 6

Recognizing (Non-)Homogeneous p-Koszulity via $A^{\mathcal{B}}_{\text{LH}}$

It is well known that quadratic Koszul algebras, which were introduced by S. Prridy ([Pr], 1970), have been widely used in topology, algebraic geometry, commutative algebra, representation theory of algebras, and quantum group theory. Because of the intimate relation with the study of curved differential graded algebras (CDG algebras), non-homogeneous 2-Koszulity for $\mathbb{N}$-filtered algebras, which, in connection with the generalized PBW property ([Pr], Theorem 5.3), has been studied further by several authors, e.g., [Pos] and [BG1]. We refer to ([PP], 2005) for more detail on this aspect. In 2001, R. Berger generalized in [Ber1] the quadratic Koszulity to $\mathbb{N}$-graded algebras defined by homogeneous relations of degree $p > 2$. Significant examples of p-Koszul algebras may be found in noncommutative projective geometry (e.g., Artin-Schelter regular cubic algebras of global dimension 3), representation thery (e.g., skew-symmetrizer killing algebras, some quiver algebras), and theoretical physics (e.g., Yang-Mills algebras), etc. In the study of some interesting algebras from symplectic reflection algebras [EG], the non-homogeneous p-Koszulity for $p > 2$ was realized in [BG2] through the p-type PBW property.

Instead of going to a general theory dealing with Koszul algebras, in this chapter we focus on *noncommutative algebras defined by relations* and intend to demonstrate how to recognize the homogeneous and non-homogeneous p-Koszulity of the algebras considered by using some computational results obtained in Ch.4 – Ch.5. Moreover, as an application of the main result (Theorem 3.1), we give some examples of (non-)homogeneous p-Koszul algebras for $p \geq 2$.

Conventions and notations are maintained as in previous chapters.

6.1 (Non-)Homogeneous p-Koszul Algebras

For the reader's convenience, in this section we recall briefly the definitions of homogeneous and non-homogeneous p-Koszul algebras.

In what follows, we let $K\langle X\rangle = K\langle X_1, \ldots, X_n\rangle$ be the free K-algebra of n generators with the natural $\mathbb{N}$-gradation, that is, each X_i has the degree 1, $1 \leq i \leq n$.

Quadratic Koszul algebra

Let J be an $\mathbb{N}$-graded ideal of $K\langle X\rangle$ and $R = K\langle X\rangle/J$. Consider the trivial $\mathbb{N}$-graded R-module $K \cong R/(\oplus_{p>0}R_p)$ and the homology groups $\mathrm{Tor}_i^R(K, K)$ of K with respect to an $\mathbb{N}$-graded R-free resolution of K. Recall from [Pr] that R is called a *Koszul algebra* if, for each $i \geq 1$, the $\mathbb{N}$-graded K-vector space $\mathrm{Tor}_i^R(K, K)$ has the property that

$$\left(\mathrm{Tor}_i^R(K, K)\right)_j = 0 \text{ for } j \neq i.$$

It is well known that the above definition is equivalent to say that K has a *linear resolution*, that is, there is an $\mathbb{N}$-graded R-free resolution of K:

$$\cdots \to L_n \xrightarrow{\delta_n} L_{n-1} \longrightarrow \cdots \longrightarrow L_1 \xrightarrow{\delta_1} L_0 \xrightarrow{\delta_0} K \to 0$$

such that for each $n \geq 0$, the $\mathbb{N}$-graded R-module $L_n = \oplus_{k\in\mathbb{Z}}(L_n)_k$ is generated in degree n in the sense that $(L_n)_k = R_{k-n}(L_n)_n$ (e.g., see [Pr], [BGS], [GV]). It is equally well known that if R is Koszul, then J is necessarily generated by $\mathbb{N}$-homogeneous elements of degree 2, namely, R is a quadratic algebra, but the converse is not necessarily true (see, e.g., [Fr], [GV]).

Homogeneous p-Koszul algebra

Let J be an $\mathbb{N}$-graded ideal $K\langle X\rangle$ and $R = K\langle X\rangle/J$. The $\mathbb{N}$-graded algebra R is said to be *homogeneous p-Koszul* in the sense of [Ber2], if the following conditions are satisfied:

(a) J is generated by $\mathbb{N}$-homogeneous elements of degree $p \geq 2$;

(b) For each $i \geq 1$, the $\mathbb{N}$-graded K-vector space $\mathrm{Tor}_i^R(K, K)$ is concentrated in degree $\zeta(i)$, or in other words,

$$\left(\mathrm{Tor}_i^R(K, K)\right)_j = 0 \text{ for } j \neq \zeta(i),$$

where $\zeta\colon \mathbb{N} \to \mathbb{N}$ is the *jump map* defined by

$$\zeta(i) = \begin{cases} mp, & \text{if } i = 2m \text{ (i.e., } i \text{ is even)}, \\ mp+1, & \text{if } i = 2m+1 \text{ (i.e., } i \text{ is odd)}. \end{cases}$$

Clearly, when $p = 2$, the above definition recovers the classical quadratic Koszul algebras.

Non-homogeneous p-Koszul algebra

Now, let the free K-algebra $K\langle X\rangle$ be equipped with the natural $\mathbb{N}$-filtration $FK\langle X\rangle$, i.e., the natural $\mathbb{N}$-grading filtration $FK\langle X\rangle = \{F_pK\langle X\rangle\}_{p\in\mathbb{N}}$ with $F_pK\langle X\rangle = \oplus_{i\le p}K\langle X\rangle_i$. For a K-subspace $V_p \subset F_pK\langle X\rangle - F_{p-1}K\langle X\rangle$, $p \ge 2$, consider the quotient algebra $A = K\langle X\rangle/\langle V_p\rangle$ and its associated $\mathbb{N}$-graded algebra $G^{\mathbb{N}}(A)$ defined by the $\mathbb{N}$-filtration FA induced by the natural $\mathbb{N}$-grading filtration $FK\langle X\rangle$. Then, A is said to be a *non-homogeneous p-Koszul algebra*, in the sense of ([Pos], [BG2]), if the following two conditions are satisfied:

(a) The $\mathbb{N}$-graded algebra $K\langle X\rangle/\langle \mathbf{LH}(V_p)\rangle$ is p-Koszul, where $\mathbf{LH}(V_p) = \{\mathbf{LH}(f) \mid f \in V_p\}$ is the set of $\mathbb{N}$-leading homogeneous elements of V_p as defined in (Ch.2, Section 3);

(b) A has the p-type PBW property in the sense of [Ber2], that is, the $\mathbb{N}$-graded K-algebra homomorphism ρ appearing in the commutative diagram

$$\begin{array}{ccc} K\langle X\rangle & \xrightarrow{\pi} & K\langle X\rangle/\langle \mathbf{LH}(V_p)\rangle \\ {\scriptstyle\phi}\big\downarrow & \swarrow{\scriptstyle\rho} & \\ G^{\mathbb{N}}(A) & & \end{array}$$

is an isomorphism (see Ch.4, Section 1).

6.2 Anick's Resolution and Homogeneous p-Koszulity

Let $(\mathcal{B}, \prec_{gr})$ be an admissible system of the free K-algebra $K\langle X\rangle = K\langle X_1,\ldots,X_n\rangle$, where $\mathcal{B}$ is the standard K-basis of $K\langle X\rangle$ and $\prec_{gr}$ is some $\mathbb{N}$-graded monomial ordering on $\mathcal{B}$ with respect to the natural $\mathbb{N}$-gradation of $K\langle X\rangle$. Given an $\mathbb{N}$-graded ideal J of $K\langle X\rangle$, let $R^{\mathcal{B}}_{\rm LH}$ be the $\mathcal{B}$-leading homogeneous algebra (or equivalently, the associated monomial algebra) of the quotient algebra $R = K\langle X\rangle/J$ with respect to $(\mathcal{B}, \prec_{gr})$, i.e.,

$R^{\mathcal{B}}_{\rm LH} = K\langle X\rangle/\langle \mathbf{LM}(J)\rangle$ (see Ch.4, Section 2). In this section we show that the homogeneous p-Koszulity of R may be determined by that of $R^{\mathcal{B}}_{\rm LH}$. The reader is referred to (Ch.5, Section 8) for related notions and notations.

To start with, let Ω be a reduced subset of $\mathcal{B}$ (see Ch.5, Section 2), and let $\Gamma_{\rm C}(\Omega)$ be the directed graph of n-chains of Ω as constructed in (Ch.5, Section 6). Considering the monomial K-algebra $\overline{A} = K\langle X\rangle/\langle\Omega\rangle$ and the Anick resolution of the trivial $\mathbb{N}$-graded $\overline{A}$-module K, then there are $\mathbb{N}$-graded K-module isomorphisms

$$\text{(1)}\qquad {\rm Tor}_i^{\overline{A}}(K,K) \cong C_{i-1},\quad i=1,2,\ldots$$

where C_{i-1} denotes the $\mathbb{N}$-graded K-vector space spanned by all $(i-1)$-chains (see Ch.5, Section 7). We say that the $\mathbb{N}$-graded K-module C_{i-1} is *generated in degree* m if every $(i-1)$-chain gives rise to a monomial of degree m. For instance, if Ω consists of quadratic monomials, then every $(i-1)$-chain of $\Gamma_{\rm C}(\Omega)$ is of the form

$$1 \longrightarrow X_{k_1} \longrightarrow X_{k_2} \longrightarrow \cdots \longrightarrow X_{k_i}$$

which yields a monomial of degree i. So, as observed in ([An3], Section 3), the above formula (1) leads to the classical Koszulity of the quadratic monomial algebra $\overline{A}$. Concerning the homogeneous p-Koszulity of $\overline{A}$, we have the following more general result.

Proposition 2.1. *For $p\ge 2$, let ζ be the associated jump map as introduced in the last section. With notation as above, if the $\mathbb{N}$-graded K-module C_{i-1} is generated in degree $\zeta(i)$, $i=1,2,\ldots$, then $\overline{A}$ is homogeneous p-Koszul.*

□

For an arbitrary $\mathbb{N}$-graded ideal J of $K\langle X\rangle$, to make use of Proposition 2.1, we first recall a result of [An3] concerning the relation between the Poincaré series $P_R(s,t)$ and the Poincaré series $P_{R^{\mathcal{B}}_{\rm LH}}(s,t)$, which is based on the Anick resolution of the trivial $\mathbb{N}$-graded module $K = R/(\oplus_{p\in\mathbb{N}}R_p)$ over $R = K\langle X\rangle/J$.

Let Ω be the reduced monomial generating set of the monomial ideal $\langle\mathbf{LM}(J)\rangle$, and for each $i=1,2,\ldots$, let C_i be the K-vector space spanned by all i-chains of the directed graph $\Gamma_{\rm C}(\Omega)$. The Anick resolution of K over R is the $\mathbb{N}$-graded R-free resolution

$$\cdots \to C_2\otimes_K R \xrightarrow{\delta_2} C_1\otimes_K R \xrightarrow{\delta_1} C_0\otimes_K R \xrightarrow{\delta_0} R \xrightarrow{\varepsilon} K \to 0$$

which is inductively constructed such that $\delta_0(x \otimes 1) = x$, and for $n \geq 1$, $u \in C_n$, if $u = u(n-1) \cdot t$ with $u(n-1) \in C_{n-1}$ and $t \in \mathcal{B} - \langle \mathbf{LM}(J) \rangle - \{1\}$, then

$$\delta_n(u \otimes 1) = u(n-1) \otimes t + f,$$

where, under taking the normal form by division by J, $\mathbf{LM}(f) \prec_{gr} u$ if $f \neq 0$. Moreover, since monomials in $\mathcal{B}$ are $\mathbb{N}$-homogeneous elements in $K\langle X \rangle$, all δ_is are $\mathbb{N}$-graded R-homomorphisms of degree zero and the element f in the expression of the image $\delta_n(u \otimes 1)$ is homogeneous. Detailed argument on this inductive procedure is given in the proof of ([An3], Theorem 1.5). The reader is also referred to [CU] for the algorithmic construction of such a resolution in terms of Gröbner basis. Now, after tensoring the above resolution over R with K we have

$$\mathrm{Tor}_i^R(K,K) = \frac{\mathrm{Ker}(\delta_{i-1} \otimes 1_K)}{\mathrm{Im}(\delta_i \otimes 1_K)} \subseteq \frac{C_{i-1} \otimes_K R \otimes_R K}{\mathrm{Im}(\delta_i \otimes 1_K)} = \frac{C_{i-1}}{\mathrm{Im}(\delta_i \otimes 1_K)}.$$

Since each derived homomorphism $\delta_i \otimes 1_K$ is graded of degree zero, and by the foregoing formula (1) the degree-j part of C_{i-1} is nothing but the degree-j part $\left(\mathrm{Tor}_i^{R_{\mathrm{LH}}^{\mathcal{B}}}(K,K)\right)_j$ of the $\mathbb{N}$-graded K-module $\mathrm{Tor}_i^{R_{\mathrm{LH}}^{\mathcal{B}}}(K,K)$, it turns out that

$$\text{(2)} \quad \dim_K \left(\mathrm{Tor}_i^R(K,K)\right)_j \leq \dim_K \left(\mathrm{Tor}_i^{R_{\mathrm{LH}}^{\mathcal{B}}}(K,K)\right)_j, \quad i \geq 1,\ j \geq 0.$$

Hence, the following result is obtained.

Lemma 2.2 ([An3], Lemma 3.4). *With notation as fixed before,*

$$P_R(s,t) \leq P_{R_{\mathrm{LH}}^{\mathcal{B}}}(s,t),$$

inequality holding coefficientwise in the sense of the formula (2) given above.

□

It follows from Proposition 2.1 and Lemma 2.2 that we are able to state the main result of this section as follows.

Theorem 2.3. *Let $R = K\langle X \rangle / J$, $R_{\mathrm{LH}}^{\mathcal{B}} = K\langle X \rangle / \langle \mathbf{LM}(J) \rangle$, and $\Gamma_{\mathrm{C}}(\Omega)$ be as fixed above, where Ω is the reduced monomial generating set of $\langle \mathbf{LM}(J) \rangle$. For $p \geq 2$, let $\zeta\colon \mathbb{N} \to \mathbb{N}$ be the jump map associated to p (see Section 1). If the $\mathbb{N}$-graded K-module C_{i-1} spanned by all $(i-1)$-chains of $\Gamma_{\mathrm{C}}(\Omega)$ is generated in degree $\zeta(i)$, $i = 1, 2, \ldots$, then $R_{\mathrm{LH}}^{\mathcal{B}}$ is homogeneous p-Koszul* *and thereby R is homogeneous p-Koszul.*

□

6.3 Working in Terms of Gröbner Bases

Combining (Ch.4, Section 2), we show in this section how to recognize the homogeneous and non-homogeneous Koszulity by means of Gröbner bases.

Let the free K-algebra $K\langle X\rangle = K\langle X_1,\dots,X_n\rangle$ of n generators be equipped with the natural $\mathbb{N}$-gradation, and let $(\mathcal{B},\prec_{gr})$ be an admissible system of $K\langle X\rangle$ with $\prec_{gr}$ an $\mathbb{N}$-graded monomial ordering on the standard basis $\mathcal{B}$. If I is an arbitrary ideal of $K\langle X\rangle$ and $A = K\langle X\rangle/I$, then as before, we have the $\mathbb{N}$-leading homogeneous algebra $A^{\mathbb{N}}_{\rm LH} = K\langle X\rangle/\langle \mathbf{LH}(I)\rangle$ of A defined by the set $\mathbf{LH}(I)$ of $\mathbb{N}$-leading homogeneous elements of I (note that if I is a graded ideal then $A = A^{\mathbb{N}}_{\rm LH}$); considering the natural $\mathbb{N}$-grading filtration $FK\langle X\rangle$ of $K\langle X\rangle$ and the associated $\mathbb{N}$-graded algebra $G^{\mathbb{N}}(A)$ of A defined by the $\mathbb{N}$-filtration FA induced by $FK\langle X\rangle$, we have $G^{\mathbb{N}}(A)\cong A^{\mathbb{N}}_{\rm LH}$; and moreover, with respect to $(\mathcal{B},\prec_{gr})$ we also have the $\mathcal{B}$-leading homogeneous algebra $A^{\mathcal{B}}_{\rm LH} = K\langle X\rangle/\langle \mathbf{LM}(I)\rangle$ of A.

Theorem 3.1. *Let the notations be as fixed above. For $p\ge 2$, let $\mathcal{G}\subset F_pK\langle X\rangle - F_{p-1}K\langle X\rangle$ be an LM-reduced Gröbner basis with respect to $(\mathcal{B},\prec_{gr})$ in the sense of (Ch.3, Section 2), and $A = K\langle X\rangle/I$ with $I = \langle\mathcal{G}\rangle$. Bearing in mind the $\mathbb{N}$-graded algebra isomorphism $G^{\mathbb{N}}(A)\cong A^{\mathbb{N}}_{\rm LH}$, the following statements hold.*
(i) *If $A^{\mathcal{B}}_{\rm LH}$ is homogeneous p-Koszul, then $A^{\mathbb{N}}_{\rm LH}$ is homogeneous p-Koszul, and hence $G^{\mathbb{N}}(A)$ is homogeneous p-Koszul.*
(ii) *If $G^{\mathbb{N}}(A)$ is homogeneous p-Koszul, then A is non-homogeneous p-Koszul.*
(iii) *Put $\Omega = \mathbf{LM}(\mathcal{G})$ and consider the directed graph $\Gamma_{\rm C}(\Omega)$ of n-chains of Ω. Let $\zeta\colon \mathbb{N}\to\mathbb{N}$ be the jump map associated to p (as introduced in Section 1). If the $\mathbb{N}$-graded K-module C_{i-1} spanned by all $(i-1)$-chains in $\Gamma_{\rm C}(\Omega)$ is generated in degree $\zeta(i)$, $i=1,2,\dots$, then $A^{\mathcal{B}}_{\rm LH}$ and $G^{\mathbb{N}}(A)$ are homogeneous p-Koszul, and thereby A is non-homogeneous p-Koszul.*

Proof. Note that by (Ch.5, Corollary 1.1) we have

$$\left(A^{\mathbb{N}}_{\rm LH}\right)^{\mathcal{B}}_{\rm LH} = A^{\mathcal{B}}_{\rm LH} = K\langle X\rangle/\langle \mathbf{LM}(\mathcal{G})\rangle.$$

If the condition of (iii) is satisfied, then (i) follows from Theorem 2.3. Moreover, by (Ch.4, Proposition 2.2, Theorem 2.4) we have

$$A^{\mathbb{N}}_{\rm LH} = K\langle X\rangle/\langle \mathbf{LH}(I)\rangle = K\langle X\rangle/\langle \mathbf{LH}(\mathcal{G})\rangle \xrightarrow[\cong]{\rho} G^{\mathbb{N}}(A),$$

which shows that A has the p-type PBW property. So, under the assumption of (iii), the homogeneous p-Koszulity of $G^{\mathbb{N}}(A)$ follows from (i) and thereby A is non-homogeneous p-Koszul by definition. □

Corollary 3.2. *Let* $\mathcal{G} \subset F_2K\langle X\rangle - F_1K\langle X\rangle$ *be an LM-reduced Gröbner basis with respect to* $(\mathcal{B}, \prec_{gr})$ *and* $A = K\langle X\rangle/I$ *with* $I = \langle \mathcal{G}\rangle$. *Then both the algebras* $A^{\mathcal{B}}_{\rm LH} = K\langle X\rangle/\langle \mathbf{LM}(I)\rangle$ *and* $A^{\mathbb{N}}_{\rm LH} = K\langle X\rangle/\langle \mathbf{LH}(I)\rangle$ *are homogeneous 2-Koszul. Hence* $G^{\mathbb{N}}(A)$ *is homogeneous 2-Koszul and consequently* A *is non-homogeneous 2-Koszul.*

Proof. This follows from the remark preceding Proposition 2.1 of Section 2 and Theorem 3.1 above. □

By Example 1 or Example 8 of (Ch.4, Section 3) and the above Corollary 3.2, we recapture the standard model of non-homogeneous 2-Koszul algebras, that is, the enveloping algebra of each finite dimensional K-Lie algebra is non-homogeneous 2-Koszul with the associated $\mathbb{N}$-graded algebra (a commutative polynomial algebra) that is homogeneous 2-Koszul.

To see more (non-)homogeneous 2-Koszul algebras, the reader is invited to revisit Examples 5 – 7 of (Ch.3, Section 5), Examples 2 – 11 of (Ch.4, Section 3). In particular, every skew polynomial algebra $\mathcal{O}_n(\lambda_{ji})$ is homogeneous 2-Koszul; every quantum grassmannian algebra in the sense of [Man] is homogeneous 2-Koszul; every Clifford algebra C is non-homogeneous 2-Koszul with the associated $\mathbb{N}$-graded algebra (an exterior algebra) that is homogeneous 2-Koszul; and every solvable polynomial algebra, as described in (Ch.4, Proposition 4.2), is homogeneous or non-homogeneous 2-Koszul.

The next two results provide (non-)homogeneous p-Koszul algebras for $p \geq 3$.

Corollary 3.3. *Let* $\mathcal{G} = \{g_1, g_2\}$ *be any Gröbner basis in the free* K*-algebra* $K\langle X\rangle = K\langle X_1, X_2\rangle$ *as described in Example 8 of (Ch.3, Section 5), that is*

$$\begin{array}{l} g_1 = X_1^2X_2 - \alpha X_1X_2X_1 - \beta X_2X_1^2 - F_1, \\ g_2 = X_1X_2^2 - \alpha X_2X_1X_2 - \beta X_2^2X_1 - F_2, \end{array} \quad \alpha, \beta \in K,\ F_1, F_2 \in K\langle X\rangle,$$

such that $\mathbf{LM}(g_1) = X_1^2X_2$, $\mathbf{LM}(g_2) = X_1X_2^2$ *with respect to the* $\mathbb{N}$*-graded lexicographic ordering* $X_2 \prec_{gr} X_1$*, where* $K\langle X\rangle$ *is equipped with the natural* $\mathbb{N}$*-gradation. Then the monomial algebra* $K\langle X\rangle/\langle \mathbf{LM}(\mathcal{G})\rangle$ *and the associated* $\mathbb{N}$*-graded algebra* $G^{\mathbb{N}}(A)$ *of* $A = K\langle X\rangle/\langle \mathcal{G}\rangle$ *defined by the natural* $\mathbb{N}$*-filtration* FA *are homogeneous 3-Koszul, and hence* A *is non-homogeneous 3-Koszul.*

In particular, every down-up algebra $A = A(\alpha,\beta,\gamma)$ *is non-homogeneous 3-Koszul with the associated* $\mathbb{N}$*-graded algebra* $G^{\mathbb{N}}(A)$ *that is homogeneous 3-Koszul.*

Proof. Let $\Omega = \{X_1^2X_2, X_1X_2^2\}$. By referring to the graph $\Gamma_{\rm C}(\Omega)$ of n-chains of Ω given in the proof of (Ch.5, Corollary 7.5), we see that

$$C_{i-1} = \begin{cases} \{X_1, X_2\}, & i = 1, \\ \{X_1X_2^2, X_1^2X_2\}, & i = 2, \\ \{X_1^2X_2^2\}, & i = 3, \\ \emptyset, & i \geq 4. \end{cases}$$

On the other hand, note that for $p = 3$, $\zeta(1) = 1$, $\zeta(2) = 3$, $\zeta(3) = 4$. Since $A_{\rm LH}^{\mathcal{B}} = K\langle X\rangle/\langle \mathbf{LM}(\mathcal{G})\rangle$, it follows from Theorem 2.3 that $A_{\rm LH}^{\mathcal{B}}$ is homogeneous 3-Koszul. Consequently, the assertion of the corollary follows from Theorem 3.1. □

Remark (i) Considering the down-up algebra $A = A(\alpha,\beta,\gamma)$ with defining relations

$$\begin{aligned} g_1 &= X_1^2X_2 - \alpha X_1X_2X_1 - \beta X_2X_1^2 - \gamma X_1, \\ g_2 &= X_1X_2^2 - \alpha X_2X_1X_2 - \beta X_2^2X_1 - \gamma X_2, \end{aligned} \quad \alpha,\beta,\gamma \in K,$$

in the case that $\beta \neq 0$, a direct proof for the homogeneous 3-Koszulity of $G^{\mathbb{N}}(A)$ and the non-homogeneous 3-Koszulity of A was given by R. Berger and V. Ginzburg in [GB2].

(ii) On the other hand, by Example 2(c) of (Ch.4, Section 3) and the foregoing discussion, in the case that K is algebraically closed every down-up algebra $A = A(\alpha,\beta,\gamma)$ is non-homogeneous 2-Koszul and its associated $\mathbb{N}$-graded algebra $G^{\mathbb{N}}(A)$ defined by the natural $\mathbb{N}$-filtration FA is homogeneous 2-Koszul. But this is not a surprise, for, the homogeneous p-Koszulity is defined with respect to the $\mathbb{N}$-gradation used for the algebra considered.

Proposition 3.4. *For every positive integer* $p \geq 2$*, there is a (non-monomial) homogeneous* p*-Koszul algebra and a non-homogeneous* p*-Koszul algebra.*

Proof. Let $K\langle X\rangle = K\langle X_1, X_2\rangle$ be the free K-algebra of two generators, and consider the $\mathbb{N}$-graded lexicographic ordering $X_1 \prec_{gr} X_2$ with respect to the natural $\mathbb{N}$-gradation of $K\langle X\rangle$. For any positive integer n, if $g = X_2^nX_1 - qX_1X_2^n + F$ with $q \in K$ and $F \in K\langle X\rangle$ such that $\mathbf{LM}(F) \prec_{gr} X_2^nX_1$, then it is easy to see that $\{g\}$ forms a Gröbner basis for the principal ideal $I = \langle g\rangle$. Put $\Omega = \{X_2^nX_1\}$. By referring to the graph $\Gamma_{\rm C}(\Omega)$ of n-chains of Ω given in the proof of (Ch.5, Corollary 7.7), we see that

$$C_{i-1} = \begin{cases} \{X_1,\ X_2\}, & i = 1, \\ \{X_2^nX_1\}, & i = 2, \\ \emptyset, & i \geq 3. \end{cases}$$

Moreover, for $p = n+1$ we have $\zeta(1) = 1$, $\zeta(2) = p = n+1$, $\zeta(3) = p+1 = n+2$. If we consider the algebra $A = K\langle X\rangle/I$, then $A^{\mathcal{B}}_{\rm LH} = K\langle X\rangle/\langle\mathbf{LM}(I)\rangle = K\langle X\rangle/\langle X_2^nX_1\rangle$. It follows from Theorem 3.1 that $A^{\mathcal{B}}_{\rm LH}$ and $A^{\mathbb{N}}_{\rm LH} = K\langle X\rangle/\langle\mathbf{LH}(I)\rangle = K\langle X\rangle/\langle\mathbf{LH}(g)\rangle \cong G^{\mathbb{N}}(A)$ are homogeneous p-Koszul, and consequently A is non-homogeneous p-Koszul in the case that $F \neq 0$ with total degree $\leq n$. □

Chapter 7

A Study of Rees Algebra by Gröbner Bases

In the study of a $\mathbb{Z}$-filtered K-algebra A with filtration $FA = \{F_nA\}_{n\in\mathbb{Z}}$, it is well known (e.g., see [Gin], [MR], [AVV], [Li0], [LVO2]) that except for the associated $\mathbb{Z}$-graded algebra $G^{\mathbb{Z}}(A) = \oplus_{n\in\mathbb{Z}} G^{\mathbb{Z}}(A)_n$ with $G^{\mathbb{Z}}(A)_n = F_nA/F_{n-1}A$, the *Rees algebra* $\widetilde{A} = \oplus_{n\in\mathbb{Z}}\widetilde{A}_n$ with $\widetilde{A}_n = F_nA$ also plays an important role, where the multiplication of $\widetilde{A}$ is induced by $F_nAF_mA \subseteq F_{n+m}A$ for all $n, m \in \mathbb{Z}$.

Let $R = \oplus_{p\in\mathbb{N}} R_p$ be an $\mathbb{N}$-graded K-algebra with an admissible system $(\mathcal{B}, \prec_{gr})$, where $\mathcal{B}$ is a skew multiplicative K-basis of R consisting of $\mathbb{N}$-homogeneous elements and $\prec_{gr}$ is an $\mathbb{N}$-graded monomial ordering on $\mathcal{B}$, and let I be an arbitrary ideal of R, $A = R/I$. Consider the $\mathbb{N}$-filtration FA of A induced by the $\mathbb{N}$-grading filtration FR of R. Through Ch.4 – Ch.6 we have seen how, in a computational way, to recognize many structural properties of A and its associated $\mathbb{N}$-graded algebra $G^{\mathbb{N}}(A)$ by using a Gröbner basis $\mathcal{G}$ of I, especially when $R = K\langle X_1, \ldots, X_n\rangle$ is a free K-algebra. In the first two sections of this chapter, we determine the defining relations of the Rees algebra $\widetilde{A}$ of A by means of the central homogenized Gröbner basis when an $\mathbb{N}$-graded algebra R as given above is considered, respectively by means of the non-central homogenized Gröbner basis when a free K-algebra $R = K\langle X_1, \ldots, X_n\rangle$ is considered, where both the central and non-central homogenized Gröbner bases are as constructed in (Ch.3, Section 6). In the latter case, (Ch.3, Proposition 6.14) enables us to see that if an algebra A is defined subject to given relations, then the Rees algebra $\widetilde{A}$ of A may be established in an algorithmic way via the defining relations of the so-called homogenized algebra $H(A)$ of A. In Section 3, with the aid of results obtained in Ch.4 – Ch.6, we prove that many structural properties of $\widetilde{A}$ may be determined in a computational way. In Section 4, we clarify the relation between Rees algebras, regular central extensions, and PBW-deformations in terms of Gröbner bases. Finally in Section 5 we show that algebras defined by dh-closed homogeneous Gröbner bases (in the sense of Ch.3, Section 7) from a polynomial algebra $K[x_1, \ldots, x_n]$ or a noncommutative free algebra $K\langle X_1, \ldots, X_n\rangle$ may be studied as Rees algebras of certain $\mathbb{N}$-filterd algebras.

Notations and conventions used in previous chapters are maintained.

7.1 Defining $\widetilde{A}$ by $\mathcal{G}^*$

In this section, we present the Rees algebra (as well as the associated graded algebra) of an $\mathbb{N}$-filtered K-algebra in terms of the central homogenized Gröbner basis introduced in (Ch.3, Section 6).

Let $R = \oplus_{p\in\mathbb{N}} R_p$ be an arbitrary $\mathbb{N}$-graded K-algebra, and $R[t]$ the polynomial ring in the commuting variable t over R which is equipped with the mixed $\mathbb{N}$-gradation (see Ch.3, Section 6), that is, $R[t] = \oplus_{p\in\mathbb{Z}} R[t]_p$ with

$$R[t]_p = \left\{ \sum_{i+j=p} a_i t^j \;\middle|\; a_i \in R_i,\ j \geq 0 \right\}, \quad p \in \mathbb{N}.$$

Consider a proper ideal I of R and the corresponding quotient algebra $A = R/I$. Taking the central (de)homogenization with respect to the commuting variable t (Ch.3, Section 6) into account, if $\langle I^*\rangle$ is the $\mathbb{N}$-graded ideal generated by the set $I^* = \{f^* \mid f \in I\}$ of homogenized elements of I in $R[t]$, then, by (Ch.3, Lemma 6.1, or Theorem 7.3) it is a good exercise to check that

(a) the image $\widetilde{t}$ of t in the quotient algebra $R[t]/\langle I^*\rangle$ is not a divisor of zero;
(b) the ideal $\langle 1-\widetilde{t}\rangle$ of $R[t]/\langle I^*\rangle$ does not contain any nonzero homogeneous element of $R[t]/\langle I^*\rangle$; and
(c) the rule of correspondence

$$\begin{array}{rcl} \alpha : R[t]/\langle I^*\rangle & \longrightarrow & A = R/I \\ F + \langle I^*\rangle & \mapsto & F_* + I, \end{array} \qquad F \in R[t]$$

defines a surjective K-algebra homomorphism with $\mathrm{Ker}\alpha = \langle 1 - \overline{t}\rangle$. Consequently,

$$A \cong R[t]/(\langle 1-t\rangle + \langle I^*\rangle). \tag{1}$$

Next, consider the $\mathbb{N}$-grading filtration

$$FR = \{F_pR\}_{p\in\mathbb{N}} \text{ with } F_pR = \oplus_{i\le p}R_i$$

of R and the $\mathbb{N}$-filtration

$$FA = \{F_pA\}_{p\in\mathbb{N}} \text{ with } F_pA = (F_pR + I)/I$$

of A induced by FR. The following result, which is quoted from [LWZ] (see also Ch.3 of [Li1]) but with a direct proof, shows that the associated $\mathbb{N}$-graded algebra $G^{\mathbb{N}}(A)$ of A and the Rees algebra $\widetilde{A}$ of A with respect to FA may be determined by $\langle I^*\rangle$.

Theorem 1.1. *Let $A = R/I$ be as fixed above. With notation as before and bearing in mind the above isomorphism* (1), *there are two $\mathbb{N}$-graded K-algebra isomorphisms:*
(i) $\widetilde{A} \cong R[t]/\langle I^*\rangle$, *and*
(ii) $G^{\mathbb{N}}(A) \cong R[t]/(\langle t\rangle + \langle I^*\rangle)$ (*compare with Theorem 3.2 of Ch.2*).

Proof. (i). Note that $\widetilde{A}_p = F_pA = (F_pR + I)/I$, and if $\sum_{i+j=p} a_it^j \in R[t]_p$ then $\sum a_i \in F_pR$. It is straightforward to check that for each $p \in \mathbb{N}$ the rule of correspondence

$$\begin{array}{rccc} \varphi_p: & R[t]_p & \longrightarrow & (F_pR + I)/I \\ & \sum\limits_{i+j=p} a_it^j & \mapsto & \sum\limits_{i} a_i + I \end{array}$$

is a surjective K-linear map with $\text{Ker}\varphi_p = R[t]_p \cap \langle I^*\rangle = \langle I^*\rangle_p$, and that furthermore all the φ_p determine the desired $\mathbb{N}$-graded K-algebra isomorphism $R[t]/\langle I^*\rangle \cong \widetilde{A}$.

(ii). Note that for each $p \in \mathbb{N}$ we have the canonical K-linear isomorphism

$$\begin{aligned} G^{\mathbb{N}}(A)_p = F_pA/F_{p-1}A &\cong (F_pR + I)/(F_{p-1}R + I) \\ &\cong F_pR/(F_{p-1}R + I \cap F_pR). \end{aligned}$$

If we write $\overline{a}_i$ for the canonical image of each $a_i \in R_i$ in $F_pR/(F_{p-1}R + I \cap F_pR)$, then it is straightforward to verify that for each $p \in \mathbb{N}$ the rule of correspondence

$$\begin{array}{rccc} \psi_p: & R[t]_p & \longrightarrow & F_pR/(F_{p-1}R + I \cap F_pR) \\ & \sum\limits_{i+j=p} a_it^j & \mapsto & \sum\limits_{i} \overline{a}_i \end{array}$$

is a surjective K-linear map with $\text{Ker}\psi_p = R[t]_p \cap (\langle t\rangle + \langle I^*\rangle) = \langle t\rangle_p + \langle I^*\rangle_p$, and that furthermore all the ψ_p determine the desired $\mathbb{N}$-graded K-algebra isomorphism $R[t]/(\langle t\rangle + \langle I^*\rangle) \cong G^{\mathbb{N}}(A)$. □

Suppose that R has an admissible system $(\mathcal{B}, \prec_{gr})$ with $\mathcal{B}$ a skew multiplicative K-basis of R consisting of $\mathbb{N}$-homogeneous elements, and $\prec_{gr}$ an $\mathbb{N}$-graded monomial ordering on $\mathcal{B}$. Then, by Theorem 1.1 above and (Ch.3, Theorem 6.3), we have the following immediate corollary.

Corollary 1.2. *With the assumption on R as above, $A = R/I$ as before, if $\mathcal{G}$ is a Gröbner basis of I with respect to the data $(\mathcal{B}, \prec_{gr})$, then we have*
(i) $\widetilde{A} \cong R[t]/\langle \mathcal{G}^* \rangle$, *and*
(ii) $G^{\mathbb{N}}(A) \cong R[t]/(\langle t \rangle + \langle \mathcal{G}^* \rangle)$,
where $\mathcal{G}^*$ is the central homogenization of $\mathcal{G}$ in $R[t]$ with respect to t.

7.2 Defining $\widetilde{A}$ by $\widetilde{\mathcal{G}}$

In this section, we present a Rees algebra by means of non-central homogenized Gröbner basis in the sense of (Ch.3, Section 6). As an immediate application of the main result (Theorem 2.1), we show, in terms of Gröbner basis, when the so-called dehomogenized algebra of a given algebra defined by relations plays the role of a Rees algebra.

Let the free K-algebra $K\langle X \rangle = K\langle X_1, \ldots, X_n \rangle$ of n generators be equipped with a fixed weight $\mathbb{N}$-gradation, say each X_i has degree $n_i > 0$, $1 \le i \le n$, and $(\mathcal{B}, \prec_{gr})$ an admissible system of $K\langle X \rangle$, where $\mathcal{B}$ is the standard K-basis of $K\langle X \rangle$ and $\prec_{gr}$ is a fixed $\mathbb{N}$-graded lexicographic ordering on $\mathcal{B}$:

$$X_{i_1} \prec_{gr} X_{i_2} \prec_{gr} \cdots \prec_{gr} X_{i_n}.$$

Then recall from (Ch.3, Section 6) that by assigning to each X_i the same degree as in $K\langle X \rangle$ and assigning to T the degree 1, the free algebra $K\langle X, T \rangle = K\langle X_1, \ldots, X_n, T \rangle$ of $n+1$ generators has the admissible system $(\widetilde{\mathcal{B}}, \prec_{_{T\text{-}gr}})$, where $\widetilde{\mathcal{B}}$ is the standard K-basis of $K\langle X, T \rangle$ and, with respect to the weight $\mathbb{N}$-gradation of $K\langle X, T \rangle$, $\prec_{_{T\text{-}gr}}$ is the $\mathbb{N}$-graded lexicographic ordering obtained by extending $\prec_{gr}$ on $\mathcal{B}$ to $\widetilde{\mathcal{B}}$:

$$T \prec_{_{T\text{-}gr}} X_{i_1} \prec_{_{T\text{-}gr}} X_{i_2} \prec_{_{T\text{-}gr}} \cdots \prec_{_{T\text{-}gr}} X_{i_n}.$$

Let I be a proper ideal of $K\langle X \rangle$ and $A = K\langle X \rangle / I$ the corresponding quotient algebra. Considering the $\mathbb{N}$-filtration FA of A induced by the $\mathbb{N}$-grading filtration $FK\langle X \rangle$ of $K\langle X \rangle = \oplus_{p \in \mathbb{N}} K\langle X \rangle_p$, and taking the non-central (de)homogenization with respect to T (Ch.3, Section 6) into account, if $\langle \widetilde{I} \rangle$ is the non-central homogenization ideal of I in $K\langle X, T \rangle$ with respect to T, where

$$\widetilde{I} = \{\widetilde{f} \mid f \in I\} \cup \{X_iT - TX_i \mid 1 \leq i \leq n\},$$

then, the following proposition, which is quoted from [LWZ], ([Li1], Ch.3), and [Li3] but with a direct proof, shows that the Rees algebra $\widetilde{A}$ of A defined by FA may be determined by $\langle \widetilde{I} \rangle$.

Theorem 2.1. *With notation and the assumption as above, there is an $\mathbb{N}$-graded K-algebra isomorphism*

$$\widetilde{A} \cong K\langle X, T\rangle / \langle \widetilde{I} \rangle.$$

Proof. Consider the weight $\mathbb{N}$-graded structure $K\langle X, T\rangle = \oplus_{p\in\mathbb{N}} K\langle X, T\rangle_p$. Let $\mathscr{C}$ be the $\mathbb{N}$-graded ideal of $K\langle X, T\rangle$ generated by $\{X_iT - TX_i \mid 1 \leq i \leq n\}$. Then by (Ch.3, Lemma 6.8), each nonzero homogeneous element $F \in K\langle X, T\rangle_p$ can be written as $F = L + H$, where $L \in \mathscr{C}$ and $H = \sum \lambda_i T^{r_i} w_i$ with $\lambda_i \in K^*$, $r_i \in \mathbb{N}$, $w_i \in \mathcal{B}$; and moreover, there exists some $r \in \mathbb{N}$ such that $T^r(H_\sim)^\sim = H$ and hence $F = L + T^r(F_\sim)^\sim$. Note that $\mathscr{C} \subset \langle \widetilde{I} \rangle$. So, if $F_\sim = \sum \lambda_i w_i \in I$ then $(F_\sim)^\sim \in \widetilde{I}$ and it turns out that $F \in K\langle X, T\rangle_p \cap \langle \widetilde{I} \rangle = \langle \widetilde{I} \rangle_p$. Thus, since $F_pA = (F_pK\langle X\rangle + I)/I$, and $\sum \lambda_i T^{r_i} w_i \in K\langle X, T\rangle_p$ implies $\sum \lambda_i w_i \in F_pK\langle X\rangle$, it is straightforward to check that for each $p \in \mathbb{N}$ the rule of correspondence

$$\begin{array}{rccc} \psi_p : & K\langle X, T\rangle_p & \longrightarrow & (F_pK\langle X\rangle + I)/I = F_pA \\ & F = L + \sum \lambda_i T^{r_i} w_i & \mapsto & \sum \lambda_i w_i + I \end{array}$$

is a surjective K-linear map with $\mathrm{Ker}\psi_p = K\langle X, T\rangle_p \cap \langle \widetilde{I} \rangle = \langle \widetilde{I} \rangle_p$, and that furthermore all the ψ_p determine the desired $\mathbb{N}$-graded K-algebra isomorphism $K\langle X, T\rangle / \langle \widetilde{I} \rangle \cong \widetilde{A}$. □

Corollary 2.2. *Let the ideal I and $A = K\langle X\rangle / I$ be as in the last proposition. If $\mathcal{G}$ is a Gröbner basis of I with respect to the data $(\mathcal{B}, \prec_{gr})$, then*

$$\widetilde{\mathcal{G}} = \{\widetilde{g} \mid g \in \mathcal{G}\} \cup \{X_iT - TX_i \mid 1 \leq i \leq n\}$$

is a homogeneous Gröbner basis for the graded ideal $\langle \widetilde{I} \rangle$ in $K\langle X, T\rangle$ with respect to the data $(\widetilde{\mathcal{B}}, \prec_{T\text{-}gr})$, and hence $\widetilde{A} \cong K\langle X, T\rangle / \langle \widetilde{\mathcal{G}} \rangle$.

Proof. This follows immediately from Theorem 2.1 and (Ch.3, Theorem 6.10). □

Let $I = \langle S\rangle$ be the ideal of $K\langle X\rangle$ generated by a subset S. Recall from the literature (e.g. [LeS], [LV]) that the *homogenized algebra* of $A = K\langle X\rangle/I$, denoted $H(A)$, is defined to be the quotient algebra

$$H(A) = K\langle X,T\rangle/\langle\widetilde{S}\rangle \text{ with}$$

$$\widetilde{S} = \{\widetilde{f} \mid f \in S\} \cup \{X_iT - TX_i \mid 1 \le i \le n\}.$$

We have observed before (Ch.3, Proposition 6.14) that in general $\langle\widetilde{S}\rangle \subsetneq \langle\widetilde{I}\rangle$, and we have also known from (Ch.3, Proposition 6.14) how to obtain algorithmically a homogeneous Gröbner basis for $\langle\widetilde{I}\rangle$ by producing a homogeneous Gröbner basis of $\langle\widetilde{S}\rangle$. So, at this stage, Theorem 2.1, Corollary 2.2 and (Ch.3, Theorem 6.10, Theorem 6.14) together yield the following assertion.

Proposition 2.3. *Let I, $A = K\langle X\rangle/I$ and $H(A)$ be as above, the following statements hold.*
(i) *Considering the $\mathbb{N}$-filtration FA of A induced by the $\mathbb{N}$-grading filtration $FK\langle X\rangle$ of $K\langle X\rangle$, the Rees algebra $\widetilde{A} = \widetilde{A} = K\langle X,T\rangle/\langle\widetilde{I}\rangle$ of A may be established in an algorithmic way via $H(A)$ in the sense of (Ch.3, Proposition 6.14).*
(ii) *If S is a Gröbner basis for the ideal $I = \langle S\rangle$, then $\langle\widetilde{S}\rangle = \langle\widetilde{I}\rangle$ and hence $H(A) = \widetilde{A}$.* □

We finish this section by looking at two typical examples.

Example 1. Let g be an n-dimensional K-Lie algebra with K-basis $\{x_1,\dots,x_n\}$, where $[x_j,x_i] = \sum_\ell \lambda_{ji}^\ell x_\ell$, $1 \le i < j \le n$, and let $U(\mathbf{g})$ be the enveloping algebra of g. Then by Theorem 2.1 and Example 1 (or Example 8) of (Ch.4, Section 3), the Rees algebra $\widetilde{U(\mathbf{g})}$ of $U(\mathbf{g})$ with respect to the natural $\mathbb{N}$-filtration $FU(\mathbf{g})$ has the set of quadratic defining relations

$$\widetilde{\mathcal{G}} = \{R_{ji} = X_jX_i - X_iX_j - \textstyle\sum_\ell \lambda_{ji}TX_\ell \mid 1 \le i < j \le n\}$$

$$\bigcup\{X_iT - TX_i \mid 1 \le i \le n\},$$

that is, $\widetilde{U(\mathbf{g})} \cong K\langle X,T\rangle/\langle\widetilde{\mathcal{G}}\rangle$, where $K\langle X,T\rangle$ is the free K-algebra generated by $\{X_1,\dots,X_n,T\}$. Note that the Rees algebra $\widetilde{U(\mathbf{g})}$ is actually the homogenized enveloping algebra $H(\mathbf{g})$ of g studied in ([LS], [LV]).

Example 2. It follows from Theorem 2.1 and Example 8 of (Ch.3, Section 5) that with respect to the natural $\mathbb{N}$-filtration FA, every down-up algebra $A = A(\alpha, \beta, \gamma)$ in the sense of ([Ben], [BR]), has its Rees algebra $\widetilde{A}$ defined by the set $\widetilde{\mathcal{G}}$ of relations

$$\begin{aligned}
&\widetilde{g}_1 = X_1^2X_2 - \alpha X_1X_2X_1 - \beta X_2X_1^2 - \gamma T^2X_1,\\
&\widetilde{g}_2 = X_1X_2^2 - \alpha X_2X_1X_2 - \beta X_2^2X_1 - \gamma T^2X_2,\\
&C_3 = X_1T - X_1T,\\
&C_4 = X_2T - TX_2,
\end{aligned}$$

that is, $\widetilde{A} \cong K\langle X_1, X_2, T\rangle/\langle\widetilde{\mathcal{G}}\rangle$. Note that the Rees algebra $\widetilde{A}$ is actually the homogenized down-up algebra proposed by [BR] and studied in [Cas].

We will see from the next section that working with a Rees algebra has more advantages than working with a homogenized algebra.

7.3 Recognizing Structural Properties of $\widetilde{A}$ via $\mathcal{G}$

Throughout this section, we let the free K-algebra $K\langle X\rangle = K\langle X_1, \ldots, X_n\rangle$ be equipped with a positive weight $\mathbb{N}$-gradation, and let I be an arbitrary proper ideal of $K\langle X\rangle$, $A = K\langle X\rangle/I$, which is endowed with the $\mathbb{N}$-filtration FA induced by the weight $\mathbb{N}$-grading filtration $FK\langle X\rangle$ of $K\langle X\rangle$. Thus, FA yields the associated graded algebra $G^{\mathbb{N}}(A)$ of A and the Rees algebra $\widetilde{A}$ of A as before. Furthermore, as in the last section, we fix an admissible system $(\mathcal{B}, \prec_{gr})$ for $K\langle X\rangle$, where $\prec_{gr}$ is some $\mathbb{N}$-graded lexicographic ordering on the standard K-basis $\mathcal{B}$ of $K\langle X\rangle$. So $\prec_{gr}$ extends to an $\mathbb{N}$-graded lexicographic ordering $\prec_{_{T\text{-}gr}}$ on the standard K-basis $\widetilde{\mathcal{B}}$ of the free K-algebra $K\langle X, T\rangle = K\langle X_1, \ldots, X_n, T\rangle$, such that $T \prec_{_{T\text{-}gr}} X_i$ for all $1 \le i \le n$.

Based on Theorem 2.1 of the last section and the results of (Ch.5, Ch.6), we show in this section that many structural properties of $\widetilde{A}$ may be determined by means of the Gröbner basis theory of I. The notations used in Section 2 are maintained.

Theorem 3.1. *With notation as fixed before, the set $N(\langle\widetilde{I}\rangle)$ of normal monomials in $\widetilde{\mathcal{B}}$ (mod $\langle\widetilde{I}\rangle$), or equivalently a K-basis for $\widetilde{A}$, is determined by the set $N(I)$ of normal monomials in $\mathcal{B}$ (mod I), that is,*

$$N(\langle\widetilde{I}\rangle) = \left\{T^r w \;\middle|\; w \in N(I),\ r \in \mathbb{N}\right\}.$$

Proof. Note that every ideal in $K\langle X\rangle$ has a Gröbner basis. This follows from Theorem 2.1 and (Ch.3, Corollary 6.11). □

Theorem 3.2. *Suppose that I has an LM-reduced finite Gröbner basis $\mathcal{G} = \{g_1, \ldots, g_s\}$ in $K\langle X\rangle$, and let $\Gamma(\mathbf{LM}(\mathcal{G}))$ be the Ufnarovski graph of $\mathcal{G}$ as defined in (Ch.5, Section 2). With notation as fixed before, the following statements hold.*

(i) *The growth of $\widetilde{A}$ is alternative, that is, $\widetilde{A}$ has exponential growth if and only if the graph $\Gamma(\mathbf{LM}(\mathcal{G}))$ has two different cycles with a common vertex; otherwise, $\widetilde{A}$ has polynomial growth of degree $d+1$, or equivalently, GK.dim$\widetilde{A} = d+1$, where d is, among all routes of $\Gamma(\mathbf{LM}(\mathcal{G}))$, the largest number of distinct cycles occurring in a single route.*

(ii) *With $\ell = \max\{l(u_i) \mid u_i \in \mathbf{LM}(\mathcal{G})\}$, if each monomial $u \in \mathcal{B} - \langle\mathbf{LM}(\mathcal{G})\rangle$ with $1 \le l(u) \le \ell$ is cyclic, then $\widetilde{A}$ is semi-prime.*

(iii) *With $\ell = \max\{l(u_i) \mid u_i \in \mathbf{LM}(\mathcal{G})\}$, if the following two conditions are satisfied:*

(a) *any $u \in \mathcal{B} - \langle\mathbf{LM}(\mathcal{G})\rangle$ with $l(u) < \ell - 1$ is a suffix of some vertex of $\Gamma(\mathbf{LM}(\mathcal{G}))$;*

(b) *for any two vertices v_i and v_j of $\Gamma(\mathbf{LM}(\mathcal{G}))$, there exists a route from v_i to v_j and vice versa,*

then $\widetilde{A}$ is prime or Noetherian prime of GK dimension 2.

Proof. (i). Let z be the homogeneous element of degree 1 in $\widetilde{A}_1 = F_1A$ represented by the identity element 1 of A (note that $1 \in F_0A$ by our fixed convention). Then it is straightforward that z is not a divisor of zero and it is contained in the center of $\widetilde{A}$; moreover, the localization of $\widetilde{A}$ at the multiplicative subset $\{z^k \mid k \in \mathbb{N}\}$ is isomorphic to $A[t, t^{-1}]$, the ring of Laurent series over A. Thus, by ([M-R], Chapter 8, Proposition 2.13, Corollary 2.15), GK.dim$\widetilde{A}$ = GK.dim$A$+1. Noticing that $\widetilde{A} \cong K\langle X, T\rangle/\langle\widetilde{I}\rangle$ and $\widetilde{\mathcal{G}}$ is a finite Gröbner basis for the ideal $\langle\widetilde{I}\rangle$, the assertion follows now from (Ch.5, Corollary 3.2).

(ii) – (iii). Let the element z be as indicated in the proof of (ii) above. Note that if $a \in \widetilde{A}$, say $a = a_p + a_{p-1} + a_{p-2} + \cdots + a_0$ with $a_j \in \widetilde{A}_j$, then for some $a' \in \widetilde{A}$,

$$a = (a_p + a_{p-1}z + a_{p-2}z^2 + \cdots + a_0z^p) + (1 - z)a'.$$

So, by sending z to 1 in A, it is easy to verify (or see a more general argument in the next section) that $A \cong \widetilde{A}/\langle 1 - z\rangle$ and the ideal $\langle 1 - z\rangle$ does not contain any nonzero homogeneous element of $\widetilde{A}$. Thus, the (semi-)

primeness of A implies the $\mathbb{N}$-graded (semi-)primeness of $\widetilde{A}$ and hence the (semi-)primeness of $\widetilde{A}$ (exercise). Therefore, the assertion follows now from (Ch.5, Theorem 5.3, Theorem 5.7). $\square$

Except for the case of Theorem 3.2(iii), the next result deals with a little more general case of recognizing the Noetherianity of $\widetilde{A}$ in an algorithmic way.

Theorem 3.3. *Let I and $A = K\langle X\rangle/I$ be as before, and let $J = \langle\Omega\rangle$ with $\Omega = \{u_1, \ldots, u_s\} \subset \mathcal{B}$ a reduced finite subset. Suppose that $J \subseteq \langle \mathbf{LM}(I)\rangle$ (for instance, $\Omega \subseteq \mathbf{LM}(I)$). If there is no edge entering (leaving) any cycle of the Ufnarovski graph $\Gamma(\Omega)$, then $\widetilde{A}$ is left (right) Noetherian of GK dimension not exceeding 2.*

Proof. By (Ch.5, Theorem 4.2), the associated $\mathbb{N}$-graded algebra $G^{\mathbb{N}}(A)$ is left (right) Noetherian of GK dimension not exceeding 1. Since the topology defined on A by the $\mathbb{N}$-filtration FA is complete, it follows from ([LVO], Ch.2, Proposition 1.2.3) that $\widetilde{A}$ is left (right) Noetherian. Finally, the proof of Theorem 3.2(i) implies that the GK dimension of $\widetilde{A}$ does not exceed 2. $\square$

Recall from ([LVO4] Ch.1, Corollary 7.6) that if R is a $\mathbb{Z}$-filtered ring with filtration FR, such that $\widetilde{R}$ is left and right Noetherian, then

$$\text{gl.dim}\widetilde{R} \leq \{1 + \text{gl.dim}R,\ 1 + \text{gl.dim}G^{\mathbb{Z}}(R)\}$$

and the equality holds if $\text{gl.dim}G^{\mathbb{Z}}(R)$ is finite.

Concerning the global dimension of the Rees algebra $\widetilde{A}$ of $A = K\langle X\rangle/I$, we have the following computational result *without the assumption on Noetherianity of $\widetilde{A}$*.

Theorem 3.4. *Let $A^{\mathcal{B}}_{\text{LH}} = K\langle X\rangle/\langle \mathbf{LM}(I)\rangle$ be the $\mathcal{B}$-leading homogeneous algebra of A defined with respect to the fixed data $(\mathcal{B}, \prec_{gr})$. The following statements hold.*

(i) *Let Ω be the reduced monomial generating set of $\langle \mathbf{LM}(I)\rangle$ and suppose $\Omega \cap X = \emptyset$. If the graph $\Gamma_{\text{C}}(\Omega)$ of n-chains of Ω does not contain any d-chain, then $\text{gl.dim}\widetilde{A} \leq d+1$.*

(ii) *If $A^{\mathcal{B}}_{\text{LH}}$ has finite global dimension and the polynomial growth of degree m, then*

$$\text{gl.dim}\widetilde{A} = m + 1,$$

and moreover, the ideal I has a finite Gröbner basis $\mathcal{G}$ such that $\widetilde{\mathcal{G}}$ is a finite homogeneous Gröbner basis in the free algebra $K\langle X, T\rangle$ and $K\langle X, T\rangle / \langle\, \widetilde{\mathcal{G}}\, \rangle \cong \widetilde{A}$.

Proof. (i). Since Ω is the reduced monomial generating set of $\langle \mathbf{LM}(I)\rangle$, it follows from (Ch.3, Theorem 4.4, Proposition 4.6) that the set

$$\mathcal{G} = \{g \in I \mid \mathbf{LM}(g) = w \text{ for some } w \in \Omega\}$$

is a Gröbner basis for I with $\mathbf{LM}(\mathcal{G}) = \Omega$. Considering the non-central homogenized Gröbner basis $\widetilde{\mathcal{G}}$ for the ideal $\langle \widetilde{I}\rangle$, by the proof of (Ch.3, Theorem 6.6) we have

$$\mathbf{LM}(\widetilde{\mathcal{G}}) = \{\mathbf{LM}(g),\ X_iT \mid g \in \mathcal{G},\ 1 \le i \le n\}.$$

Noticing $X \cap \Omega = \emptyset$ by the assumption, $\mathbf{LM}(\widetilde{\mathcal{G}})$ is reduced. Thus, it is easy to see that the graph of n-chains $\Gamma_{\text{C}}(\mathbf{LM}(\mathcal{G}))$ of $\mathcal{G}$ is a subgraph of the graph of n-chains $\Gamma_{\text{C}}(\mathbf{LM}(\widetilde{\mathcal{G}}))$ of $\widetilde{\mathcal{G}}$, and that

(a) the graph $\Gamma_{\text{C}}(\mathbf{LM}(\widetilde{\mathcal{G}}))$ of $\widetilde{\mathcal{G}}$ has no edge of the form $T \to v$ for all $v \in \widetilde{V}$, where $\widetilde{V}$ is the set of vertices of $\Gamma_{\text{C}}(\mathbf{LM}(\widetilde{\mathcal{G}}))$;

(b) if $v \in \widetilde{V}$ is of the form $v = sX_j$, $s \in \mathcal{B}$, then $\Gamma_{\text{C}}(\mathbf{LM}(\widetilde{\mathcal{G}}))$ contains the edge $v \to T$;

(c) any $(d+1)$-chain in $\Gamma_{\text{C}}(\mathbf{LM}(\widetilde{\mathcal{G}}))$ is of the form

$$1 \to X_i \to v_1 \to v_2 \to \cdots \to v_{d-1} \to T,$$

where

$$1 \to X_i \to v_1 \to v_2 \to \cdots \to v_{d-1}$$

is a d-chain in $\Gamma_{\text{C}}(\mathbf{LM}(\mathcal{G}))$.

Therefore, if the graph $\Gamma_{\text{C}}(\mathbf{LM}(\mathcal{G}))$ does not contain any d-chain, then $\text{gl.dim}\widetilde{A} \le d + 1$ by (Ch.5, Theorem 7.4(i)).

(ii). This folows from (Ch.5, Theorems 7.1 – 7.4) and the foregoing Theorem 3.2(i). □

Combining Examples 2, 3, 5 of (Ch.5, Section 3), Example 2 of (Ch.5, Section 8), (Ch.5, Corollary 7.5, Corollary 7.6, Corollary 7.7), and the proof of (Ch.6, Proposition 3.5), we have the following corollaries of the foregoing Theorems 3.1 – 3.4.

Corollary 3.5. *Let the free K-algebra $K\langle X\rangle = K\langle X_1, X_2\rangle$ be equipped with a positive weight $\mathbb{N}$-gradation, such that $d(X_1) = n_1$ and $d(X_2) = n_2$, and let $\mathcal{G} = \{g_1, g_2\}$ be any Gröbner basis with respect to some $\mathbb{N}$-graded lexicographic ordering $\prec_{gr}$, such that $\mathbf{LM}(g_1) = X_1^2X_2$ and $\mathbf{LM}(g_2) = X_1X_2^2$. Consider the Rees algebra $\widetilde{A}$ of the algebra $A = K\langle X\rangle/\langle\mathcal{G}\rangle$ determined by the $\mathbb{N}$-filtration FA induced by the weight $\mathbb{N}$-grading filtration $FK\langle X\rangle$ of $K\langle X\rangle$. Then*
(i) *$GK.dim\widetilde{A} = 4$.*
(ii) *$\widetilde{A}$ is semi-prime.*
(iii) *$gl.dim\widetilde{A} = 4$.*
(iv) *The Hilbert series of $\widetilde{A}$ is*

$$H_{\widetilde{A}}(t) = \frac{1}{(1-t^3)(1-t^2)}.$$

The above results hold for the algebras constructed in Example 8 of (Ch.3, Section 5) including every down-up algebra $A = A(\alpha, \beta, \gamma)$.

□

Corollary 3.6. *Let the free K-algebra $K\langle X\rangle = K\langle X_1, \ldots, X_n\rangle$ be equipped with a positive weight $\mathbb{N}$-gradation, $\Omega \subseteq \{X_jX_i \mid 1 \le i < j \le n\}$, or $\Omega \subseteq \{X_iX_j \mid 1 \le i < j \le n\}$, and let $\mathcal{G}$ be any Gröbner basis in $K\langle X\rangle$ with respect to some $\mathbb{N}$-graded lexicographic ordering $\prec_{gr}$, such that $\mathbf{LM}(\mathcal{G}) = \Omega$. Considering the Rees algebra $\widetilde{A}$ of the algebra $A = K\langle X\rangle/\langle\mathcal{G}\rangle$ with respect to the $\mathbb{N}$-filtration FA induced by the weight $\mathbb{N}$-grading filtration $FK\langle X\rangle$ of $K\langle X\rangle$, the following statements hold.*
(i) *$gl.dim\widetilde{A} \le n+1$.*
(ii) *$GK.dim\widetilde{A} \le n+1$.*
(iii) *If $\Omega = \{X_jX_i \mid 1 \le i < j \le n\}$, or $\Omega = \{X_iX_j \mid 1 \le i < j \le n\}$, then $gl.dim\widetilde{A} = n+1$, $GK.dim\widetilde{A} = n+1$, and the Hilbert series of $\widetilde{A}$ is*

$$H_{\widetilde{A}}(t) = \frac{1}{(1-t)^{n+1}}.$$

□

Corollary 3.7. *Let the free K-algebra $K\langle X\rangle = K\langle X_1, X_2\rangle$ be equipped with a positive weight $\mathbb{N}$-gradation, and for any fixed positive integer n, let $\mathcal{G} = \{g\}$ with $g = X_2^nX_1 - qX_1X_2^n - F$ with $q \in K$, $F \in K\langle X\rangle$. Suppose that $\mathcal{G}$ forms a Gröbner basis with respect to some $\mathbb{N}$-graded lexicographic ordering $\prec_{gr}$ on $K\langle X\rangle$ such that $\mathbf{LM}(g) = X_2^nX_1$ (for instance if both X_1 and X_2 are assigned the degree 1 and if the $\mathbb{N}$-graded monomial ordering*

$X_1 \prec_{gr} X_2$ is used). Considering the K-algebra $A = K\langle X\rangle/\langle \mathcal{G}\rangle$ and its Rees algebra $\widetilde{A}$ defined by the $\mathbb{N}$-filtration FA which is induced by the weight $\mathbb{N}$-grading filtration of $K\langle X\rangle$, then
(i) *$gl.dim\widetilde{A} = 3 = GK.dim\widetilde{A}$; and*
(ii) *the Hilbert series of $\widetilde{A}$ is*

$$H_{\widetilde{A}}(t) = \frac{1}{1 - 3t + 2t^2 + t^{n+1} - t^{n+2}},$$

which, in the case of $n \geq 2$, cannot be the form as asserted by Theorem 7.2(iii) (or Theorem 7.3(iii)).

□

Finally, concerning the Koszulity of Rees algebras, it follows from Theorem 2.1, (Ch.3, Theorem 6.6) and (Ch.6, Corollary 3.2) that the next result is clear.

Theorem 3.8. *Let the free K-algebra $K\langle X\rangle = K\langle X_1, \ldots, X_n\rangle$ be equipped with the natural $\mathbb{N}$-filtration $FK\langle X\rangle$, and let $\mathcal{G} \subset F_2K\langle X\rangle - F_1K\langle X\rangle$ be a Gröbner basis with respect to some admissible system $(\mathcal{B}, \prec_{gr})$, where $\prec_{gr}$ is an $\mathbb{N}$-graded lexicographic ordering on the standard K-basis $\mathcal{B}$ of $K\langle X\rangle$. Then the Rees algebra $\widetilde{A}$ of the algebra $A = K\langle X\rangle/\langle \mathcal{G}\rangle$ with respect to the natural $\mathbb{N}$-filtration FA is homogeneous 2-Koszul.*

□

At this point, the reader is invited to determine the Koszulity of the Rees algebras of those algebras from (Ch.3 – Ch.5) that are defined by Gröbner bases consisting of elements of $F_2K\langle X\rangle - F_1K\langle X\rangle$.

7.4 An Application to Regular Central Extensions

In this section, we clarify the relation between Rees algebras, the PBW-deformations of graded algebras (see Ch.4, Section 1) and the regular central extensions of graded algebras (cf. [LSV], [C-BH], [CS2]), which enables us to recognize a regular central extension in terms of Rees algebra.

We introduce the principle of dealing with the topic concerned in a little more general setting. Let A be a $\mathbb{Z}$-filtered K-algebra with $\mathbb{Z}$-filtration $FA = \{F_nA\}_{n\in\mathbb{Z}}$, and let $\widetilde{A} = \oplus_{n\in\mathbb{Z}}\widetilde{A}_n$ with $\widetilde{A}_n = F_nA$ be the Rees algebra of A. Note that the identity element 1 of A is contained in F_0A by our fixed convention. If we write z for the homogeneous element of degree 1 in $\widetilde{A}_1 = F_1A$ represented by 1, then it is clear that z is contained in the center of $\widetilde{A}$ and z is not a divisor of zero. Furthermore, note that if $a \in \widetilde{A}$, say $a = a_p + a_{p-1} + a_{p-2} + \cdots + a_0$ with $a_j \in \widetilde{A}_j$, then for some $a' \in \widetilde{A}$,

$$a = (a_p + a_{p-1}z + a_{p-2}z^2 + \cdots + a_0z^p) + (1-z)a'.$$

Hence, it is straightforward that the maps defined by $z \mapsto 1$, respectively $z \mapsto 0$, yield two K-algebra isomorphisms:

$$(1) \qquad \widetilde{A}/\langle 1-z\rangle \cong A, \text{ respectively } \widetilde{A}/z\widetilde{A} \cong G^{\mathbb{Z}}(A),$$

where $G^{\mathbb{Z}}(A)$ is the associated $\mathbb{Z}$-graded algebra of A with respect to FA.

Turning to an opposite direction, let $B = \oplus_{n\in\mathbb{Z}}B_n$ be a $\mathbb{Z}$-graded K-algebra and suppose B has a *central element* $z \in B_1$, i.e., z is a homogeneous element of degree 1 and z is contained in the center of B. Then we have the $\mathbb{Z}$-filtered K-algebra $A = B/\langle 1-z\rangle$ with the $\mathbb{Z}$-filtration FA induced by the $\mathbb{Z}$-grading filtration FB of B, that is, $FA = \{F_nA\}_{n\in\mathbb{Z}}$ with

$$F_nA = \left(\oplus_{i\le n}B_i + \langle 1-z\rangle\right)/\langle 1-z\rangle.$$

Inspired by the isomorphisms in (1) above, the proposition given below answers the question of when B serves as the Rees algebra of A with respect to FA, or in other words, when an analogue of the foregoing (1) holds with $\widetilde{A}$ replaced by B.

Proposition 4.1. *With notation as above, the following statements are equivalent:*
(i) *z is not a divisor of zero in B;*
(ii) *$\langle 1-z\rangle \cap B_n = \{0\}$ for all $n \in \mathbb{Z}$;*
(iii) *$B \cong \widetilde{A}$;*
(iv) *$B/zB \cong G^{\mathbb{Z}}(A)$.*

Proof. Since z is a central homogeneous element of degree 1, if $d \in B$, say $d = \sum_{i=1}^{p} d_j$ with $d_j \in B_j$, then

$$d = (d_p + d_{p-1}z + d_{p-2}z^2 + \cdots + d_0z^p) + (1-z)d'$$

for some $d' \in B$. Considering the maps defined by $z \mapsto 1$, respectively $z \mapsto 0$, it is an easy exercise to complete the proof. Otherwise, one is referred to ([Li1] Ch.1, Section 4) for finding similar argument. □

Now, let $K\langle X\rangle = K\langle X_1,\dots,X_n\rangle$ be the free K-algebra generated by $X = \{X_1,\dots,X_n\}$, and let $G = \{g_1,\dots,g_s\}$ be a finite set of $\mathbb{N}$-homogeneous elements with respect to the natural $\mathbb{N}$-gradation of $K\langle X\rangle$.

For any fixed $f_1, \ldots, f_s \in K\langle X\rangle$ satisfying $d(f_i) < d(g_i)$, $1 \le i \le s$, put

$$\mathcal{G} = \{g_i + f_i \mid 1 \le i \le s\},$$

and let $K\langle X\rangle[t]$ be the polynomial K-algebra in the commuting variable t over $K\langle X\rangle$. Then we have the following three algebras:

$$R = K\langle X\rangle/\langle G\rangle, \quad A = K\langle X\rangle/\langle \mathcal{G}\rangle, \quad B = K\langle X\rangle[t]/\langle \mathcal{G}^*\rangle,$$

where $\mathcal{G}^*$ is the central homogenization of $\mathcal{G}$ in $K\langle X\rangle[t]$ with respect to t (see Ch.3, Section 6). Recall that A is called a deformation of the $\mathbb{N}$-graded algebra R (see Ch.4, Section 1), and note also that with respect to the $\mathbb{N}$-gradation of B induced by the mixed $\mathbb{N}$-gradation of $K\langle X\rangle[t]$ (see Ch.3, Section 6), the image $\bar{t}$ of t in B is an $\mathbb{N}$-homogeneous element of degree 1 and $\bar{t}$ is contained in the center of B. In the literature (e.g., [LSV], [CS]), the $\mathbb{N}$-graded algebra B is called a *central extension of R associated to A*; and if furthermore $\bar{t}$ is not a divisor of zero, then B is called a *regular central extension.*

Proposition 4.2. *With the notation above, the following statements are equivalent.*
(i) *B is a regular central extension of R associated to A.*
(ii) *$B \cong \widetilde{A}$, where the latter is the Rees algebra of A with respect to the natural $\mathbb{N}$-filtration FA.*
(iii) *$\mathcal{G}$ is an $\mathbb{N}$-standard basis for the ideal $\langle \mathcal{G}\rangle$ in the sense of (Ch.4, Section 1).*
(iv) *A is a PBW-deformation of R, i.e., $R \cong G^{\mathbb{N}}(A)$ with respect to the natural $\mathbb{N}$-filtration FA.*

Proof. Putting $I = \langle \mathcal{G}\rangle$, note that $\mathcal{G}^* \subset I^*$. By referring to Section 1, the equivalence follows now from the foregoing Proposition 4.1 and (Ch.4, Theorem 2.4). □

Finally, combining (Ch.4, Theorem 2.4), the following result is clear.

Theorem 4.3. *With notation as before, if $\mathcal{G}$ is a Gröbner basis with respect to some admissible system $(\mathcal{B}, \prec_{gr})$ of $K\langle X\rangle$, where $\prec_{gr}$ is an $\mathbb{N}$-graded monomial ordering on the standard K-basis $\mathcal{B}$ of $K\langle X\rangle$, then the algebra $B = K\langle X\rangle[t]/\langle \mathcal{G}^*\rangle$ is a regular central extension of R associated to A.*

□

7.5 Algebras Defined by dh-Closed Homogeneous Gröbner Bases

The characterization of dh-closed graded ideals in terms of dh-closed homogeneous Gröbner bases given in (Ch.3, Section 7) indeed provides us with an effective way to study algebras defined by dh-homogeneous Gröbner bases, that is, such algebras can be studied as Rees algebras (defined by grading filtration) via studying algebras with simpler defining relations as in [Li0], [LVO2] and the previous Section 3. In this section we demonstrate details on this topic.

Notions and notations used in (Ch.3, Sections 6, 7) and present chapter are maintained.

We consider first an $\mathbb{N}$-graded K-algebra $R = \oplus_{p\in\mathbb{N}} R_p$ and the polynomial ring $R[t]$ with the mixed $\mathbb{N}$-gradation, and assume that R has an admissible system $(\mathcal{B}, \prec_{gr})$, where $\mathcal{B}$ is a skew multiplicative K-basis of R consisting of $\mathbb{N}$-homogeneous elements. Consequently, $R[t]$ has the corresponding admissible system $(\mathcal{B}^*, \prec_{t\text{-}gr})$.

Theorem 5.1. *With the convention made above, let $J = \langle \mathscr{G} \rangle$ be the graded ideal of $R[t]$ generated by a dh-closed homogeneous Gröbner basis $\mathscr{G}$, $I = \langle \mathscr{G}_* \rangle$ and $A = R/I$. By considering the $\mathbb{N}$-filtration FA of A induced by the $\mathbb{N}$-grading filtration FR of R, we have the graded K-algebra isomorphisms:*

$$G^{\mathbb{N}}(A) \cong R/\langle \mathbf{LH}(\mathscr{G}_*)\rangle, \quad \widetilde{A} \cong R[t]/\langle \mathscr{G} \rangle = R[t]/J,$$

where $\mathbf{LH}(\mathscr{G}_*) = \{\mathbf{LH}(g_*) \mid g_* \in \mathscr{G}_*\}$ is the set of $\mathbb{N}$-leading homogeneous elements of $\mathscr{G}_*$ with respect to the $\mathbb{N}$-gradation of R.

Proof. By Theorem 1.1(i) and (Ch.2, Theorem 3.2), we have

$$(1) \qquad G^{\mathbb{N}}(A) \cong R/\langle \mathbf{LH}(I)\rangle, \quad \widetilde{A} \cong R[t]/\langle I^* \rangle,$$

where $\mathbf{LH}(I) = \{\mathbf{LH}(f) \mid f \in I\}$ is the set of $\mathbb{N}$-leading homogeneous elements of I with respect to the $\mathbb{N}$-gradation of R. Since $\mathscr{G}$ is dh-closed, i.e., $(\mathscr{G}_*)^* = \mathscr{G}$, it follows from (Ch.3, Theorem 2.3, Theorem 2.5) that

$$(2) \qquad I = \langle \mathscr{G}_* \rangle = J_*, \quad \langle I^* \rangle = \langle (J_*)^* \rangle = \langle \mathscr{G} \rangle = J,$$

in which $\mathscr{G}_*$ is a Gröbner basis of I. Furthermore, by (Ch.4, Proposition 2.2) we know that $\mathbf{LH}(\mathscr{G}_*)$ is a homogeneous Gröbner basis for the graded ideal $\langle \mathbf{LH}(I)\rangle$ in R, and so we have

$$(3) \qquad \langle \mathbf{LH}(I)\rangle = \langle \mathbf{LH}(\mathscr{G}_*)\rangle.$$

Combining (1), (2) and (3), we obtain the desired graded K-algebra isomorphisms. $\square$

Thus, for a dh-closed homogeneous Gröbner basis in $R[t]$, the $\mathbb{N}$-graded algebra $R[t]/\langle\mathscr{G}\rangle = \widetilde{A}$ can be studied via studying the algebras $R/\langle\mathscr{G}_*\rangle = A$ and $R/\langle\mathbf{LH}(\mathscr{G}_*)\rangle = G^{\mathbb{N}}(A)$. For instance, $\widetilde{A}$ is semiprime (prime, a domain) if and only if A is semiprime (prime, a domain); if $G^{\mathbb{N}}(A)$ is semiprime (prime, a domain), then so are A and $\widetilde{A}$; if $G^{\mathbb{N}}(A)$ is Noetherian (artinian), then so are A and $\widetilde{A}$; if $G^{\mathbb{N}}(A)$ is of finite global dimension, then so are A and $\widetilde{A}$, etc. The reader is referred to [Li0] and [LVO1, 2, 3, 4] for more details on this topic. Here the reader is also reminded that, in many cases, the algebras A and $G^{\mathbb{N}}(A)$ may be studied via studying the monomial algebra $A^{\mathcal{B}}_{\mathrm{LH}} = R/\langle\mathbf{LM}(\mathscr{G}_*)\rangle$ (see Ch.5).

Turning to the free K-algebras $K\langle X\rangle = K\langle X_1,\ldots,X_n\rangle$ and the free K-algebra $K\langle X,T\rangle = \langle X_1,\ldots,X_n,T\rangle$, let the admissible system $(\mathcal{B},\prec_{gr})$ for $K\langle X\rangle$ and the admissible system $(\widetilde{\mathcal{B}},\prec_{_{T\text{-}gr}})$ for $K\langle X,T\rangle$ be as fixed in Section 2.

Theorem 5.2. *With the convention made above, let $\mathscr{G}$ be a dh-closed homogeneous Gröbner basis in $K\langle X,T\rangle$, and put $A = K\langle X\rangle/\langle\mathscr{G}_\sim\rangle$. By considering the $\mathbb{N}$-filtration FA of A induced by the (weight) $\mathbb{N}$-grading filtration $FK\langle X\rangle$ of $K\langle X\rangle$, we have the graded K-algebra isomorphisms:*

$$G^{\mathbb{N}}(A) \cong K\langle X\rangle/\langle\mathbf{LH}(\mathscr{G}_\sim)\rangle,\quad \widetilde{A} \cong K\langle X,T\rangle/\langle\mathscr{G}\rangle,$$

where $\mathbf{LH}(\mathscr{G}_\sim) = \{\mathbf{LH}(g_\sim) \mid g_\sim \in \mathscr{G}_\sim\}$ is the set of the $\mathbb{N}$-leading homogeneous element of $\mathscr{G}_\sim$ with respect to the fixed $\mathbb{N}$-gradation of $K\langle X\rangle$.

Proof. For convenience, let us put $J = \langle\mathscr{G}\rangle$ and $I = \langle\mathscr{G}_\sim\rangle$. By Theorem 2.1 and (Ch.2, Theorem 3.2), there are graded K-algebra isomorphisms:

$$(1)\qquad G^{\mathbb{N}}(A) \cong K\langle X\rangle/\langle\mathbf{LH}(I)\rangle,\quad \widetilde{A} \cong K\langle X,T\rangle/\langle\widetilde{I}\rangle,$$

where $\mathbf{LH}(I) = \{\mathbf{LH}(f) \mid f \in I\}$ is the set of $\mathbb{N}$-leading homogeneous elements of I with respect to the fixed $\mathbb{N}$-gradation of $K\langle X\rangle$. Since $\mathscr{G}$ is dh-closed, i.e., $(\mathscr{G}_\sim)^\sim = \mathscr{G}$, it follows from (Ch.3, Theorem 6.10, Theorem 6.12) that

$$(2)\qquad I = \langle\mathscr{G}_\sim\rangle = J_\sim,\quad \langle\widetilde{I}\rangle = \langle(J_\sim)^\sim\rangle = \langle\mathscr{G}\rangle = J,$$

in which $\mathscr{G}_\sim$ is a Gröbner basis of I. Furthermore, from (Ch.4, Proposition 2.2) we know that $\mathbf{LH}(\mathscr{G}_\sim)$ is a Gröbner basis for the graded ideal $\langle\mathbf{LH}(I)\rangle$

in $K\langle X\rangle$, and so

$$(3) \qquad \langle \mathbf{LH}(I)\rangle = \langle \mathbf{LH}(\mathscr{G}_{\sim})\rangle.$$

Combining (1), (2) and (3), we see that the desired graded K-algebra isomorphisms are established. □

Thus, for a dh-closed homogeneous Gröbner basis $\mathscr{G}$ in $K\langle X,T\rangle$, the $\mathbb{N}$-graded algebra $K\langle X,T\rangle/\langle\mathscr{G}\rangle = \widetilde{A}$ can be studied via studying the algebras $K\langle X\rangle/\langle\mathscr{G}_{\sim}\rangle = A$, $K\langle X\rangle/\langle\mathbf{LH}(\mathscr{G}_{\sim})\rangle = G^{\mathbb{N}}(A)$, and the monomial algebra $A_{\mathrm{LH}}^{\mathcal{B}} = K\langle X\rangle/\langle\mathbf{LM}(\mathscr{G}_{\sim})\rangle$, as demonstrated in the previous Section 3.

Remark Note that (Ch.3, Theorem 7.9, Theorem 7.10) has provided us with a practical stage to bring the foregoing Theorem 5.1 and Theorem 5.2 into play. Here let us say a little more about this in the commutative case. If, with respect to a fixed $x_i = t$, a dh-closed homogeneous Gröbner basis $\mathscr{G}$ for the ideal $I = \langle \mathcal{G}\rangle$ in the polynomial algebra $K[x_1,\dots,x_n]$ is considered (Ch.3, Theorem 7.9), and if K is algebraically closed, then, by referring to Corollary 1.2 and ([LVO5], Ch.3), one easily checks that the projective variety $V(\mathscr{G})$ defined by $\mathscr{G}$ in the projective space $\mathbb{P}_K^{n-1}$ is just the projective closure of the affine variety $V(\mathscr{G}_*)$ defined by $\mathscr{G}_*$ in the affine space $\mathbb{A}_K^{n-1}$, and the affine variety $V(\mathbf{LH}(\mathscr{G}_*))$ defined by $\mathbf{LH}(\mathscr{G}_*)$ in $\mathbb{A}_K^{n-1}$ corresponds to the part $V(\mathscr{G})\cap V(x_i)$ of $V(\mathscr{G})$ at infinity.

Chapter 8

Looking for More Gröbner Bases

We have seen that the realization of the mathematical principle developed through Ch.4 – Ch.7 depends on having nontrivial (finite) Gröbner bases for the defining ideals of algebras considered. Owing to such a high cost, we conclude this book by looking for more (finite) Gröbner bases in certain interesting contexts.

Notations and conventions used in previous chapters are completely maintained.

8.1 Lifting (Finite) Gröbner Bases from $\mathcal{O}_n(\lambda_{ji})$

Let $K\langle X\rangle = K\langle X_1,\ldots,X_n\rangle$ be the free K-algebra generated by $X = \{X_1,\ldots,X_n\}$ and $\mathcal{B}$ the standard K-basis of $K\langle X\rangle$. We have seen from Example 10 of (Ch.3, Section 5) that even if a two-sided ideal $\mathfrak{I}$ of $K\langle X\rangle$ is finitely generated, $\mathfrak{I}$ does not necessarily have a finite Gröbner basis with respect to any monomial ordering. However, it was proved in [EPS] that if the quotient algebra $A = K\langle X\rangle/\mathfrak{I}$ is commutative, then, after a generic linear change of variables (if necessary), $\mathfrak{I}$ has a finite Gröbner basis in $K\langle X\rangle$. This, indeed, gives another algorithmic way to study a commutative algebra via its noncommutative Gröbner representation, and its effectiveness may be illustrated, for example, by the work of [An2], [AR], [GH] and [PRS]. In [NS], the method of [EPS] was used to deal with exterior algebras in a similar way. In this section, by extending the results of [EPS] to the skew polynomial algebra $\mathcal{O}_n(\lambda_{ji}) = K\langle X\rangle/\langle\mathcal{S}\rangle$ introduced in (Ch.1, Section 1), where $\mathcal{S} = \{S_{ji} = X_jX_i - \lambda_{ji}X_iX_j \mid 1\le i<j\le n,\ \lambda_{ji}\in K^*\}$, we show that if G is a Gröbner basis in $\mathcal{O}_n(\lambda_{ji})$ and $\mathcal{I}$ is the pre-image of the ideal $I = \langle G\rangle$ under the canonical algebra epimorphism π: $K\langle X\rangle \to \mathcal{O}_n(\lambda_{ji})$, i.e.,

$\mathcal{I} = \pi^{-1}(I)$, then a Gröbner basis $\mathcal{G}$ of $\mathcal{I}$ may be constructed subject to G (Theorem 1.3). The discussion, in turn, enables us to specify several useful cases in which the Gröbner basis $\mathcal{G} = \mathcal{S} \cup \{\delta(u \cdot g) \mid u \in \mathscr{U}_{\Delta}(\mathbf{LM}(g)),\ g \in G\}$ obtained for $\mathcal{I} = \pi^{-1}(I)$ in Theorem 1.3 is always finite (Corollary 1.4), and furthermore enables us to give an effective criterion for the finiteness of $\mathcal{G}$ (Theorem 1.6).

To start with, first recall from (Ch.1, Section 1) and Example 5 of (Ch.3, Section 5) that $\mathcal{O}_n(\lambda_{ji})$ has the PBW K-basis

$$\mathscr{B} = \{x_1^{\alpha_1} \cdots x_n^{\alpha_n} \mid \alpha_1, \ldots, \alpha_n \in \mathbb{N}\}$$

which is clearly a skew multiplicative K-basis, i.e., if $x^{\alpha} = x_1^{\alpha_1} x_2^{\alpha_2} \cdots x_n^{\alpha_n}$, $x^{\beta} = x_1^{\beta_1} x_2^{\beta_2} \cdots x_n^{\beta_n} \in \mathscr{B}$ then

$$x^{\alpha} x^{\beta} = \lambda x_1^{\alpha_1+\beta_1} x_2^{\alpha_2+\beta_2} \cdots x_n^{\alpha_n+\beta_n} \text{ with } \lambda \in K^*.$$

Moreover, observe that with respect to any monomial ordering on $\mathscr{B}$ as described in Example 2 of (Ch.3, Section 1), $\mathcal{O}_n(\lambda_{ji})$ is a solvable polynomial algebra in the sense of [K-RW] (see Ch.4, Section 4). Hence, each ideal I of $\mathcal{O}_n(\lambda_{ji})$ has a finite Gröbner basis $\mathcal{G} = \{g_1, \ldots, g_s\}$ in the sense that if $f \in I$, $f \neq 0$, then $\mathbf{LM}(g_i) | \mathbf{LM}(f)$ for some $g_i \in \mathcal{G}$, i.e., $\mathbf{LM}(f) = \lambda u \mathbf{LM}(g_i) v$ for some $\lambda \in K^*$, $u, v \in \mathscr{B}$, or equivalently in the sense that $\langle \mathbf{LM}(I) \rangle = \langle \mathbf{LM}(\mathcal{G}) \rangle$, and such a Gröbner basis $\mathcal{G}$ can be produced by using the algorithm GRÖBNER proposed in [K-RW], or by using the up-to-date noncommutative system SINGULAR: PLURAL with the package twostd ([GLS], [LS]).

To make use of the Gröbner basis theory of both the free algebra $K\langle X \rangle$ and the algebra $\mathcal{O}_n(\lambda_{ji})$, from now on we use capitals $U, V, W, \ldots$ to denote monomials in $\mathcal{B}$, and use lowercase letters $u, v, w, \ldots$ to denote monomials in $\mathscr{B}$.

Let $\prec$ be a monomial ordering on $\mathscr{B}$ and I an ideal of $\mathcal{O}_n(\lambda_{ji})$. By (Ch.3, Sections 2, 4), the monomial ideal $\langle \mathbf{LM}(I) \rangle$ has the unique reduced monomial generating set

$$\Delta = \{u \in \mathbf{LM}(I) \mid \text{if } w \in \mathbf{LM}(I) \text{ and } w \neq u, \text{ then } w \nmid u\},$$

any monomial generating set of $\langle \mathbf{LM}(I) \rangle$ contains Δ (hence it is also the unique minimal monomial generating set of $\langle \mathbf{LM}(I) \rangle$), any subset $\mathcal{G} \subset I$ with $\mathbf{LM}(\mathcal{G}) = \Delta$ is a Gröbner basis for I, and if furthermore $\mathcal{G}$ satisfies the condition that $g_1, g_2 \in \mathcal{G}$ and $g_1 \neq g_2$ implies $\mathbf{LM}(g_1) \neq \mathbf{LM}(g_2)$, then $\mathcal{G}$ is a minimal Gröbner basis for I. Conversely, if G is a minimal Gröbner basis of the ideal I, then $\mathbf{LM}(\mathcal{G}) = \Delta$.

Now, let π: $K\langle X\rangle \to \mathcal{O}_n(\lambda_{ji})$ be the canonical K-algebra epimorphism, and let us fix the (left) lexicographic order $X_1 <_{lex} X_2 <_{lex} \cdots <_{lex} X_n$ on $\mathcal{B}$ (note that $<_{lex}$ is not a monomial ordering on $\mathcal{B}$). If $\prec$ is a monomial ordering on $\mathscr{B}$, then we may imitate the *lexicographic extension* used in [EPS] to get a monomial ordering $\prec_{et}$ on $\mathcal{B}$ as follows: for $U, V \in \mathcal{B}$,

$$U \prec_{et} V \text{ if } \begin{cases} \pi(U) \prec \pi(V), \text{ or} \\ \pi(U) = \pi(V) \text{ and } U <_{lex} V. \end{cases}$$

Note that since $\mathscr{B}$ is a skew multiplicative K-basis, to verify that $U \prec_{et} V$ implies $W_1UW_2 \prec_{et} W_1VW_2$, here we should use $\mathbf{LM}(\pi(W_1UW_2)) \prec \mathbf{LM}(\pi(W_1VW_2))$.

It is clear that with respect to $\prec_{et}$ the defining relations of $\mathcal{O}_n(\lambda_{ji})$ satisfy

$$\text{(1)} \qquad \mathbf{LM}(S_{ji}) = X_jX_i, \quad 1 \le i < j \le n.$$

To go further, let us make the following

Convention If I is a proper ideal of $\mathcal{O}_n(\lambda_{ji})$, then the set of canonical images $\{\overline{x}_1, \ldots, \overline{x}_n\}$ of $x_1, \ldots, x_n$ in $\mathcal{O}_n(\lambda_{ji})/I$ forms a *minimal* generating set for the K-algebra $\mathcal{O}_n(\lambda_{ji})/I$.

Thus, if we use a monomial ordering $\prec$ on $\mathscr{B}$, the convention made above amounts to saying that no x_i appears as the leading monomial of some $f \in I$, or equivalently, $\mathbf{LM}(I) \cap \{x_1, \ldots, x_n\} = \emptyset$; and in the case that a graded monomial ordering $\prec_{gr}$ with respect to $\deg x_i = 1$ $(1 \le i \le n)$ is used, it is equivalent to say that $\overline{x}_1, \ldots, \overline{x}_n$ are linearly independent over K.

We also fix the following notations:

For each monomial $w = x_1^{\alpha_1}x_2^{\alpha_2}\cdots x_n^{\alpha_n} \in \mathscr{B}$ with $r = \alpha_1 + \alpha_2 + \cdots + \alpha_n$, we rewrite w, in an obvious way, as

$$\text{(2)} \qquad w = x_{i_1}x_{i_2}\cdots x_{i_r} \text{ such that } i_1 \le i_2 \le \cdots \le i_r,$$

and if $r \ge 2$, then we write

$$\text{(3)} \qquad \frac{w}{x_{i_j}} = x_{i_1}\cdots x_{i_{j-1}}x_{i_{j+1}}\cdots x_{i_r}.$$

It turns out that

$$\text{(4)} \qquad x_{i_j}\frac{w}{x_{i_j}} = \mu_{i_j}w \text{ for some } \mu_{i_j} \in K^*.$$

Consider an arbitrary proper ideal I of $\mathcal{O}_n(\lambda_{ji})$ and its associated leading monomial ideal $\langle \mathbf{LM}(I)\rangle$ generated by $\mathbf{LM}(I) = \{\mathbf{LM}(f) \mid f \in I\}$ with respect to a fixed monomial ordering $\prec$ on $\mathscr{B}$. Let T be a monomial generating set of $\langle \mathbf{LM}(I)\rangle$. Then since $T \cap \{x_1, \ldots, x_n\} = \emptyset$ by our convention, for each $w = x_{i_1}x_{i_2}\cdots x_{i_r} \in T$ with $i_1 \le i_2 \le \cdots \le i_r$, the set of monomials

$$\mathscr{U}_T(w) = \left\{ u \in \mathscr{B} \cap K[x_{i_1+1}, \ldots, x_{i_r-1}] \;\middle|\; \begin{array}{l} u\frac{w}{x_{i_1}} \notin \langle \mathbf{LM}(I)\rangle, \\ u\frac{w}{x_{i_r}} \notin \langle \mathbf{LM}(I)\rangle \end{array} \right\}$$

makes sense, where $K[x_{i_1+1}, \ldots, x_{i_r-1}]$ is the subalgebra of $\mathcal{O}_n(\lambda_{ji})$ generated by $\{x_{i_1+1}, \cdots, x_{i_r-1}\}$.

Remark Note that if $|i_1 - i_r| \le 1$ then $K[x_{i_1+1}, \ldots, x_{i_r-1}] = K$. Moreover, it is possible that $\mathscr{U}_T(w) = \emptyset$. But if $T = \Delta$ is the unique minimal monomial generating set of $\langle \mathbf{LM}(I)\rangle$, then it is clear that at least $1 \in \mathscr{U}_\Delta(w)$ for each $w \in \Delta$. We will see that the case of $\mathscr{U}_\Delta(w) = \{1\}$ is much better for establishing our finiteness result.

We shall also use the K-linear map

$$\begin{array}{cccc} \delta: & \mathcal{O}_n(\lambda_{ji}) & \longrightarrow & K\langle X\rangle \\ & \sum \lambda_{\alpha(i)} x_1^{\alpha_{i1}} x_2^{\alpha_{i2}} \cdots x_n^{\alpha_{in}} & \mapsto & \sum \lambda_{\alpha(i)} X_1^{\alpha_{i1}} X_2^{\alpha_{i2}} \cdots X_n^{\alpha_{in}}. \end{array}$$

Using notation (2) made before, if $w = x_1^{\alpha_1}x_2^{\alpha_2}\cdots x_n^{\alpha_n} \in \mathscr{B}$ with $r = \alpha_1 + \alpha_2 + \cdots + \alpha_n$, then w has the representation $w = x_{i_1}x_{i_2}\cdots x_{i_r}$ with $i_1 \le i_2 \le \cdots \le i_r$, and hence

$$\delta(w) = X_{i_1}X_{i_2}\cdots X_{i_r} \text{ with } i_1 \le i_2 \le \cdots \le i_r. \tag{5}$$

With the preparation made above, in what follows we let I be a proper ideal of $\mathcal{O}_n(\lambda_{ji})$, $\langle \mathbf{LM}(I)\rangle$ the associated leading monomial ideal of I with respect to a fixed admissible system $(\mathscr{B}, \prec)$ of $\mathcal{O}_n(\lambda_{ji})$; and we consider the ideal $\mathcal{I} = \pi^{-1}(I)$ of $K\langle X\rangle$ as well as the associated leading monomial ideal $\langle \mathbf{LM}(\mathcal{I})\rangle$ of $\mathcal{I}$ in $K\langle X\rangle$ with respect to the data $(\mathcal{B}, \prec_{et})$ associated to $(\mathscr{B}, \prec)$.

Lemma 1.1. *Let $W = X_{i_1}X_{i_2}\cdots X_{i_r} \in \mathcal{B}$. Then $W \in \langle \mathbf{LM}(\mathcal{I})\rangle$ if and only if $\pi(W) \in \langle \mathbf{LM}(I)\rangle$ or $i_j > i_{j+1}$ in W for some j.*

Proof. Let $W = X_{i_1}X_{i_2}\cdots X_{i_r} \in \mathcal{B}$. If $W \in \langle \mathbf{LM}(\mathcal{I})\rangle$, then since $\mathcal{I} = \pi^{-1}(I)$, it follows from the definition of $\prec_{et}$ and (Ch.3, Lemma 4.5) that $\pi(W) \in \langle \mathbf{LM}(I)\rangle$. Conversely, note that the set $\mathcal{S}$ of defining relations of $\mathcal{O}_n(\lambda_{ji})$ is contained in $\mathcal{I}$, and that by the foregoing observation (1), $\mathbf{LM}(\mathcal{S}) = \{X_jX_i \mid 1 \le i < j \le n\}$. If W is a normal monomial (modulo $\mathcal{S}$), then $\pi(W) \neq 0$. By (Ch.3, Lemma 4.5), there is some $f \in I$ such that $\mathbf{LM}(f) = \pi(W)$. Writing $F = \delta(f)$, then $\pi(F) = f$ implies $F \in \mathcal{I}$, and by the definition of $\prec_{et}$ we get $W = \mathbf{LM}(F) \in \langle \mathbf{LM}(\mathcal{I})\rangle$. If W is not a normal monomial (modulo $\mathcal{S}$), then $r \ge 2$ and we may assume $i_j > i_{j+1}$ in W for some j. Thus $X_{i_j}X_{i_{j+1}} \in \mathbf{LM}(\mathcal{S}) \subset \mathbf{LM}(\mathcal{I})$ and hence $W \in \langle \mathbf{LM}(\mathcal{I})\rangle$. □

Lemma 1.2. *Let*

$$\Omega = \{V \in \mathbf{LM}(\mathcal{I}) \mid \textit{if } S \in \mathbf{LM}(\mathcal{I}) \textit{ and } S \neq V \textit{ then } S \nmid V\}$$

be the unique minimal monomial generating set of $\langle \mathbf{LM}(\mathcal{I})\rangle$, *where the division* $S|V$ *is done in* $K\langle X\rangle$ *with respect to* $\prec_{et}$, *and let* $N(\mathcal{I})$ *denote the set of normal monomials in* $\mathcal{B}$ *(modulo* $\mathcal{I}$ *with respect to* $\prec_{et}$*). Then* $\Omega = \mathbf{LM}(\mathcal{S}) \cup \Omega_1$, *where* $\mathbf{LM}(\mathcal{S}) = \{X_jX_i \mid 1 \le i < j \le n\}$ *and* $\Omega_1 =$

$$\left\{ V = X_{i_1}X_{i_2}\cdots X_{i_r} \in \mathbf{LM}(\mathcal{I}) \;\middle|\; \begin{array}{l} i_1 \le i_2 \le \cdots \le i_r, \\ X_{i_2}\cdots X_{i_r},\ X_{i_1}\cdots X_{i_{r-1}} \in N(\mathcal{I}) \end{array} \right\}.$$

Proof. Note that $\mathbf{LM}(\mathcal{S}) = \{X_jX_i \mid 1 \le i < j \le n\}$ by the foregoing observation (1), and that $\mathbf{LM}(\mathcal{S}) \subset \mathbf{LM}(\mathcal{I})$. The equality $\Omega = \mathbf{LM}(\mathcal{S}) \cup \Omega_1$ follows from Lemma 1.1. □

Theorem 1.3. *Let* $\Omega = \mathbf{LM}(\mathcal{S}) \cup \Omega_1$ *be the unique minimal monomial generating set of* $\langle \mathbf{LM}(\mathcal{I})\rangle$ *obtained in Lemma 1.2, and let* G *be a minimal Gröbner basis of* I *in* $\mathcal{O}_n(\lambda_{ji})$ *with respect to the fixed data* $(\mathscr{B}, \prec)$. *Put* $\mathbf{LM}(G) = \Delta$. *Then, with notation as before, the following statements hold.*
(i) $\Omega_1 = \{\mathbf{LM}(\delta(u \cdot \mathbf{LM}(g))) \mid u \in \mathscr{U}_\Delta(\mathbf{LM}(g)),\ g \in G\}$.
(ii) $\mathcal{G} = \mathcal{S} \cup \{\delta(u \cdot g) \mid u \in \mathscr{U}_\Delta(\mathbf{LM}(g)),\ g \in G\}$ *is a Gröbner basis for the ideal* $\mathcal{I}$ *with respect to the data* $(\mathcal{B}, \prec_{et})$.
(iii) *Let* $\mathcal{G}$ *be the Gröbner basis obtained in (ii) above. If* $\mathscr{U}_\Delta(\mathbf{LM}(g))$ *is a finite set for every* $g \in G$, *then* $\mathcal{G}$ *is a finite Gröbner basis for* $\mathcal{I}$; *and if* $\mathscr{U}_\Delta(\mathbf{LM}(g)) = \{1\}$ *for every* $g \in G$, *then* $\mathcal{G} = \mathcal{S} \cup \{\delta(g) \mid g \in G\}$ *is a finite and minimal Gröbner basis for* $\mathcal{I}$.

Proof. (i). For each $V = X_{i_1} \cdots X_{i_r} \in \Omega_1$ with $i_1 \le i_2 \le \cdots \le i_r$, by Lemma 1.1,

$$v = x_{i_1}(x_{i_2} \cdots x_{i_r}) = (x_{i_1} \cdots x_{i_{r-1}})x_{i_r} = \pi(V) \in \mathbf{LM}(I)$$

but $x_{i_2} \cdots x_{i_r} \notin \mathbf{LM}(I)$, $x_{i_1} \cdots x_{i_{r-1}} \notin \mathbf{LM}(I)$. Thus, if $\mathbf{LM}(g) \in \Delta$ such that $\mathbf{LM}(g)|v$, then $\mathbf{LM}(g)$ must contain both x_{i_1} and x_{i_r}. Let $v = \lambda s\mathbf{LM}(g)w$ with $\lambda \in K^*$ and $s, w \in \mathscr{B}$. Then $v = \lambda' u \cdot \mathbf{LM}(g)$ with $\lambda' \in K^*$ and $u \in \mathscr{B}$. Rewriting this equality as

$$x_{i_1} \cdot \frac{v}{x_{i_1}} = \lambda'' x_{i_1} u \cdot \frac{\mathbf{LM}(g)}{x_{i_1}} \text{ for some } \lambda'' \in K^*,$$

$$\frac{v}{x_{i_r}} \cdot x_{i_r} = \lambda' u \cdot \frac{\mathbf{LM}(g)}{x_{i_r}} \cdot x_{i_r},$$

we see that u satisfies $u \cdot \frac{\mathbf{LM}(g)}{x_{i_1}} \notin \langle \mathbf{LM}(I) \rangle$, $u \cdot \frac{\mathbf{LM}(g)}{x_{i_r}} \notin \langle \mathbf{LM}(I) \rangle$, and it is clear that $u \in \mathscr{B} \cap K[x_{i_1+1}, \ldots, x_{i_r-1}]$. Thus,

$$\begin{aligned} V = \delta(v) &= \mathbf{LM}(\delta(u \cdot \mathbf{LM}(g))) \\ &\in \{\mathbf{LM}(\delta(u \cdot \mathbf{LM}(g))) \mid u \in \mathscr{U}_{\Delta}(\mathbf{LM}(g)),\ g \in G\}. \end{aligned}$$

Conversely, noticing that Δ is now the unique minimal monomial generating set of $\langle \mathbf{LM}(I) \rangle$, if $\mathbf{LM}(g) = x_{i_1} x_{i_2} \cdots x_{i_r} \in \Delta$ with $i_1 \le i_2 \le \cdots \le i_r$, then $\mathscr{U}_{\Delta}(\mathbf{LM}(g)) \neq \emptyset$ (see the remark preceding Lemma 1.1). Let $u \in \mathscr{U}_{\Delta}(\mathbf{LM}(g))$. Then since $u \in \mathscr{B} \cap K[x_{i_1+1}, \ldots, x_{i_r-1}]$ and

$$u \cdot (x_{i_2} \cdots x_{i_r}) = u \cdot \frac{\mathbf{LM}(g)}{x_{i_1}} \notin \langle \mathbf{LM}(I) \rangle,$$

$$u \cdot (x_{i_1} \cdots x_{i_{r-1}}) = u \cdot \frac{\mathbf{LM}(g)}{x_{i_r}} \notin \langle \mathbf{LM}(I) \rangle,$$

but

$$x_{i_1} \cdot (u \cdot x_{i_2} \cdots x_{i_r}) = \lambda u \cdot \mathbf{LM}(g) \in \langle \mathbf{LM}(I) \rangle \text{ for some } \lambda \in K^*,$$

$$(u \cdot x_{i_1} \cdots x_{i_{r-1}}) \cdot x_{i_r} = u \cdot \mathbf{LM}(g) \in \langle \mathbf{LM}(I) \rangle,$$

it follows from Lemma 1.1 that $\mathbf{LM}(\delta(u \cdot \mathbf{LM}(g))) \in \Omega_1$. This completes the proof of (i).

(ii). Note that by the definition of δ we have $\mathcal{G} \subset \mathcal{I}$, and that by the definition of $\prec_{et}$ it is clear that $\mathbf{LM}(\delta(u \cdot g)) = \mathbf{LM}(\delta(u \cdot \mathbf{LM}(g)))$ for each

$\mathbf{LM}(\delta(u\cdot\mathbf{LM}(g))) \in \Omega_1$. Hence $\mathbf{LM}(\mathcal{G}) = \Omega$. It follows that $\mathcal{G}$ is a Gröbner basis for the ideal $\mathcal{I}$ with respect to $(\mathcal{B}, \prec_{et})$.

(iii). Since every ideal of $\mathcal{O}_n(\lambda_{ji})$ has a finite Gröbner basis, if $\mathscr{U}_\Delta(\mathbf{LM}(g))$ is a finite set or $\mathscr{U}_\Delta(\mathbf{LM}(g)) = \{1\}$ for every $g \in G$, then the finiteness and the minimality of $\mathcal{G}$ follow directly from its construction. □

From the previous discussion we may record some immediate consequences recognizing the finiteness of $\mathcal{G}$.

Corollary 1.4. *With notation as before, let $\mathcal{G}$ be the Gröbner basis of $\mathcal{I}$ obtained in Theorem 1.3. In one of the following cases, $\mathcal{G}$ is finite.*
(i) *$\mathcal{O}_n(\lambda_{ji})/I$ is finite dimensional over K (note that this property is recognizable by using a Gröbner basis G of I).*
(ii) *Each element g in a given minimal Gröbner basis G of I has the property that $\mathbf{LM}(g) = x_{i_1}x_{i_2}\cdots x_{i_r}$ with $i_1 \le i_2 \le \cdots \le i_r$ and $|i_1 - i_r| \le 1$ (note that this property is certainly recognizable for a finite G).*
(iii) *$n = 2$, that is, the algebra we are considering is the skew polynomial algebra $\mathcal{O}_2(\lambda_{ji})$ subject to the single relation $x_2x_1 - q_{21}x_1x_2 = 0$.*

Proof. (i). Since $K\langle X\rangle/\mathcal{I} \cong \mathcal{O}_n(\lambda_{ji})/I$, if $\dim_K(\mathcal{O}_n(\lambda_{ji})/I) < \infty$, then $N(\mathcal{I})$, the set of normal monomials in $\mathcal{B}$ (modulo $\mathcal{I}$ with respect to $\prec_{et}$), is finite. But from Lemma 1.2 we know that

$$\Omega_1 \subseteq \{V = X_jW \in \mathbf{LM}(\mathcal{I}) \mid 1 \le j \le n,\ W \in N(\mathcal{I})\}.$$

Hence Ω_1 is finite, and consequently $\mathcal{G}$ is finite.

(ii). If G has the property as mentioned, then by the definition of $\mathscr{U}_\Delta(w)$ and the remark preceding Lemma 1.1, $\mathscr{U}_\Delta(w) = \{1\}$ for each $w \in \Delta = \mathbf{LM}(G)$. It follows from Theorem 1.3(iii) that $\mathcal{G}$ is finite.

(iii). If $n = 2$, then obviously the property mentioned in (ii) holds. Hence $\mathcal{G}$ is finite. □

Corollary 1.4(i) enables us to realize quotient algebras of an exterior algebra as quotient algebras of free algebras defined by finite Gröbner bases.

Corollary 1.5. *Let $\mathsf{E} = K[b_1, \ldots, b_n]$ be the exterior algebra of n generators over K, that is, E is defined subject to the relations*

$$b_i^2 = 0,\ b_jb_i + b_ib_j = 0, \quad 1 \le i < j \le n.$$

Let $K\langle X\rangle = K\langle X_1,\ldots,X_n\rangle$ be the free K-algebra of n generators and π: $K\langle X\rangle \rightarrow \mathsf{E}$ the canonical K-algebra epimorphism with $\pi(X_i) = b_i$, $1 \le i \le n$. If I is an arbitrary ideal of E and $\mathcal{I} = \pi^{-1}(I)$, then $\mathcal{I}$ has a finite Gröbner basis in $K\langle X\rangle$.

Proof. Consider the skew polynomial algebra $\mathcal{O}_n(\lambda_{ji})$ with $\lambda_{ji} = -1$, $1 \le i < j \le n$, then the canonical epimorphism $\pi: K\langle X\rangle \rightarrow \mathsf{E}$ is realized by the composition of canonical epimorphisms π_1 and π_2:

$$\pi: K\langle X\rangle \xrightarrow{\pi_1} \mathcal{O}_n(\lambda_{ji}) \xrightarrow{\pi_2} \mathsf{E} \longrightarrow 0.$$

Note that $\mathrm{Ker}\pi_2$ is generated by $\{x_i^2 \mid 1 \le i \le n\}$ and $\dim_K \mathsf{E} = 2^n$. By passing to $\mathcal{O}_n(\lambda_{ji})$, it follows from Corollary 1.4(i) that $\mathcal{I}$ has a finite Gröbner basis in $K\langle X\rangle$. □

Remark For a graded ideal I of E generated in degree ≥ 2, the same result as mentioned in Corollary 1.5 was obtained in ([NS], Corollary 2.4).

Bearing in mind that every Gröbner basis in $\mathcal{O}_n(\lambda_{ji})$ is finite and if we pay a little more attention to the construction of $\mathscr{U}_{\Delta}(w)$, then the next result provides a more general criterion for the finiteness of $\mathcal{G}$ obtained in Theorem 1.3, which may be checked manually.

Theorem 1.6. *With notation as in Theorem 1.3, let G be a minimal Gröbner basis of I in $\mathcal{O}_n(\lambda_{ji})$ and $\Delta = \mathbf{LM}(G)$. The following two statements are equivalent:*

(i) *For each $w = x_{i_1}x_{i_2}\cdots x_{i_r} \in \Delta$ with $i_1 \le i_2 \le \cdots \le i_r$ and $|i_1 - i_r| \ge 2$, one of the following two conditions is satisfied:*

(a) *For each $i_1 < \ell < i_r$, there is some positive integer α_ℓ such that*

$$x_\ell^{\alpha_\ell} \cdot \frac{w}{x_{i_1}} \in \langle \mathbf{LM}(I)\rangle;$$

(b) *For each $i_1 < \ell < i_r$, there is some positive integer β_ℓ such that*

$$x_\ell^{\beta_\ell} \cdot \frac{w}{x_{i_r}} \in \langle \mathbf{LM}(I)\rangle;$$

(ii) *$\mathcal{G} = \mathcal{S} \cup \{\delta(u \cdot g) \mid u \in \mathscr{U}_{\Delta}(\mathbf{LM}(g)),\ g \in G\}$ is a finite Gröbner basis of $\mathcal{I} = \pi^{-1}(I)$ in $K\langle X\rangle$.*

If furthermore in (a), (b) of (i) above $\alpha_\ell = 1$ or $\beta_\ell = 1$ for all $i_1 < \ell < i_r$, then $\mathcal{G} = \mathcal{S} \cup \{\delta(g) \mid g \in G\}$ is a finite and minimal Gröbner basis of $\mathcal{I} = \pi^{-1}(I)$ in $K\langle X\rangle$.

Proof. (i) $\Rightarrow$ (ii). First, if $w \in \Delta$ with $|i_1 - i_r| \le 1$, then $\mathscr{U}_\Delta(w) = \{1\}$ as indicated in Corollary 1.4(ii). Otherwise, if (a) or (b) is satisfied, then by definition it is clear that $\mathscr{U}_\Delta(w)$ is finite. So in either cases $\mathcal{G}$ is finite by Theorem 1.3.

(ii) $\Rightarrow$ (i). Suppose that there is one $w = x_{i_1}x_{i_2}\cdots x_{i_r} \in \Delta$ with $i_1 \le i_2 \le \cdots \le i_r$, $|i_1 - i_r| \ge 2$, and there is some $i_1 < \ell < i_r$, such that $x_\ell^{\alpha_\ell} \cdot \frac{w}{x_{i_1}} \notin \langle \mathbf{LM}(I)\rangle$ for every positive integer α_ℓ. Then $u_\ell = x_\ell^{\alpha_\ell} \in \mathscr{U}_\Delta(w)$. Note that $\mathcal{O}_n(\lambda_{ji})$ is a domain. Thus $\mathscr{U}_\Delta(w)$ is an infinite set and it follows that $\mathcal{G}$ cannot be finite.

The last assertion is obvious by Theorem 1.3. □

Example 1. Consider the skew polynomial algebra $\mathcal{O}_3(\lambda_{ji})$ with $\lambda_{21} = \lambda_{31} = -1$, $\lambda_{32} = 1$, and let

$$G = \{g_1 = x_1^2x_3 - x_2,\ g_2 = x_1x_2x_3^2,\ g_3 = x_2^2\}.$$

Then a direct calculation shows that

$$\begin{aligned} &g_1x_1 = -x_1g_1,\ g_1x_2 = x_2g_1, \quad g_1x_3 = x_3g_1;\\ &g_2x_1 = -x_1g_2,\ g_2x_2 = -x_2g_2,\ g_2x_3 = -x_3g_2;\\ &g_3x_1 = x_1g_3, \quad g_3x_2 = x_2g_3, \quad g_3x_3 = x_3g_3. \end{aligned}$$

Hence the left ideal generated by G coincides with the two-sided ideal generated by G in $\mathcal{O}_3(\lambda_{ji})$. Furthermore, with respect to any graded monomial ordering $\prec_{gr}$ (subject to $\deg x_i = 1$, $1 \le i \le 3$) it is equally easy to check that the left S-polynomials of G are

$$S(g_1, g_2) = -x_3g_3, \quad S(g_1, g_3) = -x_2g_3, \quad S(g_2, g_3) = 0.$$

It follows that G is a left Gröbner basis for the ideal $\langle G\rangle$ (viewed as a left ideal). Therefore, by ([K-RW], Theorem 5.4), G is a two-sided Gröbner basis for $\langle G\rangle$ as well. We see that G is indeed a minimal Gröbner basis. So, putting $\Delta = \{w_1 = x_1^2x_3,\ w_2 = x_1x_2x_3^2,\ w_3 = x_2^2\}$, it is straightforward to see that $\mathscr{U}_\Delta(w_1) = \{1,\ x_2\}$, $\mathscr{U}_\Delta(w_2) = \mathscr{U}_\Delta(w_3) = \{1\}$; and applying Theorem 1.6 to $\langle G\rangle$, the ideal $\mathcal{I} = \pi^{-1}(\langle G\rangle)$ in $K\langle X_1, X_2, X_3\rangle$ has the Gröbner basis

$$\begin{aligned} \mathcal{G} &= \mathcal{S} \cup \{\delta(g_1),\ \delta(x_2 \cdot g_1),\ \delta(g_2),\ \delta(g_3)\}\\ &= \left\{\begin{matrix} X_3X_1 - q_{31}X_1X_3, & X_3X_2 - q_{32}X_2X_3, & X_2X_1 - q_{21}X_1X_2, & \\ X_1^2X_3 - X_2, & X_1^2X_2X_3 - X_2^2, & X_1X_2X_3^2, & X_2^2 \end{matrix}\right\} \end{aligned}$$

with respect to the extended monomial ordering $\prec_{et}$ on $K\langle X_1, X_2, X_3\rangle$.

Example 2. Consider in $\mathcal{O}_3(\lambda_{ji})$ the subset $G = \{g = x_1x_2x_3\}$, where the parameters $\lambda_{ji} \in K^*$ are arbitrarily chosen. Then G is a Gröbner basis with respect to any monomial ordering $\prec$. With $\Delta = \{w = x_1x_2x_3\}$ we see that $\mathscr{U}_\Delta(w) = \{1,\ x_2^n \mid n \geq 1\}$. Hence, with respect to the extended monomial ordering $\prec_{et}$ on $K\langle X_1, X_2, X_3\rangle$, the ideal $\mathcal{I} = \pi^{-1}(\langle g\rangle)$ in $K\langle X_1, X_2, X_3\rangle$ has the infinite Gröbner basis

$$\mathcal{G} = \left\{\begin{array}{ll} X_3X_1 - \lambda_{31}X_1X_3, & X_3X_2 - \lambda_{32}X_2X_3 \\ X_2X_1 - \lambda_{21}X_1X_2 & \end{array}\right\} \cup \{X_1X_2^nX_3 \mid n \geq 1\}.$$

As in the commutative case (see the beginning part of [EPS]), indeed, one may check that $\mathcal{I} = \pi^{-1}(\langle g\rangle)$ does not have a finite Gröbner basis with respect to any monomial ordering on $K\langle X_1, X_2, X_3\rangle$.

We are now in a position to state an analogue of ([EPS], Theorem 3.1) for $\mathcal{O}_n(\lambda_{ji})$. To this end, we say that a monomial ideal L of $\mathcal{O}_n(\lambda_{ji})$ is *p-Borel-fixed*, where $p = 0$ or p is a prime number, if it satisfies the following condition:

- For each monomial generator w of L, if w is divisible by x_j^t but no higher power of x_j, then $x_i^s \cdot \dfrac{w}{x_j^s} \in L$ for all $i < j$ and $s \leq_p t$, where $s \leq_p t$ if $\binom{t}{s} \not\equiv 0 \pmod{p}$, while $s \leq_0 t$ is the usual order on the natural numbers.

Theorem 1.7. *With notation as in Theorem 1.3, let G be a minimal Gröbner basis of the ideal I in $\mathcal{O}_n(\lambda_{ji})$ and $\Delta = \mathbf{LM}(G)$.*
(i) If $\langle \mathbf{LM}(I)\rangle$ is p-Borel-fixed for some prime p, then $\mathcal{G} = \mathcal{S} \cup \{\delta(u \cdot g) \mid u \in \mathscr{U}_\Delta(\mathbf{LM}(g)),\ g \in G\}$ is a finite Gröbner basis of $\mathcal{I} = \pi^{-1}(I)$ in $K\langle X\rangle$.
(ii) If $\langle \mathbf{LM}(I)\rangle$ is 0-Borel-fixed, then $\mathcal{G} = \mathcal{S} \cup \{\delta(g) \mid g \in G\}$ is a finite and minimal Gröbner basis of $\mathcal{I} = \pi^{-1}(I)$ in $K\langle X\rangle$.

Proof. By the definition of a p-Borel-fixed monomial ideal, this follows from Corollary 1.6 immediately. □

Finally, let us sketch how the machinery used in the proof of ([EPS], Corollary 1.1) may be carried to obtain a p-Borel-fixed monomial ideal of the form $\langle \mathbf{LM}(I)\rangle$ in $\mathcal{O}_n(\lambda_{ji})$.

Note that $\mathcal{O}_n(\lambda_{ji})$, as a K-vector space, coincides with the K-vector space of the commutative polynomial K-algebra of n variables, in particular, they have the same K-basis $\mathscr{B} = \{x_1^{\alpha_1} \cdots x_n^{\alpha_n} \mid \alpha_1, \ldots, \alpha_n \in \mathbb{N}\}$. So, the linear action of $GL_n(K)$ on the n-dimensional subspace $\mathscr{V} = \sum_{i=1}^{n} Kx_i$ extends to $\mathcal{O}_n(\lambda_{ji})$, that is, for $A = (a_{ij}) \in GL_n(K)$, if $A \cdot x_j = \sum_{i=1}^{n} a_{ij}x_i$, then, for any monomial $w = x_{i_1}x_{i_2}\cdots x_{i_r} \in \mathscr{B}$ with $i_1 \le i_2 \le \cdots \le i_r$,

$$A \cdot w = (A \cdot x_{i_1})(A \cdot x_{i_2}) \cdots (A \cdot x_{i_r}).$$

It follows that we may speak of a *Borel-fixed* monomial ideal L of $\mathcal{O}_n(\lambda_{ji})$, namely, the monomial ideal L satisfies $A \cdot L = L$ for every $A \in B$, where B is the Borel subgroup of $GL_n(K)$ consisting of triangular matrices. Since the multiplication of $\mathcal{O}_n(\lambda_{ji})$ is given by

$$w, v \in \mathscr{B}, \text{ then } wv = \lambda_{wv}u \text{ for some } \lambda_{wv} \in K^* \text{ and } u \in \mathscr{B},$$

as in the commutative case, it is straightforward that a monomial ideal L is Borel-fixed if and only if a monomial generator $w \in L$ is divisible by x_j then $x_i \cdot \frac{w}{x_j} \in L$ for all $i < j$. Hence, if L is Borel-fixed then it is p-Borel-fixed as described preceding Theorem 1.7. Moreover, if the ordering $x_n \prec x_{n-1} \prec \cdots \prec x_1$ is used on $\mathscr{B}$ and $w = x_1^{\alpha_1} \cdots x_n^{\alpha_n} \in \mathscr{B}$, then for $A = (a_{ij}) \in GL_n(K)$, $A \cdot w$ is a linear combination of monomials with coefficients that are polynomials in the a_{ji}, in particular, the coefficient of $\mathbf{LM}(A \cdot w)$ is in $K[a_{11}, a_{22}, \ldots, a_{nn}]$. So, for any graded ideal $L = \oplus_{p\in\mathbb{N}} L_p$ of the $\mathbb{N}$-graded algebra $\mathcal{O}_n(\lambda_{ji})$ subject to $\deg x_i = 1$, $1 \le i \le n$, by considering the linear action of $GL_n(K)$ on each finite dimensional vector space L_p by passing to the exterior product of L_p, we may also mimic the commutative results due to Galligo, Bayer-Stillman and Pardue (see [Eis], Section 15.9) and mention an analogue as follows. Since its proof is verbatim the same as that for ideals in a commutative polynomial algebra, we omit it and refer the reader to [Eis] for details.

Theorem 1.8. *Let K be an infinite field, and let I be a graded ideal of $\mathcal{O}_n(\lambda_{ji})$. Then there exists a nonempty Zariski open set U of $GL_n(K)$, such that $\langle \mathbf{LM}(A \cdot I)\rangle = \langle \mathbf{LM}(A' \cdot I)\rangle$ for each $A, A' \in U$. Writing this constant monomial ideal as J, then J is Borel-fixed.*

8.2 Lifting (Finite) Gröbner Bases from a Class of Algebras

In this section, by transferring Gröbner bases between filtered structures and graded structures as in [Li1] and Ch.4, we extend the main results 1.3 – 1.6 of Section 1 to a class of finitely generated K-algebras including enveloping algebras of finite dimensional Lie algebras, Clifford algebras, and many algebras presented in Ch.3, Ch.4.

Let K-algebra $K\langle X\rangle = K\langle X_1, \ldots, X_n\rangle$ be the free K-algebra generated by $X = \{X_1, \ldots, X_n\}$, and $\mathcal{B}$ the standard K-basis of $K\langle X\rangle$. In the current section we let $K\langle X\rangle$ be equipped with the natural $\mathbb{N}$-gradation subject to $\deg(X_i) = 1$. Then recall from Ch.2 that every ideal $\mathfrak{I}$ of $K\langle X\rangle$ is associated to the $\mathbb{N}$-graded ideal $\langle \mathbf{LH}(\mathfrak{I})\rangle$ generated by the set of $\mathbb{N}$-leading homogeneous elements $\mathbf{LH}(\mathfrak{I}) = \{\mathbf{LH}(f) \mid f \in \mathfrak{I}\}$ which defines the $\mathbb{N}$-leading homogeneous algebra $A^{\mathbb{N}}_{\mathrm{LH}} = K\langle X\rangle/\langle \mathbf{LH}(\mathfrak{I})\rangle$ of A, and furthermore, if the quotient algebra $A = K\langle X\rangle/\mathfrak{I}$ is equipped with the $\mathbb{N}$-filtration FA induced by the natural $\mathbb{N}$-grading filtration $FK\langle X\rangle$ of $K\langle X\rangle$, then (Ch.2, Theorem 3.2) yields the $\mathbb{N}$-graded algebra isomorphism $K\langle X\rangle/\langle \mathbf{LH}(\mathfrak{I})\rangle \xrightarrow{\cong} G^{\mathbb{N}}(A)$, where $G^{\mathbb{N}}(A)$ is the associated graded algebra of A deifned by FA. In addition to these known preliminaries, we also need the following analogue of (Ch.3, Proposition 2.2(i)).

Lemma 2.1. *Let h be a nonzero homogeneous element of $K\langle X\rangle$. Then $h \in \langle \mathbf{LH}(\mathfrak{I})\rangle$ if and only if $h \in \mathbf{LH}(\mathfrak{I})$.*

Proof. Since h is a homogeneous element, if $h \in \langle \mathbf{LH}(\mathfrak{I})\rangle$, then it follows that

$$h = \sum_{i,j} H_{ij}\mathbf{LH}(f_i)T_{ij},$$

where H_{ij}, T_{ij} are homogeneous elements and $f_i \in \mathfrak{I}$. If we put $f_i = \mathbf{LH}(f_i) + f_i'$, where $\deg(f_i') < \deg(f_i)$, then $f = \sum_{i,j} H_{ij}f_iT_{ij} \in \mathfrak{I}$ and

$$f = \sum_{ij} H_{ij}\mathbf{LH}(f_i)T_{ij} + \sum_{i,j} H_{ij}f_i'T_{ij} = h + \sum_{i,j} H_{ij}f_i'T_{ij}.$$

Hence it is clear that $h = \mathbf{LH}(f) \in \mathbf{LH}(\mathfrak{I})$. The converse is trivial. □

Now, as before let $\mathcal{O}_n(\lambda_{ji}) = K\langle X\rangle/\langle \mathcal{S}\rangle$ be the K-algebra with the set of defining relations

$$\mathcal{S} = \{S_{ji} = X_jX_i - \lambda_{ji}X_iX_j \mid \lambda_{ji} \in K^*,\ 1 \le i < j \le n\},$$

and let π: $K\langle X\rangle \to \mathcal{O}_n(\lambda_{ji})$ be the canonical algebra epimorphism. Note that $\mathcal{O}_n(\lambda_{ji})$ is an $\mathbb{N}$-graded algebra subject to $\deg x_i = 1$, $1 \le i \le n$, and with the PBW K-basis $\mathscr{B} = \{x_1^{\alpha_1}x_2^{\alpha_2}\cdots x_n^{\alpha_n} \mid \alpha_1, \ldots, \alpha_n \in \mathbb{N}\}$ consisting of $\mathbb{N}$-homogeneous elements. The basic result of this section is mentioned as follows.

Theorem 2.2. *Let $\mathfrak{I}$ be an ideal of $K\langle X\rangle$ and $A = K\langle X\rangle/\mathfrak{I}$ the corresponding quotient algebra. Suppose that the set $\mathcal{S}$ of defining relations of $\mathcal{O}_n(\lambda_{ji})$ is contained in $\mathbf{LH}(\mathfrak{I})$, and put $I = \langle \mathbf{LH}(\mathfrak{I})\rangle/\langle \mathcal{S}\rangle$. Keeping the convention and notation as made in Section 1, let G be a homogeneous and minimal Gröbner basis of the graded ideal I in $\mathcal{O}_n(\lambda_{ji})$ obtained by using a graded monomial ordering $\prec_{gr}$ with respect to $degx_i = 1$, $1 \le i \le n$. The following statements hold.*
(i) *Put $\Delta = \mathbf{LM}(G)$. Then the Gröbner basis $\mathcal{G} = \mathcal{S} \cup \{\delta(u \cdot g) \mid u \in \mathscr{U}_\Delta(\mathbf{LM}(g)),\ g \in G\}$ obtained for $\pi^{-1}(I)$ in Theorem 1.3 is a homogeneous Gröbner basis of $\langle \mathbf{LH}(\mathfrak{I})\rangle$ with respect to the data $(\mathcal{B}, \prec_{et})$. Moreover, $\mathcal{G} \subset \mathbf{LH}(\mathfrak{I})$.*
(ii) *Let $\mathcal{G}$ be the Gröbner basis obtained in* (i) *above. If $\mathcal{F} \subset \mathfrak{I}$ is such that $\mathbf{LH}(\mathcal{F}) = \mathcal{G}$, then $\mathcal{F}$ is a Gröbner basis for $\mathfrak{I}$ in $K\langle X\rangle$ with respect to the data $(\mathcal{B}, \prec_{et})$.*
(iii) *Let $\mathcal{F}$ be a Gröbner basis obtained in* (ii) *above with the property that $f_1, f_2 \in \mathcal{F}$, $f_1 \ne f_2$ implies $\mathbf{LH}(f_1) \ne \mathbf{LH}(f_2)$. Then $\mathcal{F}$ is finite if and only if the condition* (a) *or* (b) *mentioned in Theorem 1.6 is satisfied with respect to G.*

Proof. Before giving the proof of each part, let us first note that the canonical epimorphism π: $K\langle X\rangle \to \mathcal{O}_n(\lambda_{ji})$ does not change the degree of monomials with respect to $\deg X_i = 1$ and $\deg x_i = 1$, $1 \le i \le n$, that is, for $U, V \in \mathcal{B}$, $\pi(U), \pi(V) \in \mathscr{B}$,

$$d(U) < d(V) \text{ if and only if } d(\pi(U)) < d(\pi(V)).$$

Moreover, the lexicographic extension $\prec_{et}$ of $\prec_{gr}$ on $\mathcal{B}$ defined in Section 2 is a graded monomial ordering with respect to the natural $\mathbb{N}$-gradation of $K\langle X\rangle$.

(i). By the assumption, $\pi^{-1}(I) = \langle \mathbf{LH}(\mathfrak{I})\rangle$. Also note that every $u \in \mathscr{B}$ is a homogeneous element in the natural $\mathbb{N}$-gradation of $\mathcal{O}_n(\lambda_{ji})$. Hence, the Gröbner basis $\mathcal{G}$ obtained for $\pi^{-1}(I) = \langle \mathbf{LH}(\mathfrak{I})\rangle$ subject to G is a homogeneous Gröbner basis with respect to the data $(\mathcal{B}, \prec_{et})$. The inclusion $\mathcal{G} \subset \mathbf{LH}(\mathfrak{I})$ follows from Lemma 2.1.

(ii). Since $\mathcal{G} \subset \mathbf{LH}(\mathfrak{I})$ by (i), if $\mathcal{F} \subset \mathfrak{I}$ is such that $\mathbf{LH}(\mathcal{F}) = \mathcal{G}$, then it follows from (Ch.4, Proposition 2.2(ii)) that $\mathcal{F}$ is a Gröbner basis for $\mathfrak{I}$ in $K\langle X\rangle$ with respect to the data $(\mathcal{B}, \prec_{et})$.

(iii). Under the assumption made on $\mathcal{F}$ it is clear that the finiteness of $\mathcal{F}$ is equivalent to the finiteness of $\mathcal{G}$. So Theorem 1.6 applies to $\mathcal{G}$ and consequently to $\mathcal{F}$. □

We have seen that the realization of Theorem 2.1 depends on knowing a homogeneous Gröbner basis G for the ideal $I = \langle \mathbf{LH}(\mathfrak{I})\rangle/\langle\mathcal{S}\rangle$, but this requires knowing a generating set of $\langle\mathbf{LH}(\mathfrak{I})\rangle$ in advance. Generally speaking, even if we start with a finite generating set M of $\mathfrak{I}$, it is difficult to know whether the equality $\langle\mathbf{LH}(M)\rangle = \langle\mathbf{LH}(\mathfrak{I})\rangle$ may hold (i.e., whether M is an $\mathbb{N}$-standard basis for $\mathfrak{I}$ in the sense of (Ch.4, Section 1)). Nevertheless, (Ch.4, Proposition 2.2, Theorem 2.4) still allows us to apply Theorem 2.1 to a class of algebras effectively. More precisely, in what follows we let $\prec_{gr}$ be a graded monomial ordering on the standard K-basis $\mathcal{B}$ of $K\langle X\rangle$ with respect to the natural $\mathbb{N}$-gradation of $K\langle X\rangle$, and $R = K\langle X\rangle/\langle\mathcal{F}\rangle$, where

$$\mathcal{F} = \left\{ F_{ji} = X_jX_i - \lambda_{ji}X_iX_j - \sum_{\ell=1}^{n}\lambda_\ell X_\ell - c_{ji} \;\middle|\; \begin{array}{l} \lambda_{ji}\in K^*,\ \lambda_\ell, c_{ji}\in K, \\ 1\le i<j\le n \end{array} \right\}$$

is a Gröbner basis in $K\langle X\rangle$ with respect to the fixed data $(\mathcal{B},\prec_{gr})$ such that

$$\mathbf{LM}(F_{ji}) = X_jX_i,\ 1\le i<j\le n.$$

For convenience, let us write $R = K[a_1,\ldots,a_n]$ with $a_i = \pi(X_i)$ under the canonical epimorphism π: $K\langle X\rangle \to R$. Then, with the assumption made on $\mathcal{F}$ above, it follows from (Ch.4, Theorem 3.1) that R has the PBW K-basis $\mathscr{B} = \{a_1^{\alpha_1}\cdots a_n^{\alpha_n} \mid \alpha_1,\ldots,\alpha_n\in\mathbb{N}\}$. By giving a (two-sided) monomial ordering $\prec$ on $\mathscr{B}$, such as the lexicographic ordering or the graded (reverse) lexicographic ordering, R turns out to be a solvable polynomial algebra in the sense of [K-RW] (in [Li1] such a solvable polynomial algebra R is called a *linear solvable polynomial algebra* and is studied in some detail). Hence each ideal L of R has a finite Gröbner basis G. Furthermore, starting with a finite generating set of L, a Gröbner basis G of L can be produced by using the algorithm **GRÖBNER** proposed in [K-RW], or by employing the up-to-date noncommutative system SINGULAR: PLURAL with the package TWOSTD ([GLS], [LS]).

Let $\mathsf{G} = \{f_1,\ldots,f_s\}$ be a minimal Gröbner basis for an ideal L of R with respect to some graded monomial ordering $<_{gr}$ on $\mathscr{B}$ subject to $\deg a_i = 1$, $1\le i\le n$, and $\mathcal{L} = \pi^{-1}(L)$ in $K\langle X\rangle$. Our aim is to construct a Gröbner basis $\mathcal{G}$ for $\mathcal{L}$ subject to G, which may be finite in many cases. To this end, let us make the following

Convention The $\mathbb{N}$-filtration for each quotient algebra of $K\langle X\rangle$ considered below means the one induced by the natural $\mathbb{N}$-filtration $FK\langle X\rangle$ of $K\langle X\rangle$ as before.

Considering the $\mathbb{N}$-filtration FR of $R = K\langle X\rangle/\langle\mathcal{F}\rangle$, the $\mathbb{N}$-filtration FL of L induced by FR, and the $\mathbb{N}$-filtration $F(R/L)$ of $R/L \cong K\langle X\rangle/\mathcal{L}$ respectively, it follows from (Ch.4, Section 2) and the classical result concerning filtered rings that the following diagram of generator-preserving and degree-preserving graded ring isomorphisms is commutative:

$$\begin{array}{ccccccccc} 0 \to & \dfrac{\langle \mathbf{LH}(\mathcal{L})\rangle}{\langle \mathbf{LH}(\mathcal{F})\rangle} & \to & \dfrac{K\langle X\rangle}{\langle \mathbf{LH}(\mathcal{F})\rangle} & \to & \dfrac{K\langle X\rangle}{\langle \mathbf{LH}(\mathcal{L})\rangle} & \to 0 \\ & \cong\downarrow & & \cong\downarrow & & \cong\downarrow & \\ 0 \to & G(L) & \to & G(R) & \to & G(R/L) & \to 0 \end{array}$$

For each $f \in F_pR - F_{p-1}R$, if we write $\sigma(f)$ for the nonzero homogeneous element of degree p in $G^{\mathbb{N}}(R)_p = F_pR/F_{p-1}R$ represented by f, then by ([Li1], Ch.4, Theorem 2.1), $\sigma(\mathsf{G}) = \{\sigma(f_1), \ldots, \sigma(f_s)\}$ is a homogeneous and minimal Gröbner basis for $G(L)$ with respect to the same type of graded monomial ordering $<_{gr}$ on the K-basis $\sigma(\mathscr{B}) = \{\sigma(a_1)^{\alpha_1}\cdots\sigma(a_n)^{\alpha_n} \mid \alpha_1, \ldots, \alpha_n \in \mathbb{N}\}$ of $G^{\mathbb{N}}(R)$. So, for each $f_i = \sum \lambda_{\alpha(i)} a_1^{\alpha_{i_1}} a_2^{\alpha_{i_2}} \cdots a_n^{\alpha_{i_n}} \in \mathsf{G}$, if we write $F_i = \sum \lambda_{\alpha(i)} X_1^{\alpha_{i_1}} X_2^{\alpha_{i_2}} \cdots X_n^{\alpha_{i_n}}$ and write $\overline{\mathbf{LH}(F_i)}$ for the image of $\mathbf{LH}(F_i)$ in $\langle \mathbf{LH}(\mathcal{L})\rangle/\langle \mathbf{LH}(\mathcal{F})\rangle$, then $\pi(F_i) = f_i$, i.e., $F_i \in \mathcal{L}$, $1 \le i \le s$, and the commutative diagram above shows that $\overline{\mathsf{G}} = \{\ \overline{\mathbf{LH}(F_1)}, \cdots, \overline{\mathbf{LH}(F_s)}\ \}$ is a homogeneous and minimal Gröbner basis for the ideal $\langle \mathbf{LH}(\mathcal{L})\rangle/\langle \mathbf{LH}(\mathcal{F})\rangle$ with respect to the graded monomial ordering $<_{gr}$ on the K-basis $\mathscr{B}$ of $K\langle X\rangle/\langle \mathbf{LH}(\mathcal{F})\rangle$ (note that we have used the same $\mathscr{B}$ to denote the K-basis for both R and $K\langle X\rangle/\langle \mathbf{LH}(\mathcal{F})\rangle$). Since $\mathbf{LH}(\mathcal{F}) = \mathcal{S} = \{X_jX_i - \lambda_{ji}X_iX_j \mid 1 \le i < j \le n\}$, that is, $K\langle X\rangle/\langle \mathbf{LH}(\mathcal{F})\rangle = \mathcal{O}_n(\lambda_{ji})$, we have reached the requirement of Theorem 2.1 with $\mathfrak{I} = \mathcal{L}$ and $G = \overline{\mathsf{G}}$ here. Hence, if we use the same δ (as in Section 1) to denote the linear map

$$\begin{array}{cccc} \delta: & R & \longrightarrow & K\langle X\rangle \\ & \sum \lambda_{\alpha(i)} a_1^{\alpha_{i_1}} a_2^{\alpha_{i_2}} \cdots a_n^{\alpha_{i_n}} & \mapsto & \sum \lambda_{\alpha(i)} X_1^{\alpha_{i_1}} X_2^{\alpha_{i_2}} \cdots X_n^{\alpha_{i_n}} \end{array}$$

and view each $u \in \mathscr{U}_\Delta\left(\mathbf{LM}\left(\overline{\mathbf{LH}(F_i)}\right)\right)$ as an element in the K-basis $\mathscr{B}$ of R, then noticing that $a^\alpha = a_1^{\alpha_1}\cdots a_n^{\alpha_n}$, $a^\beta = a_1^{\beta_1}\cdots a_n^{\beta_n} \in \mathscr{B}$ yield

$$a^\alpha a^\beta = \lambda_{\alpha\beta} a^{\alpha+\beta} + \sum \lambda_i u_i \text{ with } u_i \in \mathscr{B}$$

satisfying $\deg(u_i) < \alpha_1 + \cdots + \alpha_n + \beta_1 + \cdots + \beta_n$, the next result is now clear.

Theorem 2.3. *With notation as fixed above, the following statements hold.*
(i) *Put* $\Delta = \mathbf{LM}(\overline{\mathsf{G}})$. *Then the Gröbner basis*

$$\mathcal{G} = \mathcal{S} \cup \left\{ \delta\left(u \cdot \overline{\mathbf{LH}(F_i)}\right) \;\middle|\; u \in \mathscr{U}_\Delta\left(\mathbf{LM}\left(\overline{\mathbf{LH}(F_i)}\right)\right),\ \overline{\mathbf{LH}(F_i)} \in \overline{\mathsf{G}} \right\}$$

as obtained in Theorem 1.3 is a homogeneous Gröbner basis for $\langle \mathbf{LH}(\mathcal{L})\rangle$ *with respect to the data* $(\mathcal{B}, \prec_{et})$.
(ii) $\mathscr{G} = \mathcal{F} \cup \left\{ \delta(u \cdot f_i) \;\middle|\; u \in \mathscr{U}_\Delta\left(\mathbf{LM}\left(\overline{\mathbf{LH}(F_i)}\right)\right),\ f_i \in \mathsf{G} \right\}$ *is a Gröbner basis for* $\mathcal{L} = \pi^{-1}(L)$ *in* $K\langle X\rangle$ *with respect to the data* $(\mathcal{B}, \prec_{et})$ *(note that* π *is now used to denote the canonical epimorphism* $K\langle X\rangle \to R$*). The finiteness of* $\mathscr{G}$ *is referred to Corollary 1.4 and Theorem 1.6.*

□

Summing up, to obtain the Gröbner basis $\mathscr{G}$ for $\mathcal{L} = \pi^{-1}(L)$, we need only to follow five steps:

(1) Fix a graded monomial ordering of the same type on both R and $\mathcal{O}_n(\lambda_{ji})$, that is, subject to $\deg a_i = 1$, $\deg x_i = 1$, $1 \le i \le n$;
(2) Calculate a minimal Gröbner basis $\mathsf{G} = \{f_1, \ldots, f_s\}$ for L in R;
(3) With the Gröbner basis $\mathsf{G} = \{f_1, \ldots, f_s\}$ obtained in (2), if $f_i = \sum \lambda_{\alpha(i)} a_1^{\alpha_{i_1}} a_2^{\alpha_{i_2}} \cdots a_n^{\alpha_{i_n}}$, then let $F_i = \sum \lambda_{\alpha(i)} X_1^{\alpha_{i_1}} X_2^{\alpha_{i_2}} \cdots X_n^{\alpha_{i_n}}$ and take $\overline{\mathsf{G}} = \{\ \overline{\mathbf{LH}(F_1)}, \cdots, \overline{\mathbf{LH}(F_s)}\ \}$ in $\mathcal{O}_n(\lambda_{ji})$;
(4) In $\mathcal{O}_n(\lambda_{ji})$, put $\Delta = \mathbf{LM}(\overline{\mathsf{G}})$ and calculate $\mathscr{U}_\Delta\left(\mathbf{LM}\left(\overline{\mathbf{LH}(F_i)}\right)\right)$ for each $\overline{\mathbf{LH}(F_i)} \in \overline{\mathsf{G}}$;
(5) In case the set $\mathscr{U}_\Delta\left(\mathbf{LM}\left(\overline{\mathbf{LH}(F_i)}\right)\right)$ can be worked out for every $\overline{\mathbf{LH}(F_i)} \in \overline{\mathsf{G}}$, the Gröbner basis $\mathscr{G}$ for $\mathcal{L}$ in $K\langle X\rangle$ can be written down.

Example 1. Let g be a finite dimensional K-Lie algebra and $U(\mathsf{g})$ the universal enveloping algebra of g. Then $U(\mathsf{g})$ is a linear solvable polynomial algebra of the type R studied above, and so, Theorem 2.2 applies to $U(\mathsf{g})$. For instance, let $U(s\ell_2)$ be the enveloping algebra of the 3-dimensional K-Lie algebra $s\ell_2 = Ke \oplus Kf \oplus Kh$ subject to the relations $[e,f] = h$, $[h,e] = 2e$, $[h,f] = -2f$, that is, under the canonical mapping π: $X \mapsto e$, $Y \mapsto f$ and $Z \mapsto h$, $U(s\ell_2) = K[e,f,h] \cong K\langle X,Y,Z\rangle/\langle \mathcal{F}\rangle$, where $\mathcal{F} = \{YX - XY + Z,\ ZX - XZ - 2X,\ ZY - YZ + 2Y\}$. In this case we see that $K\langle X,Y,Z\rangle/\langle \mathbf{LH}(\mathcal{F})\rangle = K[x,y,z]$, the commutative polynomial algebra in three variables. Consider the two-sided ideal $L = \langle e^3, f^3, h^3 - 4h\rangle$ of $U(s\ell_2)$. We proceed to obtain a Gröbner basis for $\mathcal{L} = \pi^{-1}(L)$ by following the above five steps as follows:

(1) Fix the graded reverse lexicographic monomial ordering $e <_{grevlex} f <_{grevlex} h$ on $U(s\ell_2)$, respectively $x <_{grevlex} y <_{grevlex} z$ on $K[x, y, z]$;

(2) The system SINGULAR: PLURAL with the package TWOSTD produced a minimal two-sided Gröbner basis for L:

$$\mathsf{G} = \left\{ \begin{array}{llll} e^3, & f^3, & h^3 - 4h, & \\ eh^2 + 2eh, & e^2h + 2e^2, & ef^2 - fh, & \\ e^2f - eh - 2e, & f^2h - 2f^2, & fh^2 - 2fh, & 2efh - h^2 - 2h, \end{array} \right\}$$

(see details at `http://www.singular.uni-kl.de/Manual/latest/`);

(3) Subject to G obtained in (2) above,

$$\overline{\mathsf{G}} = \{x^3,\ y^3,\ z^3,\ xz^2,\ yz^2,\ 2xyz,\ x^2y,\ xy^2,\ x^2z,\ y^2z\} \subset K[x, y, z];$$

(4) In $K[x, y, z]$, put $\Delta = \mathbf{LM}(\overline{\mathsf{G}})$. Then a direct verification shows that $\mathscr{U}_\Delta(w) = \{1\}$ for all $w \in \overline{\mathsf{G}}$;

(5) It follows that $\mathcal{L} = \pi^{-1}(L)$ has the minimal Gröbner basis

$$\mathscr{G} = \left\{ \begin{array}{lll} YX - XY + Z, & ZY - YZ + 2Y, & ZX - XZ - 2X, \\ X^3, & Y^3, & Z^3 - 4Z, \\ XZ^2 + 2XZ, & Y^2Z - 2Y^2, & XY^2 - YZ, \\ X^2Y - XZ - 2X, & X^2Z + 2X^2, & YZ^2 - 2YZ, \\ 2XYZ - Z^2 - 2Z. & & \end{array} \right\}$$

in $K\langle X, Y, Z\rangle$ with respect to $X \prec_{et} Y \prec_{et} Z$.

Example 2. Consider in the free K-algebra $K\langle X\rangle = K\langle X_1, \ldots, X_n\rangle$ the subset $\mathcal{F}$ consisting of

$$F_{ji} = X_jX_i + X_iX_j - q_{ji},\ 1 \le i < j \le n,\ q_{ji} \in K,$$

and let $R = K\langle X\rangle/\langle\mathcal{F}\rangle$. Using the graded lexicographic monomial ordering $X_1 \prec_{gr} X_2 \prec_{gr} \cdots \prec_{gr} X_n$ subject to $\deg X_i = 1$, $1 \le i \le n$, it is straightforward to see that $\mathcal{F}$ forms a Gröbner basis with $\mathbf{LM}(\mathcal{F}) = \{X_jX_i \mid 1 \le i < j \le n\}$, and thus R is a linear solvable polynomial algebra with respect to a given monomial ordering. Note that every Clifford algebra C of n generators over K is an epimorphic image of some algebra of type R, and moreover, C is finite dimensional over K. So, by Corollary 1.4(i) and Theorem 2.2, we may conclude that

- each quotient algebra of a Clifford algebra C of n generators is realized by a quotient algebra of the form $K\langle X\rangle/\langle\mathscr{G}\rangle$, where $\mathscr{G}$ is a finite Gröbner basis in $K\langle X\rangle$.

To avoid including too much tedious verification in the text, we invite the interested reader to work out the concrete examples in this case.

8.3 New Examples of Gröbner Basis Theory

In this section, we present new examples of (left, right, and two-sided) Gröbner basis theory. This is motivated by the wonderful fact that numerous quantum binomial algebras and their Koszul dual ([G-IV], [Laf], [G-I5]), which are not familiar algebras having a two-sided Gröbner basis theory (e.g., the most well-known solvable polynomial algebras and some of their homomorphic images), provide us with the first model of a proper left Gröbner basis theory, respectively a proper right Gröbner basis theory. Here and in what follows, a one-sided or two-sided Gröbner basis theory always means the one in the sense of (Ch.3, Definition 2.3) if it is considered with respect to a skew multiplicative K-basis, otherwise it means the one indicated before Theorem 3.9.

Let $R = K[a_1, \ldots, a_n]$ be a K-algebra generated by $\{a_1, \ldots, a_n\}$. Assume that R has a two-sided Gröbner basis theory with respect to a two-sided admissible system $(\mathcal{B}, \prec)$ in the sense of (Ch.3, Section 4), where $\mathcal{B}$ is a *skew multiplicative K-basis* of R consisting of "monomials" of the form $a_{i_1}^{\alpha_1}\cdots a_{i_s}^{\alpha_s}$ with $\alpha_j \in \mathbb{N}$, and $\prec$ is a *two-sided monomial ordering* on $\mathcal{B}$. For instance, R is a commutative polynomial algebra, a free algebra, a path algebra, or the skew polynomial algebra $\mathcal{O}_n(\lambda_{ji})$ subject to the relations $x_jx_i - \lambda_{ji}x_ix_j = 0$ with $\lambda_{ji} \in K^*$ and $1 \le i < j \le n$.

Let I be an ideal of R, and $A = R/I$. Since R has a (two-sided) Gröbner basis theory, under the canonical algebra epimorphism π: $R \to A$, the set $N(I)$ of normal monomials in $\mathcal{B}$ (modulo I) projects to a K-basis of A, that is $\{\bar{u} = u + I \mid u \in N(I)\}$ (Ch.3, Theorem 3.3). Throughout this section we use $\overline{N(I)}$ to denote this basis.

Our first result is to deal with the case where I is a monomial ideal, i.e., I is generated by elements of $\mathcal{B}$.

Proposition 3.1. *Let R and $(\mathcal{B}, \prec)$ be as fixed above. Consider a subset*

$\Omega \subset \mathcal{B}$ (where $1 \notin \Omega$ if $1 \in \mathcal{B}$) and the ideal $I = \langle \Omega \rangle$ in R. The following statements hold.
(i) *$\overline{N(I)}$ is a skew multiplicative K-basis for the quotient algebra $A = R/I$.*
(ii) *The two-sided monomial ordering $\prec$ on $\mathcal{B}$ induces a two-sided monomial ordering on $\overline{N(I)}$, again denoted $\prec$, and hence the quotient $A = R/I$ has a two-sided Gröbner basis theory.*

Proof. (i). Since $\mathcal{B}$ is a skew multiplicative K-basis of R, it follows from (Ch.3, Proposition 4.3) that $u \in \mathcal{B} \cap I$ if and only if there is some $v \in \Omega$ such that $v|u$, i.e., $u = \lambda wvs$ with $\lambda \in K^*$ and $w, s \in \mathcal{B}$. So, suppose $\overline{u_1}$, $\overline{u_2} \in \overline{N(I)}$ and $\overline{u_1 u_2} \neq 0$, then $u_1 u_2 = \mu v$ with $\mu \in K^*$ and $v \in N(I)$. Hence $\overline{u_1 u_2} = \mu \bar{v}$. This shows that $\overline{N(I)}$ is a skew multiplicative K-basis for A.

(ii). Suppose $\bar{u}, \bar{v} \in \overline{N(I)}$ such that $\bar{u} \prec \bar{v}$. Then $u \prec v$ since we are using the induced ordering $\prec$ on $\overline{N(I)}$. If $\bar{w}, \bar{s} \in \overline{N(I)}$ such that $\mathbf{LM}(\bar{w}\bar{u}\bar{s}) \notin K^* \cup \{0\}$ and $\mathbf{LM}(\bar{w}\bar{v}\bar{s}) \notin K^* \cup \{0\}$, then $\bar{w}\bar{u}\bar{s} \notin K^* \cup \{0\}$, $\bar{w}\bar{v}\bar{s} \notin K^* \cup \{0\}$, and as argued in the proof of (i) above, there are $u', v' \in N(I)$ such that $wus = \lambda u'$ and $wvs = \mu v'$, where $\lambda, \mu \in K^*$. Hence, it follows from $\mathbf{LM}(wus) = u' \prec v' = \mathbf{LM}(wvs)$ that $\mathbf{LM}(\bar{w}\bar{u}\bar{s}) = \overline{u'} \prec \overline{v'} = \mathbf{LM}(\bar{w}\bar{v}\bar{s})$. This shows that (MO1) of (Ch.3, Definition 1.1) holds. To prove that (MO2) of (Ch.3, Definition 1.1) holds as well, suppose $\bar{u}, \bar{v}, \bar{w}, \bar{s} \in \overline{N(I)}$ such that $\bar{u} = \mathbf{LM}(\bar{w}\bar{v}\bar{s})$ (where $\bar{w} \neq 1$ or $\bar{s} \neq 1$ if $1 \in \mathcal{B}$, and thereby $w \neq 1$ or $s \neq 1$ by the choice of Ω), and $wvs = \lambda v'$ with $\lambda \in K^*$ and $v' \in N(I)$. Thus, $v' = \mathbf{LM}(wvs)$ implies $v \prec v'$, and consequently, $\bar{v} \prec \overline{v'} = \mathbf{LM}(\bar{w}\bar{v}\bar{s}) = \bar{u}$, as desired. □

Next we show that there is a left Gröbner basis theory for a class of algebras defined by monomials and binomials of certain special type.

Theorem 3.2. *Let R and $(\mathcal{B}, \prec)$ be as fixed above. Suppose that the skew multiplicative K-basis $\mathcal{B}$ contains every monomial of the form $a_{\ell_1}^{\alpha_1} a_{\ell_2}^{\alpha_2} \cdots a_{\ell_n}^{\alpha_n}$ with respect to some permutation $a_{\ell_1}, a_{\ell_2}, \ldots, a_{\ell_n}$ of a_1, a_2, $\ldots$, a_n, where $\alpha_1, \ldots, \alpha_n \in \mathbb{N}$. If $\mathcal{G} = \Omega \cup G$ is a minimal Gröbner basis of the ideal $I = \langle \mathcal{G} \rangle$, where $\Omega \subset \mathcal{B}$ and G consists of the $\frac{n(n-1)}{2}$ elements*

$$g_{ji} = a_{\ell_j} a_{\ell_i} - \lambda_{ji} a_{\ell_i} a_{\ell_p},\ 1 \le i < j \le n,\ i < p \le n,\ \lambda_{ji} \in K^* \cup \{0\},$$

such that $\mathbf{LM}(g_{ji}) = a_{\ell_j} a_{\ell_i}$, $1 \le i < j \le n$, then the following statements hold for the algebra $A = R/I$.
(i) *$N(I) \subseteq \{a_{\ell_1}^{\alpha_1} a_{\ell_2}^{\alpha_2} \cdots a_{\ell_n}^{\alpha_n} \mid \alpha_1, \ldots, \alpha_n \in \mathbb{N}\}$, and $\overline{N(I)}$ is a skew multiplicative K-basis for A.*

(ii) *Let $\prec_{_I}$ denote the ordering on $\overline{N(I)}$, which is induced by the lexicographic ordering or the $\mathbb{N}$-graded lexicographic ordering on $\mathbb{N}^n$, such that*

$$\overline{a_{\ell_n}} \prec_{_I} \overline{a_{\ell_{n-1}}} \prec_{_I} \cdots \prec_{_I} \overline{a_{\ell_2}} \prec_{_I} \overline{a_{\ell_1}}.$$

If $\overline{a_{\ell_1}}^{\alpha_1}\overline{a_{\ell_2}}^{\alpha_2}\cdots\overline{a_{\ell_n}}^{\alpha_n} \neq 0$ implies $a_{\ell_1}^{\alpha_1}a_{\ell_2}^{\alpha_2}\cdots a_{\ell_n}^{\alpha_n} \in N(I)$, then $\prec_{_I}$ is a left monomial ordering on $\overline{N(I)}$, and hence A has a left Gröbner basis theory with respect to the left admissible system $(\overline{N(I)}, \prec_{_I})$.

Proof. Note that for simplicity, we may assume, without loss of generality, that $a_{\ell_1} = a_1$, $a_{\ell_2} = a_2$, ..., $a_{\ell_n} = a_n$.

(i). By the assumptions, $\mathcal{B}$ is a skew multiplicative K-basis of R, $a_1, \ldots, a_n \in \mathcal{B}$, and $\mathbf{LM}(g_{ji}) = a_j a_i$, $1 \le i < j \le n$. It follows from the division algorithm by $\mathcal{G}$ that $N(I) \subseteq \{a_1^{\alpha_1}a_2^{\alpha_2}\cdots a_n^{\alpha_n} \mid \alpha_j \in \mathbb{N}\}$. For $\bar{u}$, $\bar{v} \in \overline{N(I)}$, if $\bar{u}\bar{v} \neq 0$, then $uv \neq 0$ and $uv = \lambda w$ for some $\lambda \in K^*$ and $w \in \mathcal{B}$. If $w \in N(I)$, then $\bar{u}\bar{v} = \lambda\bar{w}$; otherwise, noticing that $\mathcal{G}$ consists of monomials and binomials, the division of the monomial w by $\mathcal{G}$ yields a Gröbner representation (as described in Theorem 3.4 of Ch.3):

$$w = \sum_{i,j} \mu_{ij} s_{ij} g_{ji} t_{ij} + \eta u',$$

where $\mu_{ij}, \eta \in K^*$, $s_{ij}, t_{ij} \in \mathcal{B}$, $u' \in N(I)$, and consequently $\bar{u}\bar{v} = \lambda\bar{w} = (\lambda\eta)\overline{u'}$. Therefore, $\overline{N(I)}$ is a skew multiplicative K-basis for the quotient algebra $A = R/I$.

(ii). Let $\prec_{_I}$ be one of the orderings on $\overline{N(I)}$ as mentioned in the theorem, that is, $\prec_{_I}$ is defined subject to the lexicographic ordering

$$\overline{a_n} \prec_{_I} \overline{a_{n-1}} \prec_{_I} \cdots \prec_{_I} \overline{a_2} \prec_{_I} \overline{a_1}.$$

If $\alpha = (\alpha_1, \ldots, \alpha_n) \in \mathbb{N}^n$, then we write $|\alpha| = \alpha_1 + \cdots + \alpha_n$.

First note that if $\overline{a_k} \in \{\overline{a_1}, \overline{a_2}, \ldots, \overline{a_n}\}$ and if $\overline{a_k}\cdot\overline{a_i}^{\alpha_i} \notin K^* \cup \{0\}$, then

$$\text{(1)} \qquad \mathbf{LM}(\overline{a_k}\cdot\overline{a_i}^{\alpha_i}) = \begin{cases} \overline{a_i}^{\alpha_i+1}, & k = i \\ \overline{a_k}\cdot\overline{a_i}^{\alpha_i}, & k < i \\ \overline{a_i}^{\alpha_i}\cdot\overline{a_{p_k}} \text{ with } i < p_{_k}, & k > i. \end{cases}$$

Let $\bar{u} = \overline{a_1}^{\alpha_1}\cdots\overline{a_n}^{\alpha_n}$, $\bar{v} = \overline{a_1}^{\beta_1}\cdots\overline{a_n}^{\beta_n} \in \overline{N(I)}$ be such that $\bar{u} \prec_{_I} \bar{v}$. Then

(2) either $|\alpha| < |\beta|$
or $|\alpha| = |\beta|$ and $\alpha_1 = \beta_1$, ..., $\alpha_{i-1} = \beta_{i-1}$ but $\alpha_i < \beta_i$ for some i.

Now, if $\bar{s} = \overline{a_1}^{\gamma_1}\cdots\overline{a_n}^{\gamma_n} \in \overline{N(I)}$, and if $\bar{s}\cdot\bar{u}$, $\bar{s}\cdot\bar{v} \notin K^* \cup \{0\}$, then, by (i) we may write $\mathbf{LM}(\bar{s}\cdot\bar{u}) = \overline{a_1}^{\eta_1}\cdots\overline{a_n}^{\eta_n}$, $\mathbf{LM}(\bar{s}\cdot\bar{v}) = \overline{a_1}^{\rho_1}\cdots\overline{a_n}^{\rho_n}$. It follows from the foregoing formula (1) that $|\eta| = |\alpha| + |\gamma|$, $|\rho| = |\beta| + |\gamma|$. Thus, $|\alpha| < |\beta|$ implies $|\eta| < |\rho|$, and hence

$$\text{(3)} \qquad \mathbf{LM}(\bar{s}\cdot\bar{u}) \prec_{_I} \mathbf{LM}(\bar{s}\cdot\bar{v})$$

provided we are using the graded lexicographic ordering. In the case that $|\alpha| = |\beta|$, we have $\alpha_1 = \beta_1, \ldots, \alpha_{i-1} = \beta_{i-1}$ and $\alpha_i < \beta_i$ for some i by the indication (2) given above. Since $\bar{s}\cdot\bar{u} \neq 0$ and $\bar{s}\cdot\bar{v} \neq 0$, $\overline{a_1}^{\alpha_1}\cdots\overline{a_{i-1}}^{\alpha_{i-1}} \in \overline{N(I)}$ by the assumption of (ii). By (i), if we put $\mathbf{LM}(\bar{s}\cdot\overline{a_1}^{\alpha_1}\cdots\overline{a_{i-1}}^{\alpha_{i-1}}) = \overline{a_1}^{\tau_1}\cdots\overline{a_n}^{\tau_n}$, then

$$\begin{aligned}\mathbf{LM}(\bar{s}\cdot\bar{u}) &= \mathbf{LM}(\overline{a_1}^{\tau_1}\cdots\overline{a_n}^{\tau_n}\cdot\overline{a_i}^{\alpha_i}\cdots\overline{a_n}^{\alpha_n}),\\ \mathbf{LM}(\bar{s}\cdot\bar{v}) &= \mathbf{LM}(\overline{a_1}^{\tau_1}\cdots\overline{a_n}^{\tau_n}\cdot\overline{a_i}^{\beta_i}\cdots\overline{a_n}^{\beta_n}).\end{aligned}$$

Applying the previously derived formula (1) to $i+1 \le k \le n$, we have $\overline{a_k}^{\tau_k}\cdot\overline{a_i}^{\alpha_i} = \overline{a_i}^{\alpha_i}\cdot\overline{a_{p_k}}^{\tau_k}$, and $\overline{a_k}^{\tau_k}\cdot\overline{a_i}^{\beta_i} = \overline{a_i}^{\beta_i}\cdot\overline{a_{p_k}}^{\tau_k}$ with $i < p_k$. This turns out that

$$\begin{aligned}\mathbf{LM}(\bar{s}\cdot\bar{u}) &= \mathbf{LM}(\overline{a_1}^{\tau_1}\cdots\overline{a_i}^{\tau_i}\cdot\overline{a_i}^{\alpha_i}\cdot\overline{a_{p_{i+1}}}^{\tau_{i+1}}\cdots\overline{a_{p_n}}^{\tau_n}\cdot\overline{a_{i+1}}^{\alpha_{i+1}}\cdots\overline{a_n}^{\alpha_n}),\\ \mathbf{LM}(\bar{s}\cdot\bar{v}) &= \mathbf{LM}(\overline{a_1}^{\tau_1}\cdots\overline{a_i}^{\tau_i}\cdot\overline{a_i}^{\beta_i}\cdot\overline{a_{p_{i+1}}}^{\tau_{i+1}}\cdots\overline{a_{p_n}}^{\tau_n}\cdot\overline{a_{i+1}}^{\beta_{i+1}}\cdots\overline{a_n}^{\beta_n}).\end{aligned}$$

Hence we obtain

$$\text{(4)} \qquad \mathbf{LM}(\bar{s}\cdot\bar{u}) \prec_{_I} \mathbf{LM}(\bar{s}\cdot\bar{v}).$$

This shows that (LMO1) of (Ch.3, Definition 1.1(ii)) holds.

Next, let $\bar{u} = \overline{a_1}^{\alpha_1}\cdots\overline{a_n}^{\alpha_n}$, $\bar{v} = \overline{a_1}^{\beta_1}\cdots\overline{a_n}^{\beta_n}$, $\bar{w} = \overline{a_1}^{\gamma_1}\cdots\overline{a_n}^{\gamma_n} \in \overline{N(I)}$ be such that $\bar{u} = \mathbf{LM}(\bar{v}\cdot\bar{w})$ (where $\bar{v} \neq 1$ if $1 \in \mathcal{B}$). If $\bar{u} \preceq_{_I} \bar{w}$, then since (LMO1) holds, we would have $\mathbf{LM}(\bar{v}\cdot\bar{u}) \preceq_{_I} \mathbf{LM}(\bar{v}\cdot\bar{w}) = \bar{u}$. But then, from the argument given before the formula (3) above we would have $|\beta| + |\alpha| \le |\beta| + |\gamma| = |\alpha|$, that is clearly impossible, for, $\bar{v} \neq 0$ (also $\bar{v} \neq 1$ if $1 \in \mathcal{B}$) implies $|\beta| \neq 0$. So we must have $\bar{w} \prec_{_I} \bar{u}$. It follows that (LMO2) of (Ch.3, Definition 1.1(ii)) holds as well. Summing up, we conclude that $\prec_{_I}$ is a left monomial ordering on $\overline{N(I)}$, and thereby A has a left Gröbner basis theory with respect to the left admissible system $(\overline{N(I)}, \prec_{_I})$. □

In a similar way, we may obtain a right Gröbner basis theory for a class of algebras defined by monomials and binomials of certain special type.

Theorem 3.3. *Under the assumptions on R and $\mathcal{B}$ as in Theorem 3.2, if $\mathcal{G} = \Omega \cup G$ is a minimal Gröbner basis of the ideal $I = \langle \mathcal{G}\rangle$, where $\Omega \subset \mathcal{B}$ and G consists of the $\frac{n(n-1)}{2}$ elements*

$$g_{ji} = a_{\ell_j}a_{\ell_i} - \lambda_{ji}a_{\ell_p}a_{\ell_j},\ 1 \le i < j \le n,\ p < j,\ \lambda_{ji} \in K^* \cup \{0\},$$

such that $\mathbf{LM}(g_{ji}) = a_{\ell_j}a_{\ell_i}$, $1 \le i < j \le n$, *then the following statements hold for the algebra* $A = R/I$.
(i) $N(I) \subseteq \{a_{\ell_1}^{\alpha_1}a_{\ell_2}^{\alpha_2}\cdots a_{\ell_n}^{\alpha_n} \mid \alpha_1, \ldots, \alpha_n \in \mathbb{N}\}$, *and* $\overline{N(I)}$ *is a skew multiplicative* K*-basis for* A.
(ii) *Let* $\prec_{_I}$ *denote the ordering on* $\overline{N(I)}$, *which is induced by the* $\mathbb{N}$*-graded reverse lexicographic ordering on* $\mathbb{N}^n$, *such that*

$$\overline{a_{\ell_n}} \prec_{_I} \overline{a_{\ell_{n-1}}} \prec_{_I} \cdots \prec_{_I} \overline{a_{\ell_2}} \prec_{_I} \overline{a_{\ell_1}}.$$

If $\overline{a_{\ell_1}}^{\alpha_1}\overline{a_{\ell_2}}^{\alpha_2}\cdots\overline{a_{\ell_n}}^{\alpha_n} \ne 0$ *implies* $a_{\ell_1}^{\alpha_1}a_{\ell_2}^{\alpha_2}\cdots a_{\ell_n}^{\alpha_n} \in N(I)$, *then* $\prec_{_I}$ *is a right monomial ordering on* $\overline{N(I)}$, *and hence* A *has a right Gröbner basis theory with respect to the right admissible system* $(\overline{N(I)}, \prec_{_I})$.

□

We give below examples to show that in general algebras of the type as described in Theorem 3.2 may not necessarily have a two-sided monomial ordering on $\overline{N(I)}$, and thereby each left monomial ordering $\prec_{_I}$ used in Theorem 3.2 is in general not a right monomial ordering on $\overline{N(I)}$; and that in general algebras of the type as described in Theorem 3.3 may not necessarily have a two-sided monomial ordering on $\overline{N(I)}$, and thereby the right monomial ordering $\prec_{_I}$ used in Theorem 3.3 is in general not a left monomial ordering on $\overline{N(I)}$.

Example 1. Consider in the free K-algebra $K\langle X\rangle = K\langle X_1, X_2, X_3\rangle$ the subset $\mathcal{G}$ consisting of

$$\begin{aligned} g_{21} &= X_2X_1 - \lambda X_1X_3 \\ g_{31} &= X_3X_1 - \mu X_1X_2 \quad \text{where } \lambda, \mu, \gamma \in K^* \cup \{0\}. \\ g_{32} &= X_3X_2 - \gamma X_2X_3 \end{aligned}$$

Then, with respect to the natural $\mathbb{N}$-graded lexicographic ordering subject to $X_1 \prec_{grlex} X_2 \prec_{grlex} X_3$, where each X_i is assigned the degree 1, it is straightforward to verify that $\mathbf{LM}(g_{ji}) = X_jX_i$, $1 \le i < j \le 3$, $\mathcal{G}$ forms a Gröbner basis of the ideal $I = \langle \mathcal{G}\rangle$ for each solution (λ, μ, γ) of the equation $\lambda\mu(\gamma^2 - 1) = 0$, and hence $\mathcal{G}$ yields $N(I) = \{X_1^{\alpha_1}X_2^{\alpha_2}X_3^{\alpha_3} \mid \alpha_1, \alpha_2, \alpha_3 \in \mathbb{N}\}$. Since $\mathcal{G}$ is of the type required by Theorem 3.2, it follows that the algebra $A = K\langle X\rangle/I$ has a left Gröbner basis theory with respect to the left admissible system $(\overline{N(I)}, \prec_{_I})$, where $\prec_{_I}$ is induced by the lexicographic ordering or the $\mathbb{N}$-graded lexicographic ordering on $\mathbb{N}^n$, such that $\overline{X_3} \prec_{_I} \overline{X_2} \prec_{_I} \overline{X_1}$. But since the algebra A has $1, \overline{X_1}, \overline{X_2}, \overline{X_3} \in \overline{N(I)}$ such that

$$\begin{aligned} \overline{X_2}\cdot\overline{X_1} &= \lambda\overline{X_1}\cdot\overline{X_3}, \\ \overline{X_3}\cdot\overline{X_1} &= \mu\overline{X_1}\cdot\overline{X_2}, \end{aligned}$$

we see that if $\lambda \neq 0$, $\mu \neq 0$, then any total ordering $<$ such that $\overline{X_3} < \overline{X_2}$ or $\overline{X_2} < \overline{X_3}$ cannot be a two-sided monomial ordering on $\overline{N(I)}$. Therefore, $\prec_I$ cannot be a right monomial ordering on $\overline{N(I)}$.

Example 2. In the free K-algebra $K\langle X\rangle = K\langle X_1, X_2, X_3\rangle$ consider the subset $\mathcal{G}$ consisting of

$$\begin{aligned} g_{21} &= X_2X_1 - \lambda X_1X_2 \\ g_{31} &= X_3X_1 - \mu X_2X_3 \quad \text{where } \lambda, \mu, \gamma \in K^* \cup \{0\}. \\ g_{32} &= X_3X_2 - \gamma X_1X_3 \end{aligned}$$

Then, with respect to the natural $\mathbb{N}$-graded lexicographic ordering subject to $X_1 \prec_{grlex} X_2 \prec_{grlex} X_3$, where each X_i is assigned the degree 1, it is straightforward to verify that $\mathbf{LM}(g_{ji}) = X_jX_i$, $1 \leq i < j \leq 3$, $\mathcal{G}$ forms a Gröbner basis of the ideal $I = \langle \mathcal{G}\rangle$ for each solution (λ, μ, γ) of the equation $(\lambda^2 - 1)\mu\gamma = 0$, and hence $\mathcal{G}$ yields $N(I) = \{X_1^{\alpha_1}X_2^{\alpha_2}X_3^{\alpha_3} \mid \alpha_1, \alpha_2, \alpha_3 \in \mathbb{N}\}$. Since $\mathcal{G}$ is of the type required by Theorem 3.3, it follows that the algebra $A = K\langle X\rangle/I$ has a right Gröbner basis theory with respect to the right admissible system $(\overline{N(I)}, \prec_I)$, where $\prec_I$ is induced by the $\mathbb{N}$-graded reverse lexicographic ordering on $\mathbb{N}^n$, such that $\overline{X_3} \prec_I \overline{X_2} \prec_I \overline{X_1}$. But since the algebra A has $1, \overline{X_1}, \overline{X_2}, \overline{X_3} \in \overline{N(I)}$ such that

$$\begin{aligned} \overline{X_3} \cdot \overline{X_1} &= \mu \overline{X_2} \cdot \overline{X_3}, \\ \overline{X_3} \cdot \overline{X_2} &= \gamma \overline{X_1} \cdot \overline{X_3}, \end{aligned}$$

we see that if $\mu \neq 0$, $\gamma \neq 0$, then any total ordering $<$ such that $\overline{X_1} < \overline{X_2}$ or $\overline{X_2} < \overline{X_1}$ cannot be a two-sided monomial ordering on $\overline{N(I)}$. Therefore, $\prec_I$ cannot be a left monomial ordering on $\overline{N(I)}$.

We also give an example to show that if the ideal $I = \langle \mathcal{G}\rangle$ has the Gröbner basis $\mathcal{G}$ consisting of binomials but is neither the type as described in Theorem 3.2 nor the type as described in Theorem 3.3, then the algebra defined by $\mathcal{G}$ may not have a two-sided monomial ordering, and each of the (left or right) monomial orderings used in both theorems cannot be a left or right monomial ordering for such an algebra.

Example 3. In the free K-algebra $K\langle X\rangle = K\langle X_1, X_2, X_3, X_4\rangle$ consider the subset $\mathcal{G}$ consisting of

$$\begin{aligned} g_{21} &= X_2X_1 - X_1X_2,\ g_{41} = X_4X_1 - X_2X_3, \\ g_{31} &= X_3X_1 - X_2X_4,\ g_{42} = X_4X_2 - X_1X_3, \\ g_{32} &= X_3X_2 - X_1X_4,\ g_{43} = X_4X_3 - X_3X_4. \end{aligned}$$

Then, with respect to the natural $\mathbb{N}$-graded lexicographic ordering subject to $X_1 \prec_{grlex} X_2 \prec_{grlex} X_3 \prec_{grlex} X_4$, where each X_i is assigned the degree 1, it is straightforward to verify that $\mathbf{LM}(g_{ji}) = X_jX_i$, $1 \le i < j \le 4$, $\mathcal{G}$ forms a Gröbner basis of the ideal $I = \langle \mathcal{G} \rangle$, and hence $\mathcal{G}$ yields $N(I) = \{X_1^{\alpha_1}X_2^{\alpha_2}X_3^{\alpha_3}X_4^{\alpha_4} \mid \alpha_1, \ldots, \alpha_4 \in \mathbb{N}\}$. Since the algebra $A = K\langle X \rangle / I$ has $1, \overline{X_1}, \overline{X_2}\ \overline{X_3}, \overline{X_4} \in \overline{N(I)}$ such that

$$\begin{cases} \overline{X_3} \cdot \overline{X_1} = \overline{X_2} \cdot \overline{X_4} \\ \overline{X_3} \cdot \overline{X_2} = \overline{X_1} \cdot \overline{X_4} \end{cases} \begin{cases} \overline{X_4} \cdot \overline{X_1} = \overline{X_2} \cdot \overline{X_3} \\ \overline{X_4} \cdot \overline{X_2} = \overline{X_1} \cdot \overline{X_3} \end{cases}$$

$$\begin{cases} \overline{X_4} \cdot \overline{X_1} = \overline{X_2} \cdot \overline{X_3} \\ \overline{X_3} \cdot \overline{X_1} = \overline{X_2} \cdot \overline{X_4} \end{cases} \begin{cases} \overline{X_3} \cdot \overline{X_2} = \overline{X_1} \cdot \overline{X_4} \\ \overline{X_4} \cdot \overline{X_2} = \overline{X_1} \cdot \overline{X_3} \end{cases}$$

it is clear that there is no two-sided monomial ordering on $\overline{N(I)}$, and that each of the (left or right) monomial orderings used in Theorem 3.2 and Theorem 3.3 cannot be a (left, right, or two-sided) monomial ordering on $\overline{N(I)}$.

In consideration of the examples given above, algebras defined by monomials and binomials of the types as described in Theorem 3.2 and Theorem 3.3, that may have a two-sided Gröbner basis theory are necessarily the type similar to the familiar skew polynomial algebra $\mathcal{O}_n(\lambda_{ji})$ subject to the relations $x_jx_i - \lambda_{ji}x_ix_j = 0$ with $\lambda_{ji} \in K^*$ and $1 \le i < j \le n$. This leads to the next result.

Theorem 3.4. *Under the assumptions on R and $\mathcal{B}$ as in Theorem 3.2, if $\mathcal{G} = \Omega \cup G$ is a minimal Gröbner basis of the ideal $I = \langle \mathcal{G} \rangle$, where $\Omega \subset \mathcal{B}$ and G consists of the $\frac{n(n-1)}{2}$ elements*

$$g_{ji} = a_{\ell_j}a_{\ell_i} - \lambda_{ji}a_{\ell_i}a_{\ell_j},\ 1 \le i < j \le n,\ \lambda_{ji} \in K^* \cup \{0\},$$

such that $\mathbf{LM}(g_{ji}) = a_{\ell_j}a_{\ell_i}$, $1 \le i < j \le n$, then the following statements hold for the algebra $A = R/I$.

(i) *$N(I) \subseteq \{a_{\ell_1}^{\alpha_1}a_{\ell_2}^{\alpha_2} \cdots a_{\ell_n}^{\alpha_n} \mid \alpha_1, \ldots, \alpha_n \in \mathbb{N}\}$, and $\overline{N(I)}$ is a skew multiplicative K-basis for A.*

(ii) *If $\overline{a_{\ell_1}}^{\alpha_1}\overline{a_{\ell_2}}^{\alpha_2} \cdots \overline{a_{\ell_n}}^{\alpha_n} \ne 0$ implies $a_{\ell_1}^{\alpha_1}a_{\ell_2}^{\alpha_2} \cdots a_{\ell_n}^{\alpha_n} \in N(I)$, then A has a two-sided Gröbner basis theory with respect to the two-sided admissible system $(\overline{N(I)}, \prec_{_I})$, where $\prec_{_I}$ is the two-sided monomial ordering on $\overline{N(I)}$ induced by one of the following orderings on $\mathbb{N}^n$: the lexicographic ordering, the $\mathbb{N}$-graded lexicographic ordering, and the $\mathbb{N}$-graded reverse lexicographic ordering, such that*

$$\overline{a_{\ell_n}} \prec_{_I} \overline{a_{\ell_{n-1}}} \prec_{_I} \cdots \prec_{_I} \overline{a_{\ell_2}} \prec_{_I} \overline{a_{\ell_1}}.$$

Proof. Note that in this case we have $g_{ji} = a_ja_i - \lambda_{ji}a_ia_j$, $1 \le i < j \le n$, $\lambda_{ji} \in K^* \cup \{0\}$. Let $\prec_{_I}$ be one of the orderings on $\overline{N(I)}$ as mentioned. If $\bar{u} = \overline{a_1}^{\alpha_1} \cdots \overline{a_n}^{\alpha_n}$, $\bar{v} = \overline{a_1}^{\beta_1} \cdots \overline{a_n}^{\beta_n}$, $\bar{w} = \overline{a_1}^{\gamma_1} \cdots \overline{a_n}^{\gamma_n}$, $\bar{s} = \overline{a_1}^{\eta_1} \cdots \overline{a_n}^{\eta_n} \in \overline{N(I)}$ satisfy $\bar{u} \prec_{_I} \bar{v}$, $\bar{w}\bar{u}\bar{s} \notin K^* \cup \{0\}$, and $\bar{w}\bar{v}\bar{s} \notin K^* \cup \{0\}$, then, the assumption that $\overline{a_1}^{\alpha_1} \cdots \overline{a_n}^{\alpha_n} \ne 0$ implies $a_1^{\alpha_1} \cdots a_n^{\alpha_n} \in N(I)$ and the feature of g_{ji} entail that

$$\begin{aligned} \mathbf{LM}(\bar{w}\bar{u}\bar{s}) &= \overline{a_1}^{\gamma_1+\alpha_1+\eta_1} \cdots \overline{a_n}^{\gamma_n+\alpha_n+\eta_n} \\ &\prec_{_I} \overline{a_1}^{\gamma_1+\beta_1+\eta_1} \cdots \overline{a_n}^{\gamma_n+\beta_n+\eta_n} = \mathbf{LM}(\bar{w}\bar{v}\bar{s}). \end{aligned}$$

This shows that (MO1) of (Ch.3, Definition 1.1(i)) holds. Similarly, (MO2) of (Ch.3, Definition 1.1(i)) holds as well. Hence $\prec_{_I}$ is a two-sided monomial ordering on $\overline{N(I)}$, and thereby A has a two-sided Gröbner basis theory with respect to the admissible system $(\overline{N(I)}, \prec_{_I})$. □

Focusing on quotient algebras of free K-algebras we have the following immediate corollary.

Corollary 3.5. *Let $R = K\langle X_1, \ldots, X_n\rangle$ be the free K-algebra of n generators, $\mathcal{B}$ the standard K-basis of R, and let $X_{\ell_1}, X_{\ell_2}, \ldots, X_{\ell_n}$ be a permutation of $X_1, X_2, \ldots, X_n$.*
(i) *If, with respect to some two-sided monomial ordering $\prec$ on $\mathcal{B}$, $\mathcal{G} = \Omega \cup G$ is a minimal Gröbner basis of the ideal $I = \langle \mathcal{G}\rangle$, where $\Omega \subset \mathcal{B}$ and G consists of the $\frac{n(n-1)}{2}$ elements*

$$g_{ji} = X_{\ell_j}X_{\ell_i} - \lambda_{ji}X_{\ell_i}X_{\ell_p},\ 1 \le i < j \le n,\ i < p \le n,\ \lambda_{ji} \in K^* \cup \{0\},$$

such that $\mathbf{LM}(g_{ji}) = X_{\ell_j}X_{\ell_i}$, $1 \le i < j \le n$, then the conditions of Theorem 3.2 are satisfied, and consequently the algebra $A = R/I$ has a left Gröbner basis theory with respect to the left admissible system $(\overline{N(I)}, \prec_{_I})$, where $\prec_{_I}$ is as described in Theorem 3.2.
(ii) *If, with respect to some two-sided monomial ordering $\prec$ on $\mathcal{B}$, $\mathcal{G} = \Omega \cup G$ is a minimal Gröbner basis of the ideal $I = \langle \mathcal{G}\rangle$, where $\Omega \subset \mathcal{B}$ and G consists of the $\frac{n(n-1)}{2}$ elements*

$$g_{ji} = X_{\ell_j}X_{\ell_i} - \lambda_{ji}X_{\ell_p}X_{\ell_j},\ 1 \le i < j \le n,\ p < j,\ \lambda_{ji} \in K^* \cup \{0\},$$

such that $\mathbf{LM}(g_{ji}) = X_{\ell_j}X_{\ell_i}$, $1 \le i < j \le n$, then the conditions of Theorem 3.3 are satisfied, and consequently the algebra $A = R/I$ has a right Gröbner basis theory with respect to the right admissible system $(\overline{N(I)}, \prec_{_I})$, where $\prec_{_I}$ is as described in Theorem 3.3.

(iii) *If, with respect to some two-sided monomial ordering $\prec$ on $\mathcal{B}$, $\mathcal{G} = \Omega \cup G$ is a minimal Gröbner basis of the ideal $I = \langle \mathcal{G} \rangle$, where $\Omega \subset \mathcal{B}$ and G consists of the $\frac{n(n-1)}{2}$ elements*

$$g_{ji} = X_{\ell_j} X_{\ell_i} - \lambda_{ji} X_{\ell_i} X_{\ell_j},\ 1 \le i < j \le n,\ \lambda_{ji} \in K^* \cup \{0\},$$

such that $\mathbf{LM}(g_{ji}) = X_{\ell_j} X_{\ell_i}$, $1 \le i < j \le n$, then the conditions of Theorem 3.4 are satisfied, and consequently the algebra $A = R/I$ has a two-sided Gröbner basis theory with respect to the two-sided admissible system $(\overline{N(I)}, \prec_{_I})$, where $\prec_{_I}$ is as described in Theorem 3.4.

□

Except for the commutative polynomial algebras, exterior algebras, and the skew polynomial algebras like $\mathcal{O}_n(\lambda_{ji})$, which are quotient algebras of free algebras defined by monomial and binomial relations, typical examples illustrating Theorem 3.4, or more precisely, illustrating Corollary 3.5(iii), are those algebras given in Example 6 of (Ch.3, Section 5) including the quantum Grassmannian (or quantum exterior) algebra introduced in ([Man], Section 3).

Remark (i) In previous results and examples, we used $K^* \cup \{0\}$ instead of K not only to emphasize that the parameters λ_{ji} may be 0, but also to indicate the consistency with Definition 1.1 of (Ch.3, Section 1). This remark is valid for the rest of this section.
(ii) Thanks to [G-I4] and [G-I5], there are numerous (left, right) Noetherian algebras of the type described through Theorem 3.2 – Corollary 3.5, in particular, many *quantum binomial algebras* and their Koszul dual studied in [G-IV], [Laf] and [G-I5] are such Noetherian algebras. For instance, the algebras discussed in the foregoing Examples (2), (3) with $\lambda, \mu, \gamma \ne 0$ and $\gamma^2 = 1$, respectively $\lambda^2 = 1$. By (Ch.3, Proposition 4.6), if the algebra $A = R/I$ involved in Theorem 3.2 – Corollary 3.5 is respectively left, right, and two-sided Noetherian, then A has respectively a finite left, right, and two-sided Gröbner basis theory.

To go further, we let $R = \oplus_{\gamma \in \Gamma} R_\gamma$ be a Γ-graded K-algebra, where Γ is a well-ordered semigroup with the *well-ordering* $<$. If I is an ideal of R and $A = R/I$, then we have the Γ-leading homogeneous algebra of A as defined in Ch.2, that is the Γ-graded algebra $A^{\Gamma}_{\rm LH} = R/\langle \mathbf{LH}(I) \rangle$, where $\mathbf{LH}(I)$ is the set of Γ-leading homogeneous elements of I. Suppose R has a Gröbner basis theory. Our consideration below is to indicate that a given

Gröbner basis $\mathcal{G}$ of I may not immediately gives rise to a Gröbner basis theory for A (for instance if it is not the type considered before), but in case $\mathbf{LH}(\mathcal{G})$ consists of a subset Ω of monomials and $\frac{n(n-1)}{2}$ elements of the form g_{ji} as in Theorem 3.2, Theorem 3.3, or Theorem 3.4, it does determine a left, right, or two-sided Gröbner basis theory for the algebra $A^{\Gamma}_{\rm LH}(A)$. The reason that we bring the object $A^{\Gamma}_{\rm LH}$ into play is, of course, due to (Ch.4, Proposition 2.2) and the picture (described in Ch.5, Section 1):

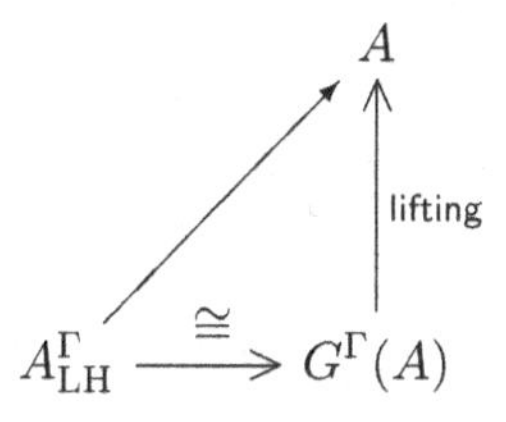

In what follows we always assume that $R = \oplus_{\gamma\in\Gamma}R_{\gamma}$ is finitely generated by Γ-homogeneous elements $a_1,\ldots,a_n$ (i.e., $R = K[a_1,\ldots,a_n]$), and that R has a Gröbner basis theory with respect to some admissible system $(\mathcal{B},\prec_{gr})$, where $\mathcal{B}$ is a skew multiplicative K-basis consisting of Γ-homogeneous elements of the form $a_{i_1}^{\alpha_1}\cdots a_{i_s}^{\alpha_s}$, and $\prec_{gr}$ is a Γ-graded monomial ordering on $\mathcal{B}$. Thus, for any ideal I of R, (Ch.4, Proposition 2.2(i)) tells us that the set $N(I)$ of normal monomials in $\mathcal{B}$ (modulo I) coincides with the set $N(\langle\mathbf{LH}(I)\rangle)$ of normal monomials in $\mathcal{B}$ (modulo $\langle\mathbf{LH}(I)\rangle$), i.e., $N(I) = N(\langle\mathbf{LH}(I)\rangle)$. So we may write $\widehat{N(I)} = \{\hat{u} = u + \langle\mathbf{LH}(I)\rangle \mid u\in N(I)\}$ for the K-basis of $A^{\Gamma}_{\rm LH}$ determined by $N(\langle\mathbf{LH}(I)\rangle)$. Note that $\widehat{N(I)}$ consists of Γ-homogeneous elements.

With notation as before, by (Ch.4, Proposition 2.2(ii)) and the foregoing Theorem 3.2 – Theorem 3.4 we are ready to mention the next three results.

Theorem 3.6. *Let $R = \oplus_{\gamma\in\Gamma}R_{\gamma}$ and $(\mathcal{B},\prec_{gr})$ be as fixed above, and assume further that R is finitely generated by Γ-homogeneous elements $a_1,\ldots,a_n$ (i.e., $R = K[a_1,\ldots,a_n]$), and that $\mathcal{B}$ contains every monomial of the form $a_{\ell_1}^{\alpha_1}a_{\ell_2}^{\alpha_2}\cdots a_{\ell_n}^{\alpha_n}$ with respect to some permutation $a_{\ell_1},a_{\ell_2},\ldots,a_{\ell_n}$ of $a_1,a_2,\ldots,a_n$, where $\alpha_1,\ldots,\alpha_n\in\mathbb{N}$. Let I be an ideal of R, $A = R/I$, and $A^{\Gamma}_{\rm LH} = R/\langle\mathbf{LH}(I)\rangle$. If $\mathcal{G}$ is a minimal Gröbner basis of I such that $\mathbf{LH}(\mathcal{G}) = \Omega\cup G$, where $\Omega\subset\mathcal{B}$ and G consists of the $\frac{n(n-1)}{2}$ elements*

$$g_{ji} = a_{\ell_j}a_{\ell_i} - \lambda_{ji}a_{\ell_i}a_{\ell_p},\ 1\le i<j\le n,\ i<p\le n,\ \lambda_{ji}\in K^*\cup\{0\},$$

satisfying $\mathbf{LM}(g_{ji}) = a_{\ell_j}a_{\ell_i}$, $1 \le i < j \le n$, *then the following statements hold.*

(i) $N(I) \subseteq \{a_{\ell_1}^{\alpha_1}a_{\ell_2}^{\alpha_2}\cdots a_{\ell_n}^{\alpha_n} \mid \alpha_1,\ldots,\alpha_n \in \mathbb{N}\}$, *and* $\widehat{N(I)}$ *is a skew multiplicative* K*-basis for* A_{LH}^{Γ}.

(ii) *Let* $\prec_{_{\mathbf{LH}(I)}}$ *denote the ordering on* $\widehat{N(I)}$, *which is induced by the lexicographic ordering or the* $\mathbb{N}$*-graded lexicographic ordering on* $\mathbb{N}^n$, *such that*

$$\widehat{a_{\ell_n}} \prec_{_{\mathbf{LH}(I)}} \widehat{a_{\ell_{n-1}}} \prec_{_{\mathbf{LH}(I)}} \cdots \prec_{_{\mathbf{LH}(I)}} \widehat{a_{\ell_2}} \prec_{_{\mathbf{LH}(I)}} \widehat{a_{\ell_1}}.$$

If $\widehat{a_{\ell_1}}^{\alpha_1}\widehat{a_{\ell_2}}^{\alpha_2}\cdots\widehat{a_{\ell_n}}^{\alpha_n} \neq 0$ *implies* $a_{\ell_1}^{\alpha_1}a_{\ell_2}^{\alpha_2}\cdots a_{\ell_n}^{\alpha_n} \in N(I)$, *then* $\prec_{_{\mathbf{LH}(I)}}$ *is a left monomial ordering on* $\widehat{N(I)}$, *and hence* A_{LH}^{Γ} *has a left Gröbner basis theory with respect to the left admissible system* $(\widehat{N(I)}, \prec_{_{\mathbf{LH}(I)}})$.

□

Theorem 3.7. *Under the assumptions in Theorem 3.6, let* I *be an ideal of* R, $A = R/I$, *and* $A_{\mathrm{LH}}^{\Gamma} = R/\langle \mathbf{LH}(I)\rangle$. *If* $\mathcal{G}$ *is a minimal Gröbner basis of* I *such that* $\mathbf{LH}(\mathcal{G}) = \Omega \cup G$, *where* $\Omega \subset \mathcal{B}$ *and* G *consists of the* $\frac{n(n-1)}{2}$ *elements*

$$g_{ji} = a_{\ell_j}a_{\ell_i} - \lambda_{ji}a_{\ell_p}a_{\ell_j},\ 1 \le i < j \le n,\ p < j,\ \lambda_{ji} \in K^* \cup \{0\},$$

satisfying $\mathbf{LM}(g_{ji}) = a_{\ell_j}a_{\ell_i}$, $1 \le i < j \le n$, *then the following statements hold.*

(i) $N(I) \subseteq \{a_{\ell_1}^{\alpha_1}a_{\ell_2}^{\alpha_2}\cdots a_{\ell_n}^{\alpha_n} \mid \alpha_1,\ldots,\alpha_n \in \mathbb{N}\}$, *and* $\widehat{N(I)}$ *is a skew multiplicative* K*-basis for* A_{LH}^{Γ}.

(ii) *Let* $\prec_{_{\mathbf{LH}(I)}}$ *denote the ordering on* $\widehat{N(I)}$, *which is induced by the* $\mathbb{N}$*-graded reverse lexicographic ordering on* $\mathbb{N}^n$, *such that*

$$\widehat{a_{\ell_n}} \prec_{_{\mathbf{LH}(I)}} \widehat{a_{\ell_{n-1}}} \prec_{_{\mathbf{LH}(I)}} \cdots \prec_{_{\mathbf{LH}(I)}} \widehat{a_{\ell_2}} \prec_{_{\mathbf{LH}(I)}} \widehat{a_{\ell_1}}.$$

If $\widehat{a_{\ell_1}}^{\alpha_1}\widehat{a_{\ell_2}}^{\alpha_2}\cdots\widehat{a_{\ell_n}}^{\alpha_n} \neq 0$ *implies* $a_{\ell_1}^{\alpha_1}a_{\ell_2}^{\alpha_2}\cdots a_{\ell_n}^{\alpha_n} \in N(I)$, *then* $\prec_{_{\mathbf{LH}(I)}}$ *is a right monomial ordering on* $\widehat{N(I)}$, *and hence* A_{LH}^{Γ} *has a right Gröbner basis theory with respect to the right admissible system* $(\widehat{N(I)}, \prec_{_{\mathbf{LH}(I)}})$.

□

Theorem 3.8. *Under the assumptions in Theorem 3.6, let* I *be an ideal of* R, $A = R/I$, *and* $A_{\mathrm{LH}}^{\Gamma} = R/\langle \mathbf{LH}(I)\rangle$. *If* $\mathcal{G}$ *is a minimal Gröbner basis of* I *such that* $\mathbf{LH}(\mathcal{G}) = \Omega \cup G$, *where* $\Omega \subset \mathcal{B}$ *and* G *consists of the* $\frac{n(n-1)}{2}$ *elements*

$$g_{ji} = a_{\ell_j}a_{\ell_i} - \lambda_{ji}a_{\ell_i}a_{\ell_j},\ 1 \le i < j \le n,\ \lambda_{ji} \in K^* \cup \{0\},$$

satisfying $\mathbf{LM}(g_{ji}) = a_{\ell_j}a_{\ell_i}$, $1 \le i < j \le n$, *then the following statements hold.*

(i) $N(I) \subseteq \{a_{\ell_1}^{\alpha_1}a_{\ell_2}^{\alpha_2}\cdots a_{\ell_n}^{\alpha_n} \mid \alpha_1,\ldots,\alpha_n \in \mathbb{N}\}$, *and* $\widehat{N(I)}$ *is a skew multiplicative* K*-basis for* $A_{\rm LH}^{\Gamma}$.
(ii) *If* $\widehat{a_{\ell_1}}^{\alpha_1}\widehat{a_{\ell_2}}^{\alpha_2}\cdots\widehat{a_{\ell_n}}^{\alpha_n} \neq 0$ *implies* $a_{\ell_1}^{\alpha_1}a_{\ell_2}^{\alpha_2}\cdots a_{\ell_n}^{\alpha_n} \in N(I)$, *then* $A_{\rm LH}^{\Gamma}$ *has a two-sided Gröbner basis theory with respect to the admissible system* $(\widehat{N(I)}, \prec_{_{\mathbf{LH}(I)}})$, *where* $\prec_{_{\mathbf{LH}(I)}}$ *is the two-sided monomial ordering on* $\widehat{N(I)}$ *induced by one of the following orderings on* $\mathbb{N}^n$: *the lexicographic ordering, the* $\mathbb{N}$*-graded lexicographic ordering, and the* $\mathbb{N}$*-graded reverse lexicographic ordering, such that*

$$\widehat{a_{\ell_n}} \prec_{_{\mathbf{LH}(I)}} \widehat{a_{\ell_{n-1}}} \prec_{_{\mathbf{LH}(I)}} \cdots \prec_{_{\mathbf{LH}(I)}} \widehat{a_{\ell_2}} \prec_{_{\mathbf{LH}(I)}} \widehat{a_{\ell_1}}.$$

□

Examples given below not only illustrate Theorem 3.6 – 3.8, but also answers why the three theorems said nothing about the existence of a (left, right, or two-sided) Gröbner basis theory for the quotient algebra $A = R/I$.

Example 4. Let $\mathcal{G}$ be the subset of the free K-algebra $K\langle X\rangle = K\langle X_1, X_2, X_3\rangle$ consisting of

$$\begin{aligned} g_{21} &= X_2X_1 - \lambda X_1X_3 + \alpha X_1, \\ g_{31} &= X_3X_1, \qquad\qquad\qquad \text{where } \lambda, \mu, \alpha \in K^* \cup \{0\}. \\ g_{32} &= X_3X_2 - \mu X_2X_3 + \alpha X_3, \end{aligned}$$

Then, with respect to the natural $\mathbb{N}$-graded lexicographic ordering subject to $X_1 \prec_{grlex} X_2 \prec_{grlex} X_3$, where each X_i is assigned the degree 1, it is straightforward to verify that $\mathbf{LM}(g_{ji}) = X_jX_i$, $1 \le i < j \le 3$, $\mathcal{G}$ forms a Gröbner basis of the ideal $I = \langle \mathcal{G}\rangle$ for any λ, μ, $\alpha \in K^* \cup \{0\}$, and thereby $N(I) = \{X_1^{\alpha_1}X_2^{\alpha_2}X_3^{\alpha_3} \mid \alpha_1, \alpha_2, \alpha_3 \in \mathbb{N}\}$. Consider the quotient algebra $A = K\langle X\rangle/I$. Since $\mathcal{G}$ is of the type required by Theorem 3.6 such that $\mathbf{LH}(\mathcal{G}) = \{X_2X_1 - \lambda X_1X_3,\ X_3X_1,\ X_3X_2 - \mu X_2X_3\}$, it follows that the algebra $A_{\rm LH}^{\mathbb{N}} = K\langle X\rangle/\langle \mathbf{LH}(\mathcal{G})\rangle$ has a left Gröbner basis theory with respect to the left admissible system $(\widehat{N(I)}, \prec_{_{\mathbf{LH}(I)}})$, where $\prec_{_{\mathbf{LH}(I)}}$ is as described in Theorem 3.6.

If $\mu = 0$ and $\alpha \neq 0$, then note that the algebra A has $\overline{X_2}$, $\overline{X_2}^2$, $\overline{X_3} \in \overline{N(I)}$ such that $\overline{X_2} \neq \overline{X_2}^2$, $\overline{X_3} \cdot \overline{X_2} = -\alpha\overline{X_3}$, and $\overline{X_3} \cdot \overline{X_2}^2 = -\alpha\overline{X_3} \cdot \overline{X_2}$. Thus, with respect to any total ordering $\prec$ on $\overline{N(I)}$, we have

$$\mathbf{LM}(\overline{X_3} \cdot \overline{X_2}^2) = \overline{X_3} = \mathbf{LM}(\overline{X_3} \cdot \overline{X_2}).$$

This shows that $\prec$ cannot be a (left, right, or two-sided) monomial ordering on $\overline{N(I)}$, and therefore, the algebra A cannot have a (left, right, and two-sided) Gröbner basis theory whenever the K-basis $\overline{N(I)}$ is used.

Example 5. Let $\mathcal{G}$ be the subset of the free K-algebra $K\langle X\rangle = K\langle X_1, X_2, X_3\rangle$ consisting of

$$\begin{aligned} g_{21} &= X_2X_1 - \lambda X_1X_2 + \alpha X_1, \\ g_{31} &= X_3X_1, \qquad\qquad\qquad\qquad \text{where } \lambda, \mu, \alpha \in K^* \cup \{0\}. \\ g_{32} &= X_3X_2 - \mu X_1X_3 + \alpha X_3, \end{aligned}$$

Then, with respect to the natural $\mathbb{N}$-graded lexicographic ordering subject to $X_1 \prec_{grlex} X_2 \prec_{grlex} X_3$, where each X_i is assigned the degree 1, it is straightforward to verify that $\mathbf{LM}(g_{ji}) = X_jX_i$, $1 \le i < j \le 3$, $\mathcal{G}$ forms a Gröbner basis of the ideal $I = \langle \mathcal{G}\rangle$ for any λ, μ, $\alpha \in K^* \cup \{0\}$, and thereby $N(I) = \{X_1^{\alpha_1}X_2^{\alpha_2}X_3^{\alpha_3} \mid \alpha_1, \alpha_2, \alpha_3 \in \mathbb{N}\}$. Consider the quotient algebra $A = K\langle X\rangle/I$. Since $\mathcal{G}$ is of the type required by Theorem 3.7 such that $\mathbf{LH}(\mathcal{G}) = \{X_2X_1 - \lambda X_1X_2,\ X_3X_1,\ X_3X_2 - \mu X_1X_3\}$, it follows that the algebra $A_{\rm LH}^{\mathbb{N}} = K\langle X\rangle/\langle \mathbf{LH}(\mathcal{G})\rangle$ has a right Gröbner basis theory with respect to the right admissible system $(\widehat{N(I)}, \prec_{_{\mathbf{LH}(I)}})$, where $\prec_{_{\mathbf{LH}(I)}}$ is as described in Theorem 3.7.

If $\lambda = 0$ and $\alpha \ne 0$, then note that the algebra A has $\overline{X_2}, \overline{X_2}^2, \overline{X_1} \in \overline{N(I)}$ such that $\overline{X_2} \ne \overline{X_2}^2$, $\overline{X_2}\cdot\overline{X_1} = -\alpha\overline{X_1}$, and $\overline{X_2}^2\cdot\overline{X_1} = -\alpha\overline{X_2}\cdot\overline{X_1}$. Thus, with respect to any total ordering $\prec$ on $\overline{N(I)}$, we have

$$\mathbf{LM}(\overline{X_2}^2\cdot\overline{X_1}) = \overline{X_1} = \mathbf{LM}(\overline{X_2}\cdot\overline{X_1}).$$

This shows that $\prec$ cannot be a (left, right, or two-sided) monomial ordering on $\overline{N(I)}$, and thereby, the algebra A cannot have a (left, right, or two-sided) Gröbner basis theory whenever the K-basis $\overline{N(I)}$ is used.

Example 6. This example provides further investigation of the Gröbner basis theory of the algebra considered in Example 2 of (Ch.4, Section 4).

Let $\mathcal{G}$ be the subset of the free K-algebra $K\langle X\rangle = K\langle X_1, X_2, X_3\rangle$ consisting of

$$\begin{aligned} g_{31} &= X_3X_1 - \lambda X_1X_3 + \gamma X_3, \\ g_{12} &= X_1X_2 - \lambda X_2X_1 + \gamma X_2, \\ g_{32} &= X_3X_2 - \omega X_2X_3 + f(X_1), \end{aligned}$$

where $\lambda, \gamma, \omega \in K^* \cup \{0\}$, and $f(X_1)$ is a polynomial in the variable X_1. If $d(f) \le 2$, we assign $d(X_1) = d(X_2) = d(X_3) = 1$; if $d(f) = n \ge$

3, we assign $d(X_1) = d(X_2) = 1$, $d(X_3) = n$. By Example 6 of (Ch.4, Section 3), if the $\mathbb{N}$-graded lexicographic ordering $X_2 \prec_{grlex} X_1 \prec_{grlex} X_3$ with respect to either $\mathbb{N}$-gradation assigned to $K\langle X\rangle$ above is used, then $\mathbf{LM}(g_{ji}) = \{X_3X_1, X_1X_2, X_3X_2\}$, $\mathcal{G}$ forms a Gröbner basis of the ideal $I = \langle \mathcal{G}\rangle$ for any λ, ω, $\gamma \in K^* \cup \{0\}$ and arbitrarily chosen $f(X_1)$, and thereby $N(I) = \{X_2^{\alpha_2}X_1^{\alpha_1}X_3^{\alpha_3} \mid \alpha_1, \alpha_2, \alpha_3 \in \mathbb{N}\}$. Furthermore, it follows from (Ch.4, Proposition 2.2(ii)) that

$$\mathbf{LH}(\mathcal{G}) = \{X_3X_1 - \lambda X_1X_3,\ X_1X_2 - \lambda X_2X_1,\ X_3X_2 - \omega X_2X_3\}$$

if $d(f) \geq 3$, respectively

$$\mathbf{LH}(\mathcal{G}) = \{X_3X_1 - \lambda X_1X_3,\ X_1X_2 - \lambda X_2X_1,\ X_3X_2 - \omega X_2X_3 + aX_1^2\}$$

if $f(x_1) = aX_1^2 + bX_1 + c$ with $a, b, c \in K$, is a Gröbner basis for the ideal $\langle \mathbf{LH}(I)\rangle$ with respect to the respective $\mathbb{N}$-gradation assigned to $K\langle X\rangle$ and the corresponding monomial ordering $\prec_{grlex}$. The existence of a Gröbner basis theory for the algebras $A = K\langle X\rangle/I$ and $A_{\mathrm{LH}}^{\mathbb{N}} = K\langle X\rangle/\langle \mathbf{LH}(\mathcal{G})\rangle$ is discussed as follows.

(1) If $\lambda \neq 0$ and $\omega \neq 0$, then, in any case (i.e., $d(f) \leq 2$ or $d(f) \geq 3$), with respect to $\overline{X_2} \prec_{grlex} \overline{X_1} \prec_{grlex} \overline{X_3}$ on $\overline{N(I)}$, respectively $\widehat{X_2} \prec_{grlex} \widehat{X_1} \prec_{grlex} \widehat{X_3}$ on $\widehat{N(I)}$, both A and $A_{\mathrm{LH}}^{\mathbb{N}}$ are solvable polynomial algebras in the sense of [K-RW]. Hence, every (left, right, or two-sided) ideal has a finite Gröbner basis.

(2) If $d(f) \geq 3$ and $\lambda = 0$ or $\omega = 0$, then by Theorem 3.8, $A_{\mathrm{LH}}^{\mathbb{N}}$ has a two-sided Gröbner basis theory with respect to the admissible system $(\widehat{N(I)}, \prec_{_{\mathbf{LH}(I)}})$, where $\prec_{_{\mathbf{LH}(I)}}$ is the monomial ordering as described in Theorem 3.8.

(3) In the case that $f(X_1) = aX_1^2 + bX_1 + c$, if $a = 0$, then, for arbitrary λ and ω, $A_{\mathrm{LH}}^{\mathbb{N}}$ has a two-sided Gröbner basis theory with respect to the admissible system $(\widehat{N(I)}, \prec_{_{\mathbf{LH}(I)}})$, where $\prec_{_{\mathbf{LH}(I)}}$ is the monomial ordering as described in Theorem 3.8.

(4) In the case that $f(X_1) = aX_1^2 + bX_1 + c$, if $a \neq 0$ and $\lambda = 0$ or $\omega = 0$, then we are not clear about the existence of a (left, right, or two-sided) Gröbner basis for $A_{\mathrm{LH}}^{\mathbb{N}}$.

(5) If $\lambda = 0$ and $\gamma \neq 0$, then, in any case (i.e., $d(f) \leq 2$ or $d(f) \geq 3$), note that $\overline{X_1}, \overline{X_1}^2, \overline{X_3} \in \overline{N(I)}$, $\overline{X_1} \neq \overline{X_1}^2$, $\overline{X_3} \cdot \overline{X_1} = -\gamma \overline{X_3}$, and $\overline{X_3} \cdot \overline{X_1}^2 = \gamma^2 \overline{X_3}$. Thus, with respect to any total ordering $\prec$ on $\overline{N(I)}$, we have

$$\mathbf{LM}(\overline{X_3} \cdot \overline{X_1}) = \overline{X_3} = \mathbf{LM}(\overline{X_3} \cdot \overline{X_1}^2).$$

This shows that $\prec$ cannot be a (left, right, or two-sided) monomial ordering on $\overline{N(I)}$, and consequently, the quotient algebra $A = R/I$ cannot have a (left, right, or two-sided) Gröbner basis theory whenever the K-basis $\overline{N(I)}$ is used.

In addition to the philosophy of studying the structural properties of A via $A^{\Gamma}_{\rm LH}$ (as may be seen from previous chapters), the natural question now is to ask if there would be any appropriate condition on A or on $A^{\Gamma}_{\rm LH}$ that enables us to obtain a Gröbner basis theory for A subject to a Gröbner basis theory of $A^{\Gamma}_{\rm LH}$.

Enlightened by (Ch.2, Theorem 3.2), the next theorem, which may be viewed as a recognizable version of ([Li1] Ch.4, Theorem 1.6), provides a solution to the question posed above. Before mentioning the next theorem, we need a little more preparation, for, the ideal I concerned by the theorem is not necessarily generated by monomials and binomials, and moreover, the K-basis $\overline{N(I)}$ of $A = R/I$, respectively the K-basis $\widehat{N(I)}$ of $A^{\Gamma}_{\rm LH}$, is not necessarily a skew multiplicative K-basis. In this case, with any fixed K-basis $\mathscr{B}$ of A, the definition of a (left, right, or two-sided) monomial ordering $\prec$ on $\mathscr{B}$ is the same as that defined in (Ch.3, Definition 1.1); for $u, v \in \mathscr{B}$, we say that u divides v, denoted $u|v$, if there are $w, s \in \mathscr{B}$ such that $v = \mathbf{LM}(wus)$; and we say that A has a (left, right, or two-sided) Gröbner basis theory (comparing with Definition 2.1 and Definition 2.3 of Ch.3) if, with respect to the (left, right, or two-sided) admissible system $(\mathscr{B}, \prec)$, every proper (left, right, or two-sided) ideal I of A has a generating set $\mathcal{G}$ such that $f \in I$ implies $\mathbf{LM}(g)|\mathbf{LM}(f)$ for some $g \in \mathcal{G}$. Similar convention is valid for $A^{\Gamma}_{\rm LH}$.

Theorem 3.9. *Let $R = \oplus_{\gamma\in\Gamma}R_{\gamma}$ and $(\mathcal{B}, \prec_{gr})$ be as in Theorem 3.6, and let I be an arbitrary proper ideal of R (i.e., I is not necessarily generated by monomials and binomials). Put $A = R/I$, $A^{\Gamma}_{\rm LH} = R/\langle \mathbf{LH}(I)\rangle$. With notation as before and the convention made above, suppose that*

(a) *$A^{\Gamma}_{\rm LH}$ is a domain; and*

(b) *$A^{\Gamma}_{\rm LH}$ has a (left, right, or two-sided) Gröbner basis theory with respect to some (left, right, or two-sided) admissible system $(\widehat{N(I)}, \prec_{\widehat{gr}})$, where $\prec_{\widehat{gr}}$ is a Γ-graded (left, right, or two-sided) monomial ordering on $\widehat{N(I)}$, such that every Γ-graded (left, right, or two-sided) ideal of $A^{\Gamma}_{\rm LH}$ has a (left, right, or two-sided) Gröbner basis consisting of Γ-homogeneous elements.*

Then the order $\prec$ on $\overline{N(I)}$ defined subject to the rule: for $\overline{u}, \overline{v} \in \overline{N(I)}$,

$$\bar{u} \prec \bar{v} \text{ whenever } \hat{u} \prec_{\widehat{gr}} \hat{v},$$

is a (left, right, or two-sided) monomial ordering on $\overline{N(I)}$, such that every (left, right, or two-sided) ideal of A has a (left, right, or two-sided) Gröbner basis.

Proof. Note that it is not difficult to check that if R has a left or right Gröbner basis theory with respect to some left or right Γ-graded monomial ordering on $\mathcal{B}$, then an analogue of (Ch.4, Proposition 2.2(ii)) holds for a left or right ideal. The proof given below will show that it is enough to conclude that A has a two-sided Gröbner basis theory whenever $A^{\Gamma}_{\rm LH}$ has a two-sided Gröbner basis theory.

Since $\prec$ is clearly a well-ordering on $\overline{N(I)}$, we first show that $\prec$ satisfies the axioms (MO1) – (MO2) of (Ch.3, Definition 1.1(i)). We start by finding the relation between $\mathbf{LM}(\hat{f}\ )$ and $\mathbf{LM}(\bar{f}\ )$ for a nonzero $f \in R$, where $\hat{f} = f + \langle \mathbf{LH}(I)\rangle \in A^{\Gamma}_{\rm LH}$, $\bar{f} = f + I \in A$. First note that $I^* = I - \{0\}$ is trivially a Gröbner basis of I, and hence $\mathbf{LH}(I^*)$ is a Gröbner basis for $\langle \mathbf{LH}(I)\rangle$ by (Ch.4, Proposition 2.2(ii)). Thus, by writing $f = \mathbf{LH}(f) + f'$ with $d(f') < d(f) = d(\mathbf{LH}(f))$ and noticing that $\prec_{gr}$ is a Γ-graded monomial ordering on $\mathcal{B}$, the division of $\mathbf{LH}(f)$ by $\mathbf{LH}(I^*)$ shows that the homogeneous element $\mathbf{LH}(f)$ has a Gröbner representation:

$$\mathbf{LH}(f) = \sum_{i,j}\lambda_{ij}w_{ij}\mathbf{LH}(f_j)v_{ij} + \sum_{\ell}\lambda_\ell u_\ell$$

with $\lambda_{ij} \in K^*$, $\lambda_\ell \in K$, $w_{ij}, v_{ij} \in \mathcal{B}$, $f_j \in I^*$, where if $\lambda_\ell \neq 0$ then $u_\ell \in N(\langle \mathbf{LH}(I)\rangle) = N(I)$, $d(u_\ell) = d(\mathbf{LH}(f)) = d(f)$, and that if $f' \neq 0$ then f' has a Gröbner representation

$$f' = \sum_{p,q}\lambda_{pq}w_{pq}\mathbf{LH}(f_q)v_{pq} + \sum_{h}\lambda_h u_h$$

with $\lambda_{pq} \in K^*$, $\lambda_h \in K$, $w_{pq}, v_{pq} \in \mathcal{B}$, $f_q \in I^*$, where if $\lambda_h \neq 0$ then $u_h \in N(\langle \mathbf{LH}(I)\rangle) = N(I)$, $d(u_h) < d(\mathbf{LH}(f)) = d(f)$. Consequently,

$$\begin{aligned} f = \mathbf{LH}(f) + f' &= \sum_{i,j}\lambda_{ij}w_{ij}\mathbf{LH}(f_j)v_{ij} + \sum_{p,q}\lambda_{pq}w_{pq}\mathbf{LH}(f_q)v_{pq} \\ &\quad + \sum_{d(u_\ell)=d(f)}\lambda_\ell u_\ell + \sum_{d(u_h)<d(f)}\lambda_h u_h. \end{aligned}$$

If we write each f_j appearing in the representation of $\mathbf{LH}(f)$ obtained above as $f_j = \mathbf{LH}(f_j) + f_j'$ with $d(f_j') < d(f_j)$, then, since Γ is an ordered semigroup, f has a representation

$$f = \sum_{i,j}\lambda_{ij}w_{ij}f_jv_{ij} + f' - \sum_{ij}\lambda_{ij}w_{ij}f_j'v_{ij} + \sum_{\ell}\lambda_\ell u_\ell$$

satisfying

$$d\left(f' - \sum_{ij}\lambda_{ij}w_{ij}f_j'v_{ij}\right) < d(f),\ u_\ell \in N(\langle \mathbf{LH}(I)\rangle) = N(I)$$

and $d(u_\ell) = d(f)$. If, furthermore, we do division by I^* with respect to $\prec_{gr}$, then the remainder of $f' - \sum_{ij}\lambda_{ij}w_{ij}f_j'v_{ij}$ must have degree $< d(f)$. Summing up and recalling that the Γ-gradation of A_{LH}^{Γ} is the one induced by the Γ-gradation of R, and that the order $\prec$ on $\overline{N(I)}$ is defined subject to the Γ-graded monomial ordering $\prec_{\widehat{gr}}$ on $\widehat{N(I)}$, we are now able to conclude that if $\mathbf{LH}(f) \notin \langle \mathbf{LH}(I)\rangle$, or equivalently, if $\widehat{\mathbf{LH}(f)} \neq 0$, then

$$\text{(1)} \qquad \mathbf{LM}(\hat{f}\ ) = \hat{u} \text{ implies } \mathbf{LM}(\bar{f}\ ) = \bar{u},$$

where $u \in N(I)$.

Next, for $\bar{u},\ \bar{v},\ \bar{w},\ \bar{s} \in \overline{N(I)}$, suppose $\bar{u} \prec \bar{v}$, $\mathbf{LM}(\bar{w}\bar{u}\bar{s}) \notin K^* \cup \{0\}$ and $\mathbf{LM}(\bar{w}\bar{v}\bar{s}) \notin K^* \cup \{0\}$. Then $\hat{u} \prec_{\widehat{gr}} \hat{v}$ by the definition of $\prec$. We also conclude that $\mathbf{LM}(\hat{w}\hat{u}\hat{s}) \notin K^* \cup \{0\}$, $\mathbf{LM}(\hat{w}\hat{v}\hat{s}) \notin K^* \cup \{0\}$, and hence

$$\text{(2)} \qquad \mathbf{LM}(\widehat{wus}) = \mathbf{LM}(\hat{w}\hat{u}\hat{s}) \prec_{\widehat{gr}} \mathbf{LM}(\hat{w}\hat{v}\hat{s}) = \mathbf{LM}(\widehat{wvs}).$$

To see this, put $f = wus$, $g = wvs$. Then, noticing that $w, u, v, s \in \mathcal{B}$ are Γ-homogeneous elements of R, we have $f = \mathbf{LH}(f) = wus$, $g = \mathbf{LH}(g) = wvs$, and thereby $\hat{f} = \widehat{\mathbf{LH}(f)} = \widehat{wus} \neq 0$, $\hat{g} = \widehat{\mathbf{LH}(g)} = \widehat{wvs} \neq 0$, since A_{LH}^{Γ} is a domain by the assumption (a). Let $\mathbf{LM}(\hat{f}\ ) = \widehat{u_1}$ and $\mathbf{LM}(\hat{g}\ ) = \widehat{v_1}$ with $u_1,\ v_1 \in N(I)$. Then it follows from the assertion (1) above that

$$\text{(3)} \qquad \begin{aligned} \overline{u_1} &= \mathbf{LM}(\bar{f}\ ) = \mathbf{LM}(\overline{wus}) = \mathbf{LM}(\bar{w}\bar{u}\bar{s}), \\ \overline{v_1} &= \mathbf{LM}(\bar{g}\ ) = \mathbf{LM}(\overline{wvs}) = \mathbf{LM}(\bar{w}\bar{v}\bar{s}). \end{aligned}$$

Thus $\mathbf{LM}(\hat{w}\hat{u}\hat{s}) \notin K^* \cup \{0\}$, $\mathbf{LM}(\hat{w}\hat{v}\hat{s}) \notin K^* \cup \{0\}$ and (2) follows. Consequently, (2) and (3) give rise to

$$\mathbf{LM}(\bar{w}\bar{u}\bar{s}) = \overline{u_1} \prec \overline{v_1} = \mathbf{LM}(\bar{w}\bar{v}\bar{s}).$$

This proves that $\prec$ satisfies (MO1). It remains to show that $\prec$ satisfies (MO2) as well. Suppose that $\bar{u},\ \bar{w},\ \bar{v},\ \bar{s} \in \overline{N(I)}$ satisfy $\bar{u} = \mathbf{LM}(\bar{w}\bar{v}\bar{s})$ (where $\bar{w} \neq \bar{1}$ or $\bar{s} \neq \bar{1}$ if $1 \in \mathcal{B}$, hence $w \neq 1$ or $s \neq 1$ in $N(I)$ and $\hat{w} \neq \hat{1}$ or $\hat{s} \neq \hat{1}$ in $\widehat{N(I)}$). Since A_{LH}^{Γ} is a domain, $\mathbf{LM}(\hat{w}\hat{v}\hat{s}) \neq 0$. So, if $\mathbf{LM}(\hat{w}\hat{v}\hat{s}) = \mathbf{LM}(\widehat{wvs}) = \widehat{v_1}$ with $v_1 \in N(I)$, then, as argued above, it follows from the foregoing assertion (1) that $\bar{u} = \mathbf{LM}(\bar{w}\bar{v}\bar{s}) = \mathbf{LM}(\overline{wvs}) = \overline{v_1}$. Thus, $\hat{v} \prec_{\widehat{gr}} \widehat{v_1}$ implies that $\bar{v} \prec \bar{u}$, as desired.

Finally, we prove that if $\overline{J} = J/I$ is a proper ideal of $A = R/I$, where J is an ideal of R containing I, then $\overline{J}$ has a Gröbner basis $\overline{\mathcal{G}}$ with respect to the admissible system $(\prec, \overline{N(I)})$ obtained above. To this end, let us first note that if $f \in R - I$, then since Γ is ordered by a well-ordering and $\langle \mathbf{LH}(I)\rangle$ is a Γ-graded ideal of R, there is some $f' \in R$ such that $\mathbf{LH}(f') \notin \langle \mathbf{LH}(I)\rangle$ and $\bar{f} = \overline{f'}$. So, without loss of generality, we may always assume that

$$f \in R - I \text{ implies } \mathbf{LH}(f) \notin \langle \mathbf{LH}(I)\rangle. \tag{4}$$

Consequently, since the Γ-gradation of A^{Γ}_{LH} is induced by the Γ-gradation of R, and since $\prec_{\widehat{gr}}$ is a Γ-graded monomial ordering on $\widehat{N(I)}$, if we write $f = \mathbf{LH}(f) + f_1$ then

$$\begin{aligned} &\mathbf{LH}(\hat{f}) = \widehat{\mathbf{LH}(f)}, \\ &\mathbf{LM}(\hat{f}) = \mathbf{LM}(\mathbf{LH}(\hat{f})) = \mathbf{LM}\left(\widehat{\mathbf{LH}(f)}\right). \end{aligned} \tag{5}$$

Now, noticing that $\mathbf{LH}(I) \subset \mathbf{LH}(J)$ and thus $\widehat{J} = \langle \mathbf{LH}(J)\rangle / \langle \mathbf{LH}(I)\rangle$ is a Γ-graded ideal of A^{Γ}_{LH}, we have $\widehat{J} \neq A^{\Gamma}_{\mathrm{LH}}$ since J is a proper ideal. Hence, with respect to the data $(\widehat{N(I)}, \prec_{\widehat{gr}})$, $\widehat{J}$ has a Gröbner basis $\mathscr{G}$ consisting of Γ-homogeneous elements. Once again since the Γ-gradation of A^{Γ}_{LH} is induced by the Γ-gradation of R, and thus $\widehat{\mathbf{LH}(J)} = \{\widehat{\mathbf{LH}(h)} \mid h \in J\}$ is a generating set of $\widehat{J}$ consisting of Γ-homogeneous elements, it is straightforward to check (or see Lemma 5.2 in Section 5 below) that $\mathscr{G} \subset \widehat{\mathbf{LH}(J)}$. Hence there is a subset $\mathcal{G} \subset J - I$ such that $\widehat{\mathbf{LH}(\mathcal{G})} = \mathscr{G}$. Put $\overline{\mathcal{G}} = \{\bar{g} \mid g \in \mathcal{G}\}$. If $\bar{f} \in \overline{J} - \{0\}$, then by the above (4) and (5), there is some $g \in \mathcal{G}$ such that $\mathbf{LM}(\widehat{\mathbf{LH}(g)}) | \mathbf{LM}(\bar{f})$, and consequently $\mathbf{LM}(\hat{g}) | \mathbf{LM}(\hat{f})$, i.e., there are $\hat{u}, \hat{v} \in \widehat{N(I)}$ such that $\mathbf{LM}(\hat{f}) = \mathbf{LM}(\hat{u}\mathbf{LM}(\hat{g})\hat{v})$. It follows from the foregoing (1) and the argument given before the formula (3) that

$$\mathbf{LM}(\bar{f}) = \mathbf{LM}(\bar{u}\mathbf{LM}(\bar{g})\bar{v}).$$

This shows that $\overline{\mathcal{G}}$ is a Gröbner basis for $\overline{J}$. □

Remark From the proof given above it is clear that if every Γ-graded (left, right, or two-sided) ideal of A^{Γ}_{LH} has a finite (left, right, or two-sided) Gröbner basis consisting of Γ-homogeneous elements. Then every (left, right, or two-sided) ideal of A has a finite (left, right, or two-sided) Gröbner basis.

There are examples of algebras other than algebras of the solvable type in the sense of [K-RW] and illustrating Theorem 3.9.

Example 7. If in Corollary 3.5(i) – (ii) we have $\Omega = \emptyset$ and all the $\lambda_{ji} \neq 0$, then it follows from the proof of Theorem 3.2 (also a similar proof of Theorem 3.3, though it is not mentioned there) that the quadratic algebra considered in (i) – (ii) respectively is a domain. Hence, by (Ch.4, Proposition 2.2(ii)), if $\mathcal{G}$ is a Gröbner basis of the ideal $I = \langle \mathcal{G} \rangle$ in $K\langle X_1, \ldots, X_n \rangle$ such that $\mathbf{LH}(\mathcal{G})$ is of the form as described in Corollary 3.5(i) – (ii) respectively but with $\Omega = \emptyset$ and all the $\lambda_{ji} \neq 0$, then the quotient algebra $K\langle X_1, \ldots, X_n \rangle / I$ will provide us with a practical algebra required by Theorem 3.9. For instance, consider in $K\langle X_1, X_2, X_3, X_4 \rangle$ the subset $\mathcal{G}$ consisting of

$$\begin{aligned} g_{43} &= X_4X_3 - \lambda_{43}X_3X_4 - \alpha X_2, & g_{32} &= X_3X_2 - \lambda_{32}X_2X_3 - \gamma X_4, \\ g_{42} &= X_4X_2 - \lambda_{42}X_2X_4 + \beta X_3, & g_{31} &= X_3X_1 - \lambda_{31}X_1X_4, \\ g_{41} &= X_4X_1 - \lambda_{41}X_1X_2, & g_{21} &= X_2X_1 - \lambda_{21}X_1X_3, \end{aligned}$$

where $\lambda_{ji}, \alpha, \beta, \gamma \in K^*$. Then, $\mathcal{G}$ is a Gröbner basis for each solution of the system of equations (provided kindly by an anonymous referee of the paper [Li6] to correct a wrong system given by the author):

$$\begin{aligned} &\lambda_{31}^2 = 1, \quad \beta = \alpha\lambda_{21}\lambda_{31}^{-1}, \ \gamma = \alpha\lambda_{21}^2, \quad \lambda_{32} = \lambda_{31}\lambda_{21}, \\ &\lambda_{41} = \lambda_{21}^{-1}\lambda_{31}, \ \lambda_{42} = \lambda_{21}^{-1}\lambda_{31}, \ \lambda_{43} = \lambda_{31}\lambda_{21}, \end{aligned}$$

where the monomial ordering used is the $\mathbb{N}$-graded lexicographic monomial ordering $\prec_{grlex}$ with respect to $d(X_i) = 1$, $1 \leq i \leq 4$, such that $X_1 \prec_{grlex} X_2 \prec_{grlex} X_3 \prec_{grlex} X_4$.

8.4 Skew 2-Nomial Algebras

In Section 3 we listed some practical algebras defined by elements of the form $u - \lambda v$ and w, where u, v, w are monomials and $\lambda \in K^*$, which are not the familiar types of algebras manipulated by classical Gröbner basis theory, such as the Gröbner basis theory for solvable polynomial algebras or some of their homomorphic images ([AL], [K-RW], [HT], [Li1], [BGV], [Lev]), but all may have a left Gröbner basis theory or a right Gröbner basis theory. Inspired by such a fact, we introduce more general skew 2-nomial algebras in this section, and, as the first step of having a (one-sided or two-sided) Gröbner basis theory, we establish the existence of a skew multiplicative K-basis for such algebras. As to the existence of a (one-sided or two-sided) monomial ordering, examples of Section 3 show that it has to be a matter of examining concrete individual class of algebras.

In this section R denotes a K-algebra with a *skew multiplicative K-basis* $\mathcal{B}$. Moreover, here and in what follows, a skew multiplicative K-basis is in the sense of (Ch.3, Definition 4.1), and a Gröbner basis theory is in the sense of (Ch.3, Definition 2.3). All notations used before are maintained.

We start by generalizing the sufficiency result of the following theorem.

Theorem ([Gr4], Theorem 2.3). *Suppose that R is a K-algebra with multiplicative K-basis $\mathcal{C}$ (in the sense that $u, v \in \mathcal{C}$ implies $uv = 0$ or $uv \in \mathcal{C}$). Let I be an ideal in R and π: $R \to R/I$ be the canonical surjection. Let $\pi(\mathcal{C})^* = \pi(\mathcal{C}) - \{0\}$. Then $\pi(\mathcal{C})^*$ is a multiplicative K-basis for R/I if and only if I is a 2-nomial ideal (i.e., I is generated by elements of the form $u - v$ and w, where $u, v, w \in \mathcal{C}$).*

□

Let I be an arbitrary ideal of R. Then we may define a relation $\sim_I$ on $\mathcal{B} \cup \{0\}$ as follows: if $v, u \in \mathcal{B} \cup \{0\}$, then

$$v \sim_I u \Longleftrightarrow \text{there exists a } \lambda \in K^* \text{ such that } v - \lambda u \in I.$$

Lemma 4.1. *$\sim$ is an equivalence relation on $\mathcal{B} \cup \{0\}$.*

Proof. Note that K is a field, and $v \in I$ if and only if $v \sim_I 0$. The verification of reflexivity, symmetry, and transitivity for $\sim_I$ is straightforward.

□

Recall that elements of $\mathcal{B}$ are called monomials. If $h = v - \lambda u \in R$ with $v, u \in \mathcal{B}$ and $\lambda \in K^*$, then we call h a *skew 2-nomial* in R; and so, we may call the equivalence relation $\sim_I$ obtained above a *skew 2-nomial relation* on $\mathcal{B} \cup \{0\}$ determined by the ideal I.

Definition 4.2. An ideal I of R is said to be a *skew 2-nomial ideal* if I is generated by monomials and skew 2-nomials. If I is a skew 2-nomial ideal of R, then we call the corresponding quotient algebra $A = R/I$ a *skew 2-nomial algebra.*

Remark In comparison with other references, for instance [GI-1], [Laf], or [ES], we used the term "skew 2-nomial ideal" to distinguish it from the term "2-nomial ideal" used in [Gr4] (see also a remark given after Theorem 4.4).

Lemma 4.3. *Let I be a skew 2-nomial ideal of R and $\sim_I$ the skew 2-nomial* *relation on $\mathcal{B}\cup\{0\}$ determined by I. If $f=\sum_j\lambda_jw_j$ with $\lambda_j\in K^*$ and $w_j\in\mathcal{B}$, then $f\in I$ if and only if for each equivalence class $[w]$ of $\sim_I$, $\sum_{w_j\in[w]}\lambda_jw_j\in I$.*

Proof. The sufficiency is clear. To get the necessity, suppose $f=\sum_j\lambda_jw_j\in I$. Then since $\mathcal{B}$ is a skew multiplicative K-basis and I is an ideal, f has a representation of the form

$$f=\sum_j\lambda_jw_j=\sum_p\mu_p(v_p-\gamma_pu_p)+\sum_q\eta_qs_q$$

with $v_p-\gamma_pu_p, s_q\in I$, $\mu_p,\gamma_p,\eta_q\in K^*$, but $v_p,u_p\not\in I$ for every p. Note that for v, $s\in\mathcal{B}$, if $s\in I$ but $v\not\in I$, then $v\not\sim_I s$. So, comparing both sides of the above equality subject to the equivalence relation $\sim_I$, if $[w]$ is any equivalence class of $\sim_I$, then either

$$\sum_{w_j\in[w]}\lambda_jw_j=\sum_{v_p,u_p\in[w]}\mu_p(v_p-\gamma_pu_p)\in I,$$

or

$$\sum_{w_j\in[w]}\lambda_jw_j=\sum_{s_q\in[w]}\eta_qs_q\in I,$$

as required. □

In what follows we use the unified notation as before: if I is an ideal of R and π: $R\rightarrow R/I$ is the canonical algebra epimorphism, then, for $f\in R$, we write $\bar{f}$ for the canonical image of f in R/I, i.e., $\bar{f}=f+I$.

Theorem 4.4. *Let I be a skew 2-nomial ideal of R, $A=R/I$, and let $\sim_I$ be the skew 2-nomial relation on $\mathcal{B}\cup\{0\}$ determined by I. Considering the image $\overline{\mathcal{B}}$ of $\mathcal{B}$ under the canonical algebra epimorphism π: $R\rightarrow A$, if we put $\overline{\mathcal{B}}^*=\overline{\mathcal{B}}-\{0\}$, then the following statements hold.*
(i) *With respect to the inclusion order on subsets of $\overline{\mathcal{B}}^*$, $\overline{\mathcal{B}}^*$ contains a maximal subset, denoted $\overline{\mathcal{B}}^*_{\sim_I}$, with the property that*

$$\overline{u_1},\overline{u_2}\in\overline{\mathcal{B}}^*_{\sim_I},\ \overline{u_1}\neq\overline{u_2}\ \textit{implies}\ u_1\not\sim_I u_2.$$

(ii) *The maximal subset $\overline{\mathcal{B}}^*_{\sim_I}$ of $\overline{\mathcal{B}}^*$ obtained in* (i) *above forms a skew multiplicative K-basis for the skew 2-nomial algebra A.*

Proof. (i). The existence of $\overline{\mathcal{B}}^*_{\sim_I}$ with respect to the inclusion order on subsets of $\overline{\mathcal{B}}^*$ follows from Zorn's Lemma.

(ii). Since $\mathcal{B}$ is a skew multiplicative K-basis of R, if $u, v \in \mathcal{B}$, then either $uv = 0$ or $uv = \lambda w$ for some $\lambda \in K^*$, $w \in \mathcal{B}$. Hence, for $\bar{u}$, $\bar{v} \in \overline{\mathcal{B}}^*_{\sim_I}$, either $\bar{u}\bar{v} = 0$, or $\bar{u}\bar{v} \neq 0$ and so $\bar{u}\bar{v} = \overline{uv} = \lambda\bar{w}$ with $\lambda \in K^*$ and $\bar{w} \in \overline{\mathcal{B}}^*$. If $\bar{w} \notin \overline{\mathcal{B}}^*_{\sim_I}$, then by the definition of $\overline{\mathcal{B}}^*_{\sim_I}$, there is some $\bar{s} \in \overline{\mathcal{B}}^*_{\sim_I}$ such that $s \sim_I w$. Consequently $\bar{w} = \mu\bar{s}$ for some $\mu \in K^*$ and thereby $\bar{u}\bar{v} = \lambda\mu\bar{s}$ with $\lambda\mu \in K^*$. Thus, the skew multiplication property of $\overline{\mathcal{B}}^*_{\sim_I}$ is proved. Furthermore, we proceed to show that $\overline{\mathcal{B}}^*_{\sim_I}$ forms a K-basis of A. Again by the definition of $\overline{\mathcal{B}}^*_{\sim_I}$, if $\bar{w} \in \overline{\mathcal{B}}^*$ and $\bar{w} \notin \overline{\mathcal{B}}^*_{\sim_I}$, then $\bar{w} = \lambda\bar{v}$ for some $\lambda \in K^*$ and $\bar{v} \in \overline{\mathcal{B}}^*_{\sim_I}$. It follows that $\overline{\mathcal{B}}^*_{\sim_I}$ spans A. To see that elements of $\overline{\mathcal{B}}^*_{\sim_I}$ are linearly independent over K, suppose that $\sum_{j=1}^{\ell} \lambda_j\overline{w_j} = 0$ and that the $\overline{w_j}$ are distinct, where $\lambda_j \in K$ and $\overline{w_j} \in \overline{\mathcal{B}}^*_{\sim_I}$. Then $f = \sum_{j=1}^{\ell} \lambda_j w_j \in I$. By Lemma 4.3, $\sum_{w_j \in [u]} \lambda_j w_j \in I$ holds for every equivalence class $[u]$ of $\sim_I$. But $w_j \in [u]$ implies $w_j \sim_I u$. Noticing that $\sim_I$ is an equivalence relation and that all the $\overline{w_j}$ are distinct, it follows from the definition of $\overline{\mathcal{B}}^*_{\sim_I}$ that different w_j's are contained in different equivalence classes. So $\lambda_j w_j \in I$ and $\lambda_j\overline{w_j} = 0$. Since $\overline{w_j} \neq 0$, it turns out that $\lambda_j = 0$ for all j. This completes the proof. □

Remark Let R be a K-algebra with K-basis $\mathcal{B}$. Recall from ([Gr4], Section 2) that an ideal I of R is said to be a 2-*nomial ideal* if I can be generated by elements of the form $b_1 - b_2$ and b, where $b_1, b_2, b \in \mathcal{B}$. By ([Gr4], Theorem 2.3), if $\mathcal{B}$ is a *multiplicative K-basis* of R (i.e., $u, v \in \mathcal{B}$ implies $uv = 0$ or $uv \in \mathcal{B}$) and I is a 2-nomial ideal of R, then, under the canonical algebra epimorphism π: $R \to R/I$, the set $\overline{\mathcal{B}}^* = \overline{\mathcal{B}} - \{0\}$ forms a multiplicative K-basis of R/I. Since a multiplicative K-basis is trivially a skew multiplicative K-basis and any 2-nomial ideal is trivially a skew 2-nomial ideal, it is easy to see that in the case that $\mathcal{B}$ is a multiplicative K-basis and I is a 2-nomial ideal we must have $\overline{\mathcal{B}}^* = \overline{\mathcal{B}}^*_{\sim_I}$, where the latter is the set obtained in Theorem 4.4(i) above. That is, actually the preceding Theorem 4.4 generalizes the sufficiency result of ([Gr2], Theorem 2.3).

Next, suppose further that there is a monomial ordering $\prec$ on $\mathcal{B}$, that is, $(\mathcal{B}, \prec)$ forms an admissible system $(\mathcal{B}, \prec)$ of R and hence R has a Gröbner basis theory (note that the transitivity of the division of monomials in $\mathcal{B}$ follows from the assumption that $\mathcal{B}$ is now a skew multiplicative K-basis of R). Moreover, in this case if I is an ideal of R, and $A = R/I$, then it follows from (Ch.3, Proposition 4.8) that the set of normal monomials in

$\mathcal{B}$ (mod I) is given by $N(I) = \mathcal{B} - \mathbf{LM}(I)$. Since the image $\overline{N(I)}$ of $N(I)$ under the canonical algebra epimorphism π: $R \to A$ forms a K-basis of A and $\overline{N(I)} \subseteq \overline{\mathcal{B}}^*$, by ([Gr4], Theorem 2.3), we first note the following fact.

Proposition 4.5. *Let I be a 2-nomial ideal of R in the sense of* [Gr4]. *Then $\overline{N(I)} = \overline{\mathcal{B}}^*$ and hence $\overline{N(I)}$ is a multiplicative K-basis for the quotient algebra $A = R/I$.*

□

Turning to the case of Theorem 4.4, we shall prove, under the assumption that R has an admissible system, that $\overline{N(I)}$ forms a skew multiplicative K-basis for the skew 2-nomail algebra $A = R/I$.

Lemma 4.6. *Suppose that R has an admissible system $(\mathcal{B}, \prec)$. Let I be a skew 2-nomial ideal of R, $A = R/I$, and let $N(I)$ be the set of normal monomials in $\mathcal{B}$ (mod I). With notation as in Theorem 4.4, the following statements hold.*

(i) *For each $\bar{w} \in \overline{N(I)}$, there is a unique $\lambda \in K^*$ and a unique $\bar{u} \in \overline{\mathcal{B}}^*_{\sim_I}$ such that $\bar{w} = \lambda\bar{u}$; and if $\overline{w_1}$, $\overline{w_2} \in \overline{N(I)}$ with $\overline{w_1} = \lambda_1\overline{u_1}$ and $\overline{w_2} = \lambda_2\pi(u_2)$ respectively, then $\overline{w_1} \neq \overline{w_2}$ implies $\pi(u_1) \neq \overline{u_2}$.*

(ii) *For each $\bar{u} \in \overline{\mathcal{B}}^*_{\sim_I}$, there is some $\bar{w} \in \overline{N(I)}$ such that $\bar{u} = \mu\bar{w}$ with $\mu \in K^*$.*

Proof. To prove the assertions mentioned in (i), (ii), let us bear in mind that $\overline{\mathcal{B}}^*_{\sim_I}$ is a K-basis of A by Theorem 4.4, and that $\overline{N(I)}$ is also a K-basis of A by the Gröbner basis theory of R.

(i). Note that $\overline{N(I)} \subseteq \overline{\mathcal{B}}^*$. Let $\bar{w} \in \overline{N(I)}$. Then it follows from the definition of $\overline{\mathcal{B}}^*_{\sim_I}$ that $\bar{w} = \lambda\bar{u}$ for some $\lambda \in K^*$ and $\bar{u} \in \overline{\mathcal{B}}^*_{\sim_I}$. Since $\overline{\mathcal{B}}^*_{\sim_I}$ is a K-basis, the expression obtained is unique. If $\overline{w_1}, \overline{w_2} \in \overline{N(I)}$ and $\overline{w_1} = \lambda_1\overline{u_1}$ and $\overline{w_2} = \lambda_2\overline{u_2}$, then $\overline{u_1} = \overline{u_2}$ would clearly imply the linear dependence of $\overline{w_1}$ and $\overline{w_2}$. Hence the second assertion of (i) follows from the fact that $\overline{N(I)}$ is a K-basis.

(ii). Since $\overline{N(I)}$ is a K-basis for A, each $\bar{u} \in \overline{\mathcal{B}}^*_{\sim_I}$ has a unique linear expression $\bar{u} = \sum_{p=1}^m \mu_p\overline{w_p}$ with $\mu_p \in K^*$ and $\overline{w_p} \in \overline{N(I)}$. By (i), $\bar{u} = \sum_{p=1}^m \mu_p\lambda_p\overline{u_p}$ with $\lambda_p \in K^*$ and $\overline{u_p} \in \overline{\mathcal{B}}^*_{\sim_I}$, in which $\overline{w_p} = \lambda_p\overline{u_p}$ and all the $\overline{u_p}$ are distinct. If, without loss of generality, $\bar{u} = \overline{u_1}$, then $\bar{u} = \lambda_1^{-1}\overline{w_1}$, as desired. □

Theorem 4.7. *With notation as before, let the skew 2-nomial algebra $A = R/I$ be as in Theorem 4.4, and suppose that R has an admissible system $(\mathcal{B}, \prec)$. Then $\overline{N(I)}$ is a skew multiplicative K-basis for A.*

Proof. If $\overline{w_1}, \overline{w_2} \in \overline{N(I)}$ and $\overline{w_1 w_2} \neq 0$, then by Lemma 4.6 and the fact that $\overline{\mathcal{B}}^*_{\sim_I}$ is a skew multiplicative K-basis of A (Theorem 4.4(ii)), there are $\lambda_1, \lambda_2, \lambda_3, \mu \in K^*$, $\overline{u_1}, \overline{u_2} \in \overline{\mathcal{B}}^*_{\sim_I}$ and $\overline{w_3} \in \overline{N(I)}$ such that

$$\begin{aligned} \overline{w_1} \cdot \overline{w_2} &= \lambda_1 \overline{u_1} \lambda_2 \overline{u_2} \\ &= \lambda_1 \lambda_2 \lambda_3 \overline{u_3} \\ &= \lambda_1 \lambda_2 \lambda_3 \mu \overline{w_3}. \end{aligned}$$

It follows that the K-basis $\overline{N(I)}$ of A is a skew multiplicative K-basis. □

As we will see below, indeed, a better proof of Theorem 4.7 may be obtained by employing the division by a Gröbner basis of I, which involves the following interesting result (a noncommutative analogue of ([ES], Proposition 1.1)).

Proposition 4.8. *Suppose that R has an admissible system $(\mathcal{B}, \prec)$, and let I be a skew 2-nomial ideal of R. Then I has a Gröbner basis $\mathcal{G}$ consisting of monomials and skew 2-nomials, i.e., elements of the form $v - \lambda u$, s with $\lambda \in K^*$ and $v, u, s \in \mathcal{B}$.*

Proof. Let G be any Gröbner basis of I. Then, since I is generated by skew 2-nomials and monomials, as in the proof of Lemma 4.3, each $g \in G$ has a representation of the form

$$g = \sum_p \mu_p (v_p - \gamma_p u_p) + \sum_q \eta_q s_q$$

with $v_p - \gamma_p u_p, s_q \in I$, $\mu_p, \gamma_p, \eta_q \in K^*$, but $v_p, u_p \notin I$ for every p. Let $g = \sum_j \lambda_j w_j$, where $\lambda_j \in K^*$ and $w_j \in \mathcal{B}$. If $\mathbf{LM}(g) \notin I$, then we claim that there is some w_j appearing in $\sum_j \lambda_j w_j$ such that $w_j \neq \mathbf{LM}(g)$, $\mathbf{LM}(g) - \mu w_j \in I$, and $\mathbf{LM}(\mathbf{LM}(g) - \mu w_j) = \mathbf{LM}(g)$, where $\mu \in K^*$. Otherwise, considering the skew 2-nomial relation $\sim_I$ on $\mathcal{B} \cup \{0\}$ determined by I, and letting $[\mathbf{LM}(g)]$ be the equivalence class of $\sim_I$ represented by $\mathbf{LM}(g)$, by Lemma 3.3 we would have

$$\sum_{w_j \in [\mathbf{LM}(g)]} \lambda_j w_j = \mathbf{LC}(g)\mathbf{LM}(g) = \sum_{v_p, u_p \in [\mathbf{LM}(g)]} \mu_p (v_p - \gamma_p u_p) \in I,$$

and consequently $\mathbf{LM}(g) \in I$, a contradiction. Thus, if we put

$$\mathcal{G} = \left\{ v - \lambda u,\ s \;\middle|\; \begin{array}{l} v, u \in \mathcal{B} - I,\ \lambda \in K^*,\ v - \lambda u \in I,\ s \in \mathcal{B} \cap I, \\ \text{such that } \mathbf{LM}(v - \lambda u) = \mathbf{LM}(g_i) \text{ and} \\ s = \mathbf{LM}(g_j) \text{ for some } g_i, g_j \in G \end{array} \right\},$$

then $\mathbf{LM}(\mathcal{G}) = \mathbf{LM}(G)$ and hence $\mathcal{G}$ is a Gröbner basis for I. □

For the convenience of statement, we call the Gröbner basis $\mathcal{G}$ obtained in Lemma 4.8 a *skew* 2-*nomial Gröbner basis* of the skew 2-nomial ideal I.

Lemma 4.9. *Suppose that R has an admissible system $(\mathcal{B}, \prec)$. Let I be a skew 2-nomial ideal of R and $\mathcal{G}$ a skew 2-nomial Gröbner basis of I. If $u \in \mathcal{B}$ satisfying $u \notin I$ and $u \notin N(I)$, where $N(I)$ is the set of normal monomials in $\mathcal{B}$ (mod I), then u has a representation*

$$u = \sum_{i,j} \lambda_{ij} w_{ij} g_j v_{ij} + \lambda u'$$

with $\lambda_{ij}, \lambda \in K^*$, $w_{ij}, v_{ij} \in \mathcal{B}$, $g_j \in \mathcal{G}$ and $u' \in N(I)$ satisfying $u' \prec u$.

Proof. It is simply a consequence of doing division by $\mathcal{G}$. As $u \notin I$ and $u \notin N(I)$, there is some $g_i = u_i - \mu_i s_i \in \mathcal{G}$, such that $u = \lambda_i w_i \mathbf{LM}(g_i) v_i$ for some $\lambda_i \in K^*$, $w_i, v_i \in \mathcal{B}$. Suppose $\mathbf{LM}(g_i) = u_i$. Then since $\mathcal{B}$ is a skew multiplicative K-basis,

$$u - \lambda_i w_i g_i v_i = \lambda_i \mu_i w_i s_i v_i = \gamma_i u_1$$

with $\gamma_i \in K^*$ and $u_1 \in \mathcal{B}$ satisfying $u_1 \prec u$. Note that $u \notin I$ implies $u_1 \notin I$. If $u_1 \notin N(I)$, then repeat the division procedure for u_1. By the well-ordering property of $\prec$, the proof is finished after a finite number of reductions. □

Now, let $\pi\colon\ R \to A = R/I$ be the canonical algebra epimorphism as before, and suppose that R has an admissible system $(\mathcal{B}, \prec)$. If $N(I)$ is the set of normal monomials in $\mathcal{B}$ (mod I) and $\mathcal{G}$ is a skew 2-nomial Gröbner basis of I, then, since $\mathcal{B}$ is a skew multiplicative K-basis, for any u, $v \in N(I)$, it follows from Lemma 4.9 that either

$$\bar{u}\bar{v} = 0,$$

or

$$\bar{u}\bar{v} = \overline{uv} = \lambda \bar{w} \text{ with } \lambda \in K^*,\ w \in N(I).$$

This shows that $\overline{N(I)}$ is a skew multiplicative K-basis for the skew 2-nomial algebra A, that is, Theorem 4.7 is recaptured.

In connection with the results obtained in Section 1, we mention especially the following example.

Example 1. Let R be the polynomial K-algebra $K[x_1, \ldots, x_n]$ or the skew polynomial K-algebra $\mathcal{O}_n(\lambda_{ji})$, and let G be a minimal Gröbner basis of the ideal $I = \langle G \rangle$ in R with respect to some monomial ordering $\prec$ on the standard K-basis $\mathscr{B}$ of R. Suppose that G consists of elements of the form w, $t - \lambda s$ with $\lambda \in K^*$ and $w, t, s \in \mathscr{B}$. Then by Theorem 4.7, the quotient algebra R/I is a skew 2-nomial algebra with the skew multiplicative K-basis $\overline{N(I)}$. Furthermore, writing $\mathcal{I}$ for the pre-image of I under the canonical algebra epimorphism π: $K\langle X \rangle = K\langle X_1, \ldots, X_n \rangle \rightarrow R$, then by Theorem 1.3, the ideal $\mathcal{I}$ has the lifted Gröbner basis

$$\mathcal{G} = \mathcal{S} \cup \{\delta(u \cdot g) \mid u \in \mathscr{U}_{\Delta}(\mathbf{LM}(g)),\ g \in G\}$$

with respect to the monomial ordering $\prec_{et}$ which is the lexicographic extension of $\prec$ on the standard K-basis $\mathcal{B}$ of $K\langle X \rangle$. By the assumption on G, the definition of $\mathcal{S}$ and the definition of δ, it is clear that $\mathcal{G}$ consists of monomials and skew 2-nomials. It follows from Theorem 4.7 that the quotient algebra $K\langle X \rangle/\mathcal{I}$ is a skew 2-nomial algebra with the skew multiplicative K-basis $\overline{N(\mathcal{I})}$.

Finally, concerning the existence of a Gröbner basis theory for a skew 2-nomial algebra, let us mention the following conclusion.

Theorem 4.10. *Let R be a K-algebra with the skew multiplicative K-basis $\mathcal{B}$, and let I be a skew 2-nomial ideal of R, $A = R/I$.*
(i) *With notation as in Theorem 4.4, if there is a (left, right, or two-sided) monomial ordering $\prec$ on $\overline{\mathcal{B}}^*_{\sim_I}$, then the skew 2-nomial algebra A has a (left, right, or two-sided) Gröbner basis theory.*
(ii) *With notation and the assumption as in Theorem 4.7, if there is a (left, right, or two-sided) monomial ordering $\prec$ on $\overline{N(I)}$, then the skew 2-nomial algebra A has a (left, right, or two-sided) Gröbner basis theory.*

8.5 Almost Skew 2-Nomial Algebras

Naturally, the results of Sections 3, 4, combined with the working principle of Ch.4 – Ch.7, motivate us to introduce the class of algebras with the associated graded algebra which is a skew 2-nomial algebra (such as the algebras given in Examples 4 – 7 of Section 3), namely the almost skew 2-nomial algebras defined below. As in previous sections, we retain the convention of a skew multiplicative K-basis in the sense of (Ch.3, Definition 4.1), and that a one-sided or two-sided Gröbner basis theory means the one in the sense of (CH.3, Section 4) or it is the one indicated before Theorem 3.9, depending on whether a skew multiplicative K-basis is used or not. All notations used before are maintained.

Let $R = \oplus_{\gamma\in\Gamma}R_\gamma$ be a Γ-graded K-algebra, where Γ is a totally ordered semigroup with the total ordering $<$, and we assume that R has an admissible system $(\mathcal{B}, \prec)$, where $\mathcal{B}$ is a *skew multiplicative K-basis* of R consisting of *Γ-homogeneous elements*, and $\prec$ is a monomial ordering on $\mathcal{B}$. Hence R has a Gröbner basis theory with respect to $(\mathcal{B}, \prec)$.

Let I be an ideal of R, $A = R/I$, and $A^{\Gamma}_{\rm LH} = R/\langle \mathbf{LH}(I)\rangle$ the Γ-leading homogeneous algebra of A as defined in (Ch.2, Section 3), where $\mathbf{LH}(I)$ is the set of Γ-leading homogeneous elements of I.

Definition 5.1. With notation as above, if $\langle \mathbf{LH}(I)\rangle$ is a skew 2-nomial ideal, or equivalently, if $A^{\Gamma}_{\rm LH} = R/\langle \mathbf{LH}(I)\rangle$ is a skew 2-nomial algebra, then we call the quotient algebra $A = R/I$ an *almost skew 2-nomial algebra.*

Before mentioning the main result of this section, let us demonstrate how to obtain an almost skew 2-nomial algebra in terms of Gröbner bases in R. We begin by stating a more general version of Lemma 2.1 for $\mathbf{LH}(I)$ in the Γ-graded algebra R.

Lemma 5.2. *With notation as above, let h be a nonzero Γ-homogeneous element of R. Then $h \in \langle \mathbf{LH}(I)\rangle$ if and only if $h \in \mathbf{LH}(I)$.*

□

Proposition 5.3. *Let J be a Γ-graded ideal of R. If J is a skew 2-nomial ideal, then J has a skew 2-nomial Gröbner basis $\mathcal{G}$ (as obtained in Proposition 4.8) but consisting of Γ-homogeneous elements.*

Proof. For convenience, let $\mathrm{H}^\Gamma(J)$ denote the set of Γ-homogeneous elements of J. Noticing that J is a Γ-graded ideal, if $f = \sum_j r_{\gamma_j} \in R$ with $r_{\gamma_j} \in R_{\gamma_j}$, then $f \in J$ if and only if $r_{\gamma_j} \in J$ for all j (Ch.1, Proposition 2.1). Thus, since the division of monomials in $\mathcal{B}$ with respect to $\prec$ is transitive by the assumption on R, exactly as in (Ch.3, Section 2) we obtain a minimal homogeneous Gröbner basis for J as follows

$$G = \left\{ h \in \mathrm{H}^\Gamma(J) \;\middle|\; \begin{array}{l} \text{if } h' \in \mathrm{H}^\Gamma(J) \text{ and } \mathbf{LM}(h')|\mathbf{LM}(h), \\ \text{then } \mathbf{LM}(h) = \mathbf{LM}(h') \end{array} \right\}.$$

Furthermore, noticing that J is also a skew 2-nomial ideal, actually as in the proof of Proposition 4.8, subject to the G above we obtain a Gröbner basis of J in the desired form

$$\mathcal{G} = \left\{ v - \lambda u,\ s \;\middle|\; \begin{array}{l} v, u \in \mathcal{B} - J,\ \lambda \in K^*,\ v - \lambda u \in \mathrm{H}^\Gamma(J),\ s \in \mathcal{B} \cap J, \\ \text{such that } \mathbf{LM}(v - \lambda u) = \mathbf{LM}(h_i) \text{ and} \\ s = \mathbf{LM}(h_j) \text{ for some } h_i, h_j \in G \end{array} \right\}.$$

□

Corollary 5.4. *Suppose that R has an admissible system $(\mathcal{B}, \prec_{gr})$ in which $\prec_{gr}$ is a Γ-graded monomial ordering on $\mathcal{B}$, and let $A = R/I$ be an almost skew 2-nomial algebra in the sense of Definition 5.1. Then I has a Gröbner basis $\mathscr{G}$ such that $\mathbf{LH}(\mathscr{G})$ is a homogeneous skew 2-nomial Gröbner basis for the ideal $\langle \mathbf{LH}(I) \rangle$ with respect to $(\mathcal{B}, \prec_{gr})$.*

Proof. Writing $J = \langle \mathbf{LH}(I) \rangle$, it follows from Proposition 5.3 and Lemma 5.2 that J has a skew 2-nomial Gröbner basis $\mathcal{G} \subset \mathbf{LH}(I)$. Since $\prec_{gr}$ is a Γ-graded monomial ordering on $\mathcal{B}$, by (Ch.4, Proposition 2.2(ii)) the desired Gröbner basis $\mathscr{G}$ for I is given by

$$\mathscr{G} = \{ f \in I \mid \mathbf{LH}(f) = g,\ g \in \mathcal{G} \}.$$

□

It follows from Corollary 5.4, Theorem 3.9, Theorem 4.7, and Theorem 4.10 that we can now write down the main result of this section.

Theorem 5.5. *Suppose that R has an admissible system $(\mathcal{B}, \prec_{gr})$ in which $\prec_{gr}$ is a Γ-graded monomial ordering on $\mathcal{B}$. Let I be an ideal of R, $A = R/I$, and $A^\Gamma_{\mathrm{LH}} = R/\langle \mathbf{LH}(I) \rangle$ the Γ-leading homogeneous algebra of A. If $I = \langle \mathscr{G} \rangle$ is generated by a Gröbner basis $\mathscr{G}$ such that $\mathbf{LH}(\mathscr{G})$ consists of monomials and Γ-homogeneous skew 2-nomials, then the following statements hold.*

(i) $\langle \mathbf{LH}(I)\rangle = \langle \mathbf{LH}(\mathscr{G})\rangle$ *is a* Γ*-graded skew 2-nomial ideal,* $A^{\Gamma}_{\rm LH} = R/\langle \mathbf{LH}(I)\rangle$ *is a skew 2-nomial algebra, and hence* A *is an almost skew 2-nomial algebra.*

(ii) *Let* $N(I)$ *be the set of normal monomials in* $\mathcal{B}$ *(mod* I*), and write*

$$\widehat{N(I)} = \{\hat{u} = u + \langle \mathbf{LH}(I)\rangle \mid u \in N(I)\}$$

for the K*-basis of* $A^{\Gamma}_{\rm LH}$ *(as in Section 3). Then* $\widehat{N(I)}$ *is a skew multiplicative* K*-basis of* $A^{\Gamma}_{\rm LH}$*. If furthermore there is a (left, right, or two-sided) monomial ordering* $\prec_{\rm LH}$ *on* $\widehat{N(I)}$*, then* $A^{\Gamma}_{\rm LH}$ *has a (left, right, or two-sided) Gröbner basis theory with respect to the admissible system* $(\widehat{N(I)}, \prec_{\rm LH})$*.*

(iii) *If* $A^{\Gamma}_{\rm LH}$ *is a domain and* $\prec_{\widehat{gr}}$ *is a* Γ*-graded monomial ordering on* $\widehat{N(I)}$*, then* A *has a (left, right, or two-sided) Gröbner basis theory with respect to the (left, right, or two-sided) admissible system* $(\overline{N(I)}, \prec)$*, where*

$$\overline{N(I)} = \{\overline{u} = u + I \mid u \in N(I)\}$$

is the K*-basis of* A *determined by* $N(I)$ *(as in Section 4), and the monomial ordering* $\prec$ *on* $\overline{N(I)}$ *is defined subject to the rule: for* $\overline{u}, \overline{v} \in \overline{N(I)}$*,* $\bar{u} \prec \bar{v}$ *whenever* $\hat{u} \prec_{\widehat{gr}} \hat{v}$*.*

□

Remark and open problems

As one may see from previous sections, the (left, right, or two-sided) Gröbner basis theory we discussed for (almost) skew 2-nomial algebras covers certain new classes of algebras, such as two subclasses of quantum binomial algebras and their Koszul dual, down-up algebras (see Ch.4, Section 3, Example 6(c)), and algebras with the associated graded algebra of such types, which are beyond the scope of algebras treated by the well-known existed Gröbner basis theory. However, the existing results concerning (almost) skew 2-nomial algebras are obviously far from being complete. For instance, even for an algebra of the type as described in Examples 1, respectively Example 2 of Section 3 (which is a Noetherian quantum binomial algebra whenever $\lambda, \mu, \gamma \in K^*$ and $\gamma^2 = 1$, respectively $\lambda^2 = 1$), we do not know if there is an implementable analogue of the Buchberger Algorithm for producing a finite (left or right) Gröbner basis. Moreover, proper subclasses of practical (almost) skew 2-nomial algebras, that are not the types we have considered in Section 3 but have a (left, right, or two-sided) monomial ordering, need to be studied further. We end this chapter by summarizing several open problems related to the preceding sections.

1. Under the Noetherian assumption, is there an analogue of the Buchberger Algorithm for the algebras described in Corollary 3.5(i) – (iii) respectively? Note that if all the parameters $\lambda_{ji} \neq 0$, then the algebras described in Corollary 3.5(iii) belong to the class of GR-algebras which have been studied in an algorithmic way in [Lev].

2. For the algebras described in Corollary 3.5(i) – (iii), especially for those described in (i) – (ii) (including many quantum binomial algebras and their Koszul dual), is there a computational module theory (representation theory) in terms of (left, right, or two-sided) Gröbner bases as that developed in [Gr1] and [Gr2]?

3. Find more subclasses of skew 2-nomial algebras, in which each algebra is not the type as described in Theorem 3.2 and Theorem 3.3 respectively, but has a (left, right, or two-sided) monomial ordering.

4. Find more subclasses of almost skew 2-nomial algebras, in which the Γ-leading homogeneous algebra $A^{\Gamma}_{\rm LH}$ of each algebra A is not the type as described in Theorem 3.7 and Theorem 3.8 respectively, but satisfies the conditions of Theorem 3.9.

5. For algebras as described in Example 7 of Section 3, if the Gröbner basis theory of $A^{\mathbb{N}}_{\rm LH}$ is algorithmically realizable, is it possible to realize the lifted Gröbner basis theory of A in an algorithmic way?

6. In Example 6 of Section 3, if $\lambda = 0$ or $\omega = 0$ and $f(X_1) = aX_1^2 + bX_1 + c$ with $a \neq 0$, does $A^{\mathbb{N}}_{\rm LH}$ have a (left, right, or two-sided) Gröbner basis theory? In view of Example 6(c) of (Ch.4, Section 3), this problem applies to down-up algebras of the type $A(\alpha, 0, \gamma)$.

7. Let $\mathcal{G}$ be a Gröbner basis of the ideal $I = \langle \mathcal{G} \rangle$ in the free K-algebra $K\langle X \rangle = K\langle X_1, \ldots, X_n \rangle$ with respect to some $\mathbb{N}$-graded monomial ordering, such that $\mathbf{LH}(\mathcal{G})$ is of the type as described in Corollary 3.5(i) or Corollary 3.5(ii), but with $\Omega = \emptyset$ and all the $\lambda_{ji} \neq 0$, and such that the $\mathbb{N}$-leading homogeneous algebra $A^{\mathbb{N}}_{\rm LH} = K\langle X \rangle / \langle \mathbf{LH}(\mathcal{G}) \rangle$ of $A = K\langle X \rangle / I$ is a quantum binomial algebra in the sense of [Laf] and [G-I5]. With reference to [Laf], [G-I2] and previous Ch.4 – Ch.7, explore the structural properties of A and $\widetilde{A}$ via $A^{\mathbb{N}}_{\rm LH}$ and $A^{\mathcal{B}}_{\rm LH}$.

Bibliography

[An1] D. J. Anick, Non-commutative graded algebras and their Hilbert series, *J. Alg.*, 1(78)(1982), 120–140.

[An2] D. J. Anick, Diophantine equations, Hilbert series and undecidable spaces, *Ann. Math.II Ser.*, 122(1985), 87–112.

[An3] D. J. Anick, On Monomial algebras of finite global dimension, *Trans. Amer. Math. Soc.*, 1(291)(1985), 291–310.

[An4] D. J. Anick, On the homology of associative algebras, *Trans. Amer. Math. Soc.*, 2(296)(1986), 641–659.

[AK] J. Apel and U. Klaus, FELIX, a Special Computer Algebra System for the Computation in Commutative and Non-commutative Rings and Modules (1998). `http://felix.hgb-leipzig.de/`.

[AR] D. J. Anick and G.-C. Rota, Higher-order syzygies for the bracket algebra and for the ring of coordinates of the Grassmannian, *Proc. Nat. Acad. Sci. U.S.A.*, 88(1991), 8087–8090.

[AL] J. Apel and W. Lassner, An extension of Buchberger's algorithm and calculations in enveloping fields of Lie algebras, *J. Symbolic Comput.*, 6(1988), 361–370.

[AVV] M.J. Asensio, M. Van den bergh, and F. Van Oystaeyen, A new algebraic approach to microlocalization of filtered rings, *Trans. Amer. math. Soc.*, 2(316)(1989), 537–555.

[BCP] W. Bosma, J. Cannon and C. Playoust, The Magma algebra system I: The user language. *J. Symbolic Comput.*, 24(1997), 235–265.

[BP] J. P. Bell and P. Pekcagliyan, Primitivity of finitely presented monomial algebras, *Journal of Pure and Applied Algebra*, 7(2009), 1299-1305.

[Ben] G. Benkart, Down-up algebras and Witten's deformations of the universal enveloping algebra of sl_2, *Contemp. Math.*, 224(1999), 29–45.

[BR] G. Benkart and T. Roby, Down-up algebras, *J. Alg.*, 209(1998), 305 –344. Addendum: *J. Alg.*, 213(1999), 378.

[Ber1] R. Berger, The quantum Poincaré-Birkhoff-Witt theorem, *Comm. Math. Physics*, 143(1992), 215–234.

[Ber2] R. Berger, Koszulity for nonquadratic algebras, *J. Alg.*, 239(2001), 705–734.

[Ber3] G. Bergman, The diamond lemma for ring theory, *Adv. Math.*, 29(1978), 178–218.

[BG1] A. Braverman and D. Gaitsgory, Poincarè-Birkhoff-Witt theorem for quadratic algebras of Koszul type, *J. Alg.*, 181(1996), 315–328.

[BG2] R. Berger and V. Ginzburg, Higher symplectic reflection algebras and nonhomogeneous N-Koszul property, *J. Alg.*, 1(304)(2006), 577–601.

[BGV] J. L. Bueso, José Gómez-Torrecillas and A. Verschoren, *Algorithmic methods in non-commutative algebra: applications to quantum groups*, Kluwer Academic Publishers, 2003.

[Bj] J-E. Björk, *Rings of Differential Operators*, North-Holland Math. Library, Vol. 21, 1979.

[Bok] L. A. Bokut, Imbeddings into simple associative algebrs, *Algebra i Logika*, 2(15)(1976), 117–142.

[BM] L. A. Bokut and P. Malcolmson, Gröbner-Shirshov basis for quantum enveloping algebras, *Israel Journal of Mathematics*, 96(1996), 97–113.

[Bor] V. V. Borisenko, On matrix representations of finitely presented algebras defined by a finite set of words, *Vest. Mosk. Univ.*, 4(1985), 75–77. (In Russian)

[Bu1] B. Buchberger, *Ein Algorithmus zum Auffinden der Basiselemente des Restklassenringes nach einem nulldimensionalen polynomideal*, PhD thesis, University of Innsbruck, 1965.

[Bu2] B. Buchberger, Gröbner bases: An algorithmic method in polynomial ideal theory. In: *Multidimensional Systems Theory* (Bose, N.K., ed.), Reidel Dordrecht, 1985, 184–232.

[BW] T. Becker and V. Weispfenning, *Gröbner Bases*, Springer-Verlag, 1993.

[CG] A. M. Cohen and D. A. H. Gijsbers, GBNP, a Noncommutative Gröbner Bases Package for GAP 4 (2003). `http://www.win.tue.`

nl/~amc/pub/grobner/.

[CM] P. Carvalho and M. Musson, Down-up algebras and their representation theory, *J. Alg.*, 228(2000), 286–310. Journal

[CS1] T. Cassidy and B. Shelton, Basic properties of generalized down-up algebras, *J. Alg.*, 279(2004), 402-421

[CS2] T. Cassidy and B. Shelton, PBW-deformation theory and regular central extensions, *J. für die reine und angewandte Mathematik*, 610(2007), 1–12.

[CBH] W. Crawley-Boevey and M. P. Holland, Noncommutative deformations of Kleinian singularities, *Duke Math. J.*, 3(92)(1998), 605–635.

[CSS] Y. Chen, H. Shao and K. P. Shum, Rosso-Yamane Theorem on PBW basis of $U_q(A_N)$, *CUBO, A Mathematical Journal*, 10(3)(2008), 171–194. arXiv:0804.0954v1,http://arXiv.org

[CU] S. Cojocaru and V. Ufnarofski, BERGMAN under MS-DOS and Anick's resolution, *Discrete Mathematics and Theoretical Computer Science*, 1(1997), 139–147.

[Eis] D. Eisenbud, *Commutative Algebra with a View Toward to Algebraic Geometry*, GTM 150, Springer, New York, 1995.

[ES] D. Eisenbud and B. Sturmfels, Binomial ideals, *Duke Math. J.*, 1(84)(1996), 1–45.

[EPS] D. Eisenbud, I. Peeva and B. Sturmfels, Non-commutative Gröbner bases for commutative algebras, *Proc. Amer. Math. Soc.*, 126(1998), 687-691.

[EG] P. Etingof, and V. Ginzburg, Symplectic reflection algebras, Calogero -Moser space, and deformed Harish-Chandra homomorphism, *Invent. Math.*, 2(147)(2002), 243–346.

[FFG] D. Farkas, C. Feustel and E. Green, Synergy in the theories of Gröbner bases and path algebras, *can. J. Math.*, 45(1993), 727–739.

[FV] G. Floystad and J. E. Vatne, PBW-deformations of N-Koszul algebras, *J. Alg.*, 1(302)(2006).

[GAP] The GAP Group, GAP – Groups, Algorithms, and Programming, Version 4.4.12. http://www.gap-system.org (2008).

[GCB] The Gröbner Basis Calculator Bergman. http://servus.math.su.se/bergman/.

[GG] W. L. Gan and V. Ginzburg, Deformed preprojective algebras and symplectic reflection algebras for wreath products, *J. Alg.*, 1(283)(2005), 350–363.

[GHK] E. Green, S. Heath and J. Keller, Opal: A System for Computing Noncommutative Gröbner Bases, *Proceedings of the 8th International Conference on Rewriting Techniques and Applications*, Lecture Notes in Computer Science Vol. 1232, Springer-Verlag, 1997, 331-334.

[G-I1] T. Gateva-Ivanova, Algorithmic determination of the Jacobson radical of monomial algebras, in: *Proc. EUROCAL'85*, LNCS Vol. 378, Springer-Verlag, 1989, 355–364.

[G-I2] T. Gateva-Ivanova, Global dimension of associative algebras, in: *Proc. AAECC-6*, LNCS, Vol.357, Springer-Verlag, 1989, 213–229.

[G-I3] T. Gateva-Ivanova, On the noetherianity of some associative finitely presented algebras, *J. Alg.*, 138(1991).

[G-I4] T. Gateva-Ivanova, Noetherian properties of skew polynomial rings with binomial relations, *Trans. Amer, Math. Sco.*, 1(343)(1994), 203 - 219

[G-I5] T. Gateva-Ivanova, Binomial skew polynomial rings, Artin-Schelter regularity, and binomial solutions of the Yang-Baxter equation. `arXiv:0909.4707,http://arXiv.org`

[G-IL] T. Gateva-Ivanova and V. Latyshev, On recognizable properties of associative algebras, *J. Symbolic Computation*, 6(1988), 371–388.

[G-IV] T. Gateva-Ivanova and M. Van den Bergh, Semigroups of I type, *J. Alg.*, 206(1998), 97–112.

[Gin] V. Ginsburg, Characteristic varieties and vanishing cycles, *Invent. Math.*, 84(1986), 327–402.

[Gol] E. S. Golod, Standard bases and homology, in: *Some Current Trends in Algebra*, Proceedings, Varna, 1986, LNM, Vol. 1352, Springer-Verlag, 1988, 88–95.

[GS] E. S. Golod and I. R. Shafarevich, On the class-field tower, *Izv. Akad. Nauk.* SSSR, Ser. Mat. 28(1964). (in Russian)

[Gov] V.E. Govorov, Graded algebras, *Math. Notes*, 12(1972), 552 - 556.

[GPS] G.-M. Greuel, G. Pfister, and H. Schönemann, SINGULAR 3.1. A Computer Algebra System for Polynomial Computations. Center for Computer Algebra, University of Kaiserslautern. `http://www.singular.uni-kl.de` (2009).

[Gr1] E. Green, An introduction to noncommutative Gröbner bases, in: *Computational Algebra*, Proceedings of the fifth meeting of the Mid-Atlantic Algebra Conference, 1993, K.G. Fischer, P. Loustaunau, J. Shapiro, E.L. Green, and D. Farkas eds., Lecture Notes in Pure and Applied Mathematics, Vol. 151, Marcel Dekker, 1994, 167–190.

[Gr2] E. Green, *Noncommutative Gröbner Bases, A Computational* and *Theoretical Tool*, Lecture notes, Dec. 15, 1996. www.math.unl.edu/~shermiller2/hs/green2.ps.

[Gr3] E.L. Green, Noncommutative Gröbner bases and projective resolutions, in: *Proceedings of the Euroconference Computational Methods for Representations of Groups and Algebras, Essen, 1997*, (Michler, Schneider, eds), Progress in Mathematics, Vol. 173, Basel, Birkhaüser Verlag, 1999, 29 - 60.

[Gr4] E. Green, Multiplicative bases, Gröbner bases, and right Gröbner bases, *J. Symbolic Computation*, 29(2000), 601–623.

[GH] E. Green and R. Q. Huang, Projective resolutions of straightening closed algebras generated by minors, *Adv. in Math.*, 110(1995), 314–333.

[GL] K.R. Goodearl, E.S. Letzter, Prime and primitive spectra of multiparameter quantum affine spaces, in: *Trends in Ring Theory* (MisKolc, 1996, V. Dlab, L. Marki (eds.)), Canad. Math. Soc. Conf. Proc. Series 22, 1998, 39 - 58.

[GLS] G.-M. Greuel, V. Levandovskyy and H. Schönemann, PLURAL, a SINGULAR 3.0 Subsystem for Computations with noncommutative polynomial algebras, University of Kaiserslautern. http://www.singular.uni-kl.de (2006).

[GZ] E. Green and D. Zacharia, The cohomology ring of a monomial algebra, *Manuscripta Math.*, 1(85)(1994), 11 - 23.

[HS] J. W. Helton and M. Stankus, NCGB 3.1, a Noncommutative Gröbner Basis Package for Mathematica (2001). http://www.math.ucsd.edu/~ncalg/.

[HT] D. Hartley and P. Tuckey, Gröbner Bases in Clifford and Grassmann Algebras, *J. Symb. Comput.*, 20(1995), 197–205.

[Hu] J. E. Humphreys, *Introduction to Lie Algebras and Representation Theory*, GTM 9, Springer, 1972.

[Jac] N. Jacobson, *Lie Algebras*, New York, Wiley, 1962.

[Jat] V.A. Jategaonkar, A multiplicative analogue of the Weyl algebra, *Comm. Alg.*, 12(1984), 1669–1688.

[JBS] A. Jannussis, G. Brodimas, and D. Sourlas, Remarks on the q-quantization, *Lett. Nuovo Cimento*, 30(1981), 123–127. [Kass] C. Kassel, *Quantum Groups*, Springer-Verlag, 1995. [KL] G. Krause and T.H. Lenagan, *Growth of Algebras and Gelfand-Kirillov Dimension*, Research Notes in Math. 116, Pitman, London, 1985.

[KMP] E. Kirkman, I.M. Musson, and D.S. Passman, Noetherian down-up algebras, *Proc. Amer. Math. Soc.*, 127(1999), 2821–2827.

[KRW] A. Kandri-Rody and V. Weispfenning, Non-commutative Gröbner bases in algebras of solvable type, *J. Symbolic Comput.*, 9(1990), 1–26.

[Kur] M.V. Kuryshkin, Opérateurs quantiques généralisés de création et d'annihilation, *Ann. Fond. L. de Broglie*, 5(1980), 111–125.

[Laf] G. Laffaille, Quantum binomial algebras, Colloquium on Homology and Representation Theory (Spanish) (Vaquerfas, 1998). Bol. Acad. Nac. Cienc. (Córdoba) 65(2000), 177–182.

[LL] R. La Scala and V. Levandovskyy, Letterplace ideals and non-commutative Gröbner bases, *J. Symbolic Comput.* 44(2009), 1374–1393.

[Lat] V. N. Latyshev On the equality algorithm in Lie-nilpotent associative algebras, *Vyisn. Kiyiv. Univ., Mat. Meth.* 27(1985). (in Ukrainian)

[Le] L. Le Bruyn, Conformal sl_2 enveloping algebras, *Comm. Alg.*, 23(1995), 1325–1362. [LeS] L. Le Bruyn and S.P. Smith, Homogenized sl(2), *Proc. Amer. Math. Soc.*, 3(118)(1993), 725–730.

[Lev] V. Levandovskyy, *Noncommutative Computer Algebra for Polynomial Algebras, Gröbner Bases, Applications and Implementation*, Dissertation, Vom Fachbereich Mathematik der Universität Kaiserslautern, 2005.

[LS] V. Levandovskyy and H. Schönemann, Plural-a computer algebra system for noncommutative polynomial algebras, In: *Proc. of the International Sumposium and Algebraic Computation* (ISSAC'03, ed. J. Sendra), ACM Press, 2003, 176–183.

[LSV] L. Le Bruyn, S. P. Smith and M. Van den Bergh, Central extensions of three dimensional Artin-Schelter regular algebras, *Math. Zeitschrift*, 222(1)(1996), 171-212.

[LV] L. Le Bruyn and M. Van den Bergh, On quantum spaces of Lie algebras, *Proc. Amer. Math. Soc.*, 2(119)(1993), 407–414.

[Li0] H. Li, *Noncommutative Zariskian Filtered Rings*, Ph.D Thesis, Antwerp University, 1989.

[Li1] H. Li, *Noncommutative Gröbner Bases and Filtered-Graded Transfer*, LNM, 1795, Springer-Verlag, 2002.

[Li2] H. Li, The general PBW property, *Alg. Colloquium*, 4(14)(2007), 541–554. `arXiv:math.RA/0609172,http://arXiv.org`

[Li3] H. Li, Γ-leading homogeneous algebras and Gröbner bases, in: *Recent Developments in Algebra and Related Areas* (F. Li and C. Dong eds.), Advanced Lectures in Mathematics, Vol. 8, International Press & Higher Education Press, Boston-Beijing, 2009, 155 – 200. arXiv:math.RA/0609583,http://arXiv.org

[Li4] H. Li, On the calculation of gl.dim$G^{\mathbb{N}}(A)$ and gl.dim$\widetilde{A}$ by using Gröbner bases, *Alg. Colloquium*, 2(16)(2009), 181–194. arXiv:math.RA/0805.0686,http://arXiv.org

[Li5] H. Li, Lifting Gröbner bases from a class of algebras, *Comm. Alg*, 38(6)(2010), 2282–2299. arXiv:math.RA/0701120,http://arXiv.org

[Li6] H. Li, Looking for Gröbner basis theory for (almost) skew 2-nomial algebras, *J. Symbolic Computation*, 45(2010), 918–942. arXiv:0808.1477,http://arXiv.org

[Li7] H. Li, Algebras defined by monic Gröbner bases over rings, *Algebra and Discrete Mathematics*, to appear. arXiv:0906.4396,http://arXiv.org

[Li8] H. Li, Valuation extensions of algebras defined by monic Gröbner bases, *Algebras and Representation Theory*, 5(2011), 1–15. arXiv:1011.2860,http://arXiv.org

[LS] H. Li and C. Su, On (de)homogenized Gröbner bases, *Journal of Algebra, Number Theory: Advances and Applications*, 3(1)(2010), 35–70. arXiv:0907.0526,http://arXiv.org

[LVO1] H. Li and F. Van Oystaeyen, Zariskian Filtrations, *Comm. Alg.*, 17(1989), 2945–2970.

[LVO2] H. Li and F. van Oystaeyen, Global dimension and Auslander regularity of Rees rings, *Bull. Soc. Math. Belg.*, 43(1991), 59–87.

[LVO3] H. Li and F. Van Oystaeyen, Dehomogenization of gradings to Zariskian filtrations and applications to invertible ideals, *Proc. Amer. Math. Soc.*, 115(1)(1992), 1–11.

[LVO4] H. Li and F. Van Oystaeyen, *Zariskian Filtrations*, K-Monograph in Math., Vol.2, Kluwer Academic Publishers, 1996.

[LVO5] H. Li and F. Van Oystaeyen, *A Primer of Algebraic Geometry*, Marcel Dekker, Inc., New York·Basel, 2000.

[LVO6] H. Li and F. Van Oystaeyen, Reductions and global dimension of quantized algebras over a regular commutative domain, *Comm. Alg.*, 26(4)(1998), 1117–1124.

[LW] H. Li and Y. Wu, Filtered-graded transfer of Gröbner basis computation in solvable polynomial algebras, *Comm. Alg.*, 28(1)(2000),

15–32.

[LWZ] H. Li, Y. Wu and J. Zhang, Two applications of noncommutative Gröbner bases, *Ann. Univ. Ferrara - Sez. VII - Sc. Mat.*, XLV(1999), 1–24.

[LSX] J. Liu, J. Shen and Y. Xiao, The Gröbner bases of two-sided ideals in Clifford and Grassmann algebras, *Acta Math. Sci.*, 25A(2)(2005), 171–175. (in chinese)

[Man] Yu. I. Manin, *Quantum Groups and Noncommutative Geometry*, Les Publ. du Centre de Récherches Math., Universite de Montreal, 1988.

[Mat] H. Matsumura, *Commutative Algebra*, (Second edition), The Benjamin/Cummings Publishing Company, Inc., 1980.

[MP] J. C. McConnell and J.J. Pettit, Crossed products and multiplicative analogues of Weyl algebras, *J. London Math. Soc.*, 2(38)(1988), 47–55.

[MR] J. C. McConnell and J. C. Robson, *Noncommutative Noetherian Rings*, John Wiley & Sons, 1987.

[Mor1] T. Mora, Gröbner bases for noncommutative polynomial rings, *Proc. AAECC3* (Grenoble, 1985), LNCS, Springer-Verlag, 1986.

[Mor2] T. Mora, An introduction to commutative and noncommutative Gröbner Bases, *Theoretic Computer Science*, 134(1994), 131–173.

[NCA] NCAlgebra homepage. `http://www.math.ucsd.edu/~ncalg/`.

[NVO] C. Năstăsescu and F. Van Oystaeyen, *Graded Ring Theory*, Math. Library 28, North Holland, Amsterdam, 1982.

[NS] A. Nilsson and J. Snellman, Lifting Gröbner bases from the exterior algebra, *Comm. Alg.* 131(12)(2003), 5715–5725.

[Nor1] P. Nordbeck, On some basic applications of Gröbner bases in noncommutative polynomial rings, in: *Gröbner Bases and Applications*, Vol. 251 of *LMS Lecture Note Series*, Cambridge Univ. Press, Cambridge, 1998, 463–472.

[Nor2] P. Nordbeck, On the finiteness of Gröbner bases computation in quotients of the free algebra, *Comm. and Comp.*, 11(3)(2001), 157–180.

[Ok] J. Okninski, On monomial algebras, *Arch. Math.*, 50(1988), 417–423.

[Pie] R. S. Pierce, *Associative Algebras*, Springer-Verlag, 1982.

[Pos] L. Positselski, Nonhomogeneous quadratic duality and curvature, *Funct. Anal. Appl.*, 3(27)(1993), 197–204.

[PP] A. Polishchuk and L. Positselski, *Quadratic Algebras*, AMS University Lecture Series, Vol. 37, Providence, Rhode Island, 2005.

[PRS] I. Peeva, V. Reiner and B. Sturmfels, How to shell a monoid, *Math. Ann.*, 310(1998), 379–393.

[Pr] S. Priddy, Koszul resolutions, *Trans. Amer. Math. Soc.*, 152(1970), 39–60.

[Rai] T. Rai, *Infinite Gröbner Bases and Noncommutative Polly Cracker Cryptosystems*, Dissertation, Virginia, Polytechnic Institute and State University, Blacksburg, 2004.

[Rin] C. M. Ringel, PBW-bases of quantum groups, *J. Reine Angew.* Math. 470 (1996), 51–88.

[Ros] M. Rosso, An analogue of the Poincare-Birkhoff-Witt theorem and the universal R-matrix of $U_q(sl(N+1))$, *Comm. Math. Phys.*, 124(2)(1989), 307–318.

[Rot] J.J. Rotman, *An Introduction to Homological Algebra*, Academic Press, 1979.

[Sh] J.B. Shearer, A graded algebra with non-rational Hilbert series, *J. Alg.*, 62(1980), 228 – 231.

[Sh1] A. I. Shirshov, Some algorithm problems for Lie algebras, *Sibirsk, Mat. Ž.*, 3(1962), 292–296.

[Sh2] A. I. Shirshov, Selected works, *Nauka, Novosibirsk*, 1984. (in Russian)

[Sj] G. Sjödin, On filtered modules and their associated graded modules, *math. Scand.*, 33(1973), 229–240.

[SSW] L.W. Small, J.T. Stafford and R.B. Warfield, Affine algebras of Gelfand-Kirillov dimension one are PI, *Math. Proc. Cambridge Phil. Soc.*, 97(1985), 407–414.

[Sm1] S. P. Smith, Quantum groups: an introduction and survey for ring theorists, in: *Noncommutative rings*, S. Montgomery and L. Small eds., MSRI Publ. 24(1992), Springer-Verlag, New York, 131–178.

[Sm2] S. P. Smith, A class of algebras similar to the enveloping algebra of sl(2), *Trans. Amer. Math. Soc.*, 332(1990), 285–314.

[Uf1] V. Ufnarovski, A growth criterion for graphs and algebras defined by words, *Mat. Zametki*, 31(1982), 465–472 (in Russian); English translation: *Math. Notes*, 37(1982), 238–241.

[Uf2] V. Ufnarovski, On the use of graphs for computing a basis, growth and Hilbert series of associative algebras, (in Russian 1989), *Math. USSR Sbornik*, 11(180)(1989), 417-428.

[Uf3] V. Ufnarovski, Introduction to noncommutative Gröbner basis theory, in: *Gröbner Bases and Applications* (Linz, 1998), London Math. Soc. Lecture Notes Ser., 251, Cambridge Univ. Press, Cambridge, 1998, 259–280.

[Uf4] V. Ufnarovski, Combinatorial and asymptotic methods in algebra, in: *Algebra, VI*, Springer-Verlag, 1995, 1–196.

[Vill] R. H. Villarreal, *Monomial Algebras*, Marcel Dekker, Inc., 2001.

[Wor] S.L. Woronowicz, Twisted SU(2) group. An example of a noncommutative differential calculus. *Publ. RIMS*, 23(1987), 117-181.

[Yam] I. Yamane, A Poincare-Birkhoff-Witt theorem for quantized universal enveloping algebras of type A_N, Publ., *RIMS. Kyoto Univ.*, 25(3)(1989), 503–520.

Index